W9-BPN-387

THOMPSON & THOMPSON

Genetics IN MEDICINE

FOURTH EDITION

James S. Thompson, M.D.

Department of Anatomy
University of Toronto

Margaret W. Thompson, Ph.D.

Departments of Medical Genetics and Pediatrics
University of Toronto and
The Hospital for Sick Children, Toronto

Revised by

Margaret W. Thompson

W. B. SAUNDERS COMPANY

Philadelphia London Toronto Mexico City
Rio de Janeiro Sydney Tokyo Hong Kong

W. B. SAUNDERS COMPANY
Harcourt Brace Jovanovich, Inc.

The Curtis Center
Independence Square West
Philadelphia, PA 19106

Library of Congress Cataloging in Publication Data

Thompson, James S. (James Scott), 1919–1982

Thompson & Thompson Genetics in medicine.

Rev. ed. of: Genetics in medicine. 3rd ed. 1980.
Bibliography: p.
Includes index.

1. Medical genetics. I. Thompson, Margaret W. (Margaret Wilson), 1920– II. Title. III. Title: Genetics in medicine. [DNLM: 1. Genetics, Medical. QZ 50 T473g]

RB155.T5 1986 616′.042 85–18429

ISBN 0–7216–8854–3

Editor: Dana Dreibelbis
Designer: Karen O'Keefe
Production Manager: Bill Preston
Illustration Coordinator: Walt Verbitski
Page Layout Artist: Meg McCaffery Jolly

Listed here are the latest translated editions of this book together with the language of the translation and the publisher.

Italian (*3rd Edition*)—UTET (Unione Tipografica Editrice Torinese) Turin, Italy
Spanish (*2nd Edition*)—Salvat Editores, S. A., Barcelona, Spain
Portuguese (*3rd Edition*)—Discos CBS Industria E Commercio LTDA, Rio de Janeiro, Brazil
French (*2nd Edition*)—Doin Editeurs, Paris, France
Japanese (*3rd Edition*)—Kodansha Ltd., Tokyo, Japan
Slovak (*4th Edition*)—Osveta, Martin Czechoslovakia

Genetics in Medicine ISBN 0–7216–8854–3

Last digit is the print number: 9 8 7 6 5 4

To the memory of
James Scott Thompson,
1919–1982,
with love

PREFACE

The fourth edition of Genetics in Medicine comes at a time when medical genetics is undergoing rapid development in two important ways. On the scientific level, the generation of new knowledge of the human genome, unprecedented in its nature and extent, is bringing about redefinition of the scope of genetics in the analysis, prediction and prevention of disease. On a different plane, medical genetics is now recognized officially as a clinical and laboratory specialty area, thus fostering awareness within the medical profession of its expanding significance in medical education and practice.

The primary objective of this book is to introduce medical students to the principles, language and methods of human genetics and to indicate some of its actual and potential medical significance. Because few students have more than an elementary background in genetics, little or no previous knowledge of the subject is assumed. With each new edition, the problem of deciding what to include has become more acute. Regrettably, many important developments in laboratory and clinical genetics have had to be omitted from this edition. However, it is hoped that the book and its references will help to make the literature in medical genetics accessible to readers requiring further information.

The preparation of the previous editions benefited from the deep experience of the senior author in the administrative and practical aspects of medical education. His judgment concerning the content, presentation and appropriate level of discussion was invaluable, and his collaboration in all phases of the project is greatly missed.

I thank the many people who have contributed to the compiling of this book. In particular, I thank my university and hospital colleagues in Toronto; without exception, they have given me encouragement and practical help, but for this edition Diane Cox, Marie Crookston, Elaine Hutton, Ikuko Teshima and Hunt Willard must be singled out for special thanks. Many friends outside Toronto have also helped with suggestions, papers in press and illustrations, and among these I must make special mention of Peter Byers, Eloise Giblett, Robert Gorlin, Terry Hassold, Phyllis McAlpine and Victor McKusick. Also, I would like to acknowledge my gratitude for being able to make use of the outstanding illustrations prepared by the late David Smith. My thanks and apologies go to the numerous others who cannot be individually named here. Most of the new illustrations for this edition were prepared by Rasa Skudra of the Medical Art Department, Toronto General Hospital. Allison Foster, Pauline Kowal, Eva McGrath, Irina Oss and Baba Torres have all helped in the preparation of the manuscript, and once again Irene Jeryn of the Hospital for Sick Children Medical Library and her staff have been unfailingly helpful. Finally, I wish to thank the W. B. Saunders Company and its medical editor Dana Dreibelbis for their continuing encouragement and good will.

MARGARET W. THOMPSON

CONTENTS

1 INTRODUCTION

The place of genetics in medicine was not always as obvious as it is today. Though the significance of genetics both for the conceptual basis of medicine and for clinical practice is now generally appreciated, not many years ago the subject was thought to be concerned only with the inheritance of trivial, superficial and rare characteristics; the fundamental role of the gene in basic life processes was not understood. The discovery of the principles of heredity by the Austrian monk Gregor Mendel in 1865 received no recognition at all from medical scientists and virtually none from other biologists. Instead, his reports lay unnoticed in the scientific literature for 35 years. Charles Darwin, whose great book *The Origin of Species* (published in 1859) emphasized the hereditary nature of variability among members of a species as an important factor in evolution, had no idea how inheritance operated. At that time heredity was thought to involve blending of the traits of the two parents, and Lamarck's idea of the inheritance of acquired characteristics was still accepted. Mendel's work could have clarified Darwin's concept of the mechanism of inheritance of variability, but although Mendel was aware of Darwin's work, Darwin seems never to have known of the significance of Mendel's research or even of its existence. Darwin's cousin Francis Galton, one of the great figures of early medical genetics, also remained ignorant of Mendel's work despite its relevance to his own studies of "nature and nurture." Mendel himself, perhaps discouraged by the results of later, less favorably designed experiments, eventually abandoned experimental research, though his interest in biological science remained strong throughout his life.

Mendel's laws, which form the cornerstone of the science of genetics, were derived from his experiments with garden peas, in which he crossed pure lines differing in one or more clear-cut characteristics and followed the progeny of the crosses for at least two generations. The three laws he derived from the results of his experiments may be stated as follows:

1. **Unit inheritance.** Before Mendel's time, the characteristics of the parents were believed to blend in the offspring. Mendel clearly stated that blending did not occur and the characteristics of the parents, though they might not be expressed in the first-generation offspring, could reappear quite unchanged in a later generation. Modern teaching in genetics places little stress upon this law, but in Mendel's time it was an entirely new concept.

2. **Segregation.** The two members of a single pair of genes are never found in the same gamete but instead always segregate and pass to different gametes. In exceptional circumstances, when the members of a chromosome pair fail to segregate normally, this rule is broken. The typical consequence of such a failure is severe abnormality.

3. **Independent assortment.** Members of different gene pairs assort to the gametes independently of one another. In other words, there is random recombination of the paternal and maternal chromosomes in the gametes.

With the dawn of the new century, the rest of the scientific community was ready to catch up with Mendel. By a curious coincidence, three workers (de Vries in Holland, Correns in Germany and Tschermak in Austria) independently and simultaneously rediscovered Mendel's laws. The development of genetics as a science dates not from Mendel's own paper but from the papers that reported the rediscovery of his laws.

The universal nature of Mendel's laws was soon recognized; as early as 1902 Garrod, who ranks with Galton as a founder of medical genetics, could report in alcaptonuria the first example of what is now often called Mendelian inheritance in man. In his paper Garrod generously admitted his debt to the biologist Bateson, who had seen the genetic significance of consanguineous marriage in the parents of some patients with what Garrod called "inborn errors of metabolism." This is the first clear evidence of interaction in research between medical and nonmedical geneticists, which has continued to the present day and has contributed greatly to the rapid development of the field.

A growing understanding of the universal nature of the biochemical structure and functioning of living organisms has brought about an awareness of the crucial role of genes in life processes. The work of Garrod foreshadowed this knowledge, though in the early years of genetics its fundamental significance was not apparent. The concept was formulated clearly by Beadle and Tatum in 1941 as the "one gene–one enzyme" hypothesis. Human biochemical genetics, the analysis of human variation both in normal biochemical attributes and in hereditary metabolic diseases, has become one of the major themes in medical genetics.

In the late nineteen-fifties, the scientific study of human chromosomes became possible and the role of chromosome abnormalities as causes of malformation, retardation, infertility and reproductive failure began to be explored. More recently, the mapping of many human genes to their chromosomal locations has been accomplished, and exploration of the human gene map is proceeding rapidly.

Since the middle of the nineteen-seventies, a quite different approach to problems in medical genetics has developed, aided by powerful new technologies for the manipulation and analysis of DNA. The field of human molecular genetics obviously has great potential for research into the human genome and for application to prevention of disease, perhaps even to gene therapy. Molecular geneticists speculate that perhaps within a few years the molecular structure of the entire human genome will have been defined.

The expansion and applications of genetic knowledge have had fruitful consequences for clinical medicine. It is estimated that today at least one-third of the children in pediatric hospitals are there because of genetic disorders. This is a great change from the early years of the century, and even from the preantibiotic era. Before the days of immunization, improved nutrition and antibiotics, most hospitalized children had infectious diseases or nutritional disorders such as rickets. Today some of those with infections have genetic defects that impair their resistance and, at least in developed countries, most cases of rickets arise not from faulty nutrition but from deleterious genes. Lifesaving advances in clinical techniques for the management of medical emergencies (such as transfusion, tube feeding and maintenance of body fluids by intravenous infusion) play parts in increasing the probability of survival, thus raising the prevalence of genetic defects. Improvements in surgical procedures have also contributed to an increase in the prevalence of genetic defects during the twentieth century.

Though it grew up in close association with pediatrics, medical genetics is also relevant to many other branches of medicine. One of its most recent applications has been in obstetrics, where prenatal diagnosis of certain genetic defects has become an important aspect of antenatal care. In adult medicine it is increasingly obvious that many common conditions, such as coronary heart disease, hypertension and diabetes mellitus, have important genetic components and that preventive medicine could be much more efficient if it could be directed toward special high-risk groups rather than toward the general population.

Classification of Genetic Disorders

In medical practice, the chief significance of genetics is its role in the etiology of a large number of disorders. Virtually any trait is the result of the combined action of genetic and environmental factors, but it is convenient to distinguish between those disorders in which defects in the genetic information are of prime importance, those in which environmental hazards (including those of the intrauterine environment) are chiefly to blame and those in which a combination of genetic constitution and environment is responsible. In the first category, conditions in which the genetic information is faulty, three main types are recognized:

1. Single-gene disorders
2. Chromosome disorders
3. Multifactorial disorders

Single-gene defects are caused by mutant genes. The mutation may be present on only one chromosome of a pair (matched with a normal gene on the homologous chromosome) or on both chromosomes of the pair. In either case, the cause of the defect is a single major error in the genetic information. Single-gene disorders usually exhibit obvious and characteristic pedigree patterns. Most such disorders are rare, with a frequency of 1 in 2000 or less.

In **chromosome disorders**, the defect is not due to a single mistake in the genetic blueprint but to developmental confusion arising from an excess or deficiency of whole chromosomes or chromosome segments, which upsets the normal balance of the genome. For example, the presence of a specific extra chromosome, chromosome 21, produces a characteristic disorder, Down syndrome, even though all the genes on the extra chromosome may be quite normal. As a group, chromosome disorders are very common, affecting about seven individuals per thousand live births and accounting for about half of all spontaneous first-trimester abortions.

Multifactorial inheritance is seen in a number of common disorders, especially developmental disorders resulting in congenital malformation. Here again there is no one major error in the genetic information but rather a combination of small variations that together can produce a serious defect. Environmental factors may also be involved. Multifactorial disorders tend to recur in families but do not show the characteristic pedigree patterns of single-gene traits.

Not all disorders that affect more than one member of a family are genetic. On the contrary, occasionally a clearly definable environmental cause (for example, an infection or teratogen) may affect more than one member of a family at a time. Neel and Schull (1954) provided a useful list of indications that a condition has a genetic rather than environmental etiology:

1. The occurrence of the disease *in definite proportions* among persons related by descent, when environmental causes can be ruled out.

2. The failure of the disease to appear in unrelated lines (e.g., in spouses or in-laws).

3. A characteristic onset age and course, in the absence of known precipitating factors.

4. Greater concordance in monozygotic than in dizygotic twins.

The foregoing list was prepared some years before the role of chromosomal disorders was known. Now a fifth criterion can be added:

5. The presence in the patient of a characteristic phenotype, often including mental retardation, and a demonstrable chromosomal abnormality, with or without a family history of the same or related disorders.

The Family History

Taking a comprehensive history is an important first step in the analysis of any disorder, even if it is not obviously genetic. According to Childs (1982), "to fail to take a good family history is bad medicine and someday will be criminal negligence." Why is a comprehensive family history necessary, not only to the clinical geneticist but to clinicians in other fields as well?

1. The family history can be helpful in diagnosis.

2. It can show whether the disorder is genetic in origin.

3. It can provide information about the natural history of the disease and about variation in its expression.

4. It can clarify the pattern of inheritance, indicate which other family members are at risk and allow the risk of recurrence in those persons to be estimated.

What constitutes an adequate family history? Beginning with the patient, information should be recorded about his or her relatives in the various family branches at least as far as grandparents, great-uncles and great-aunts, and first cousins. The information should include names, dates of birth and death, and present and past medical conditions. Early infant deaths, stillbirths and abortions should be noted. Consanguinity of the parents or grandparents and geographic or ethnic origin should be documented. If the patient is a child, information about the mother's pregnancy should be recorded, with particular attention to very early events—for example, maternal infections and metabolic disorders, medications and use of alcohol. It may be necessary to obtain medical records, a procedure which requires consent from the family members whose records are pertinent.

It is useful to summarize the family history as a pedigree (see Chapter 4), which is essentially a method of recording genetic information in a form that can be rapidly and unambiguously interpreted.

The Genetic Point of View

By the early nineteen-eighties, a course in medical genetics was part of the curriculum of almost all North American medical schools. Though in most schools the time commitment is modest (about 0.5 percent of the total hours), it should be enough to introduce students to the genetic point of view.

Genetics has immediate relevance to clinical medicine. Because genetic disorders are so numerous and varied, practitioners in every area of medicine will inevitably be involved with the diagnosis, management and prevention of such conditions and will have to be aware of the potential and limitations of the genetic approach. Moreover, there are many diseases of adult life in which genetic variants are factors that play a part in determining individual susceptibility to a particular condition. The relationship of HLA to autoimmune disorders is an example. With the present-day trend toward a lower birth rate and an aging population, such disorders will require an increasing share of medical attention.

Genetics deals with variation; a fundamental aspect of the genetic approach to disease is an appreciation of human variation: its nature and extent, its origin and maintenance, its distribution in families and populations, its interaction with environment and its consequences for normal development and homeostasis. These matters, which are "basic science" rather than clinical aspects of genetics, bear on the development of public health policy as well as on the delivery of health care at the individual level.

Perhaps the most significant current theme in medical genetics is the new molecular technology, which already has shown itself to have major practical benefits for preventive medicine, especially in prenatal diagnosis. Here again, some understanding of the nature, potential, limitations and pitfalls of the new genetics is important both for application to individuals and for public health policy.

During the professional years of today's medical students, it is likely that the human genome will be completely mapped, more and better genetic methods of prevention will be developed and the replacement of some defective genes will become possible. Susceptibility to common disorders, including malignancy, may become understood and thus these conditions may become predictable or even preventable. All these advances could follow from current research activity in medical genetics. For these reasons, the language and concepts of genetics and the genetic perspective of health and disease are essential components of medical education.

GENERAL REFERENCES

Bodmer WF, Cavalli-Sforza LL. Genetics, evolution, and man. San Francisco: W. H. Freeman, 1976.

Cavalli-Sforza LL, Bodmer WF. The genetics of human populations. San Francisco: W. H. Freeman, 1971.

Connor JM, Ferguson-Smith MA. Essential medical genetics. Oxford: Blackwell Scientific Publications, 1984.

Emery AEH. Elements of medical genetics, 6th ed. Edinburgh: Churchill Livingstone, 1983.

Emery AEH, Rimoin DL, eds. Principles and practice of medical genetics. Edinburgh: Churchill Livingstone, 1983.

Harper PS. Practical genetic counselling, 2nd ed. Bristol: Wright and Littleton: Wright, PSG, 1984.

Harris H. The principles of human biochemical genetics, 3rd ed. Amsterdam, New York, Oxford: Elsevier/North–Holland Biomedical Press, 1980.

Harris H, Hirschhorn K, eds. Advances in human genetics, Vols. 1–14, 1970–1985.

McKusick VA. Mendelian inheritance in man. Catalogs of autosomal dominant, autosomal recessive and X-linked phenotypes, 6th ed. Baltimore: Johns Hopkins, 1983.

Morton NE. Outline of genetic epidemiology. Basel: Karger, 1982.

Nora JL, Fraser FC. Medical genetics, principles and practice, 2nd ed. Philadelphia: Lea & Febiger, 1981.

Novitski E. Human genetics, 2nd ed. New York: Macmillan, 1982.

Sutton HE. An introduction to human genetics, 3rd ed. Philadelphia: Saunders College, 1980.

Vogel F, Motulsky AG. Human genetics, problems and approaches. New York, Heidelberg, Berlin: Springer-Verlag, 1979.

Weatherall DJ. The new genetics and clinical practice. London: Nuffield Provincial Hospitals Trust, 1982.

2

THE CHROMOSOMAL BASIS OF HEREDITY

When a cell divides, the nuclear material (chromatin) loses the relatively homogeneous appearance characteristic of nondividing cells and condenses to form a number of rod-shaped organelles, which are called **chromosomes** (*chroma*, color; *soma*, body) because they stain deeply with certain biological dyes. Units of genetic information (**genes**) are encoded in the deoxyribonucleic acid (**DNA**) of the chromosomes.

Each species has a characteristic chromosome constitution (**karyotype**) with respect to chromosome number and morphology. The gene map for the chromosome set is also characteristic of the species. The genes are arranged along the chromosomes in linear order, each gene having a precise position or **locus**. Genes that have their loci on the same chromosome are said to be **linked** or, more precisely, **syntenic** (in synteny). Alternative forms of a gene that can occupy the same locus are called **alleles**. Any one chromosome bears only a single allele at a given locus, though in the population as a whole there may be multiple alleles, any one of which can occupy that locus.

The **genotype** of an individual is his genetic constitution; the term is usually used with reference to a single locus. The **phenotype** is the expression of the genotype as a morphological, biochemical or physiological trait. The term **genome** refers to the full DNA content of the chromosome set.

Little was known about human cytogenetics until 1956 when Tjio and Levan developed effective techniques for chromosome study and found the normal human chromosome number to be 46, not 48 as had been previously believed. Since that time much has been learned about human chromosomes, their molecular composition, their numerous and varied abnormalities and the locations of the genes they contain. Because of the importance of cytogenetics in diagnosis, chromosome laboratories have been established in many major hospitals.

The Human Chromosomes

The 46 chromosomes of normal human somatic cells constitute 23 homologous pairs. The members of a homologous pair match with respect to the genetic information each carries; that is, they have the same gene loci in the same sequence, though at any one locus they may have either the same or different alleles. One member of each chromosome pair is inherited from the father, the other from the mother, and one of each pair is transmitted to each child. Twenty-two pairs are alike in males and females and hence are called **autosomes.** The two **sex chromosomes,** the remaining pair, differ in males and females and are

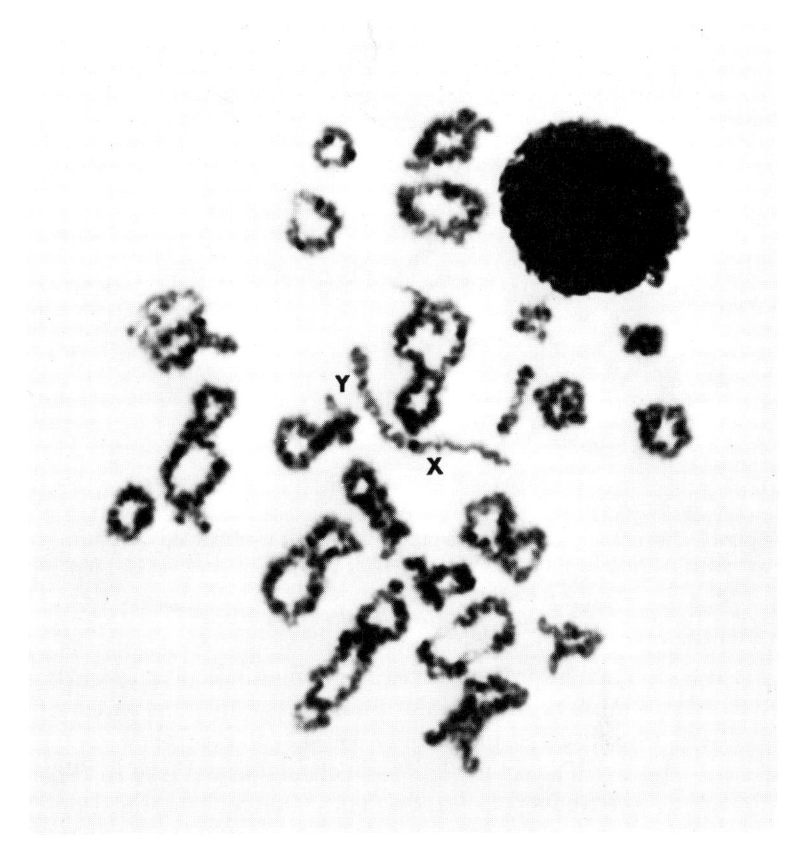

Figure 2–1. Meiosis in the human male. Note 23 chromosome pairs (bivalents), chiasmata in the bivalents, centromeres of individual chromosomes and terminal association of the X and Y. Photomicrograph courtesy of A. Chen.

of major importance in sex determination. Normally the members of a pair of autosomes are microscopically indistinguishable, and the same is true of the female sex chromosomes, the X chromosomes. In the male, the sex chromosomes are different from one another. One is an X, identical to the Xs of the female, inherited from the mother and passed on to daughters; the other, the Y chromosome, is inherited from the father and passed on to sons. Because the human X and Y pair at meiosis through a short region at the tip of the short arm of each (Fig. 2–1), it is inferred that they possess a short homologous segment.

There are two kinds of cell division—mitosis and meiosis. **Mitosis** is ordinary cell division by which the body grows and replaces dead or injured cells. It results in two daughter cells that are precisely identical to the parent cell in chromosome complement and genetic information. **Meiosis** occurs only once in a life cycle and results in the production of reproductive cells (**gametes**), each of which has a complement of 23 chromosomes. Somatic cells are said to have the **diploid** or 2n chromosome complement (*diploos*, double), whereas gametes have the **haploid** or n chromosome complement (*haploos*, single). Though a few specialized cell types are polyploid, and abnormal chromosome numbers can arise both in somatic cells and gametes by accidents of cell division, the general rule is that somatic cells are diploid and gametes are haploid. The terms haploid genome or diploid genome are also used to refer, respectively, to the full set of genes in a haploid or diploid chromosome set.

CHROMOSOME CLASSIFICATION

When prepared for analysis, the chromosomes of a human metaphase cell appear under the microscope as a **chromosome spread** (Fig. 2–2). To analyze

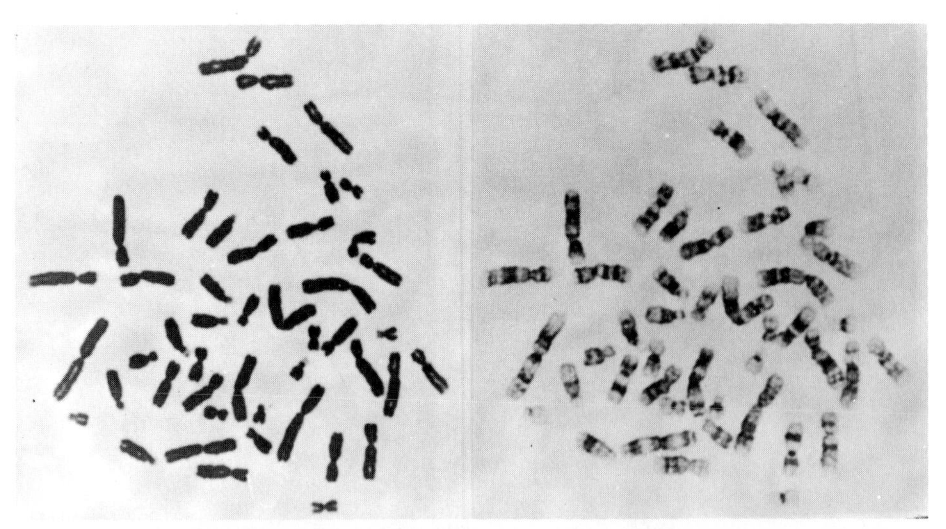

Figure 2–2. A chromosome spread prepared from a lymphocyte culture. The same cell is shown with solid staining (left) and Giemsa banding (right). Photomicrograph courtesy of R. G. Worton.

such a spread, the chromosomes are cut out from a photomicrograph and arranged in pairs in a standard classification. This process is called **karyotyping,** and the completed picture is a karyotype.

The original classification was devised in 1960 at a meeting of cytogeneticists in Denver, Colorado. The "Denver classification" distinguished seven chromosome groups, identified by the letters A through G on the basis of their overall length and centromere position. With more advanced staining methods the chromosomes can be individually identified; classification by groups alone is little used today.

The location of the **centromere,** or primary constriction, is a constant feature of each chromosome. Human chromosomes can be classified by centromere position into three types: metacentric if the centromere is central, submetacentric if it is somewhat off-center, and acrocentric if it is near one end. A fourth type, telocentric, which has a terminal centromere, does not occur in man.

The human acrocentric chromosomes (those of the D and G groups) have small masses of chromatin known as **satellites** attached to their short arms by narrow stalks (secondary constrictions). The stalks contain the genes for 18S and 28S ribosomal RNA (rRNA), each chromosome containing about 40 copies of ribosomal genes in tandem array. The 18S and 28S rRNA transcribed from these genes, together with 5S rRNA transcribed from elsewhere in the genome, is used in the nucleolus for the synthesis of ribosomes. Ribosomes are then transferred to the cytoplasm, where they play an essential role in protein synthesis.

Since the Denver Conference there have been several other conferences on standardization of chromosome classification. The nomenclature has been extended, each chromosome being identified unequivocally by its banding pattern and each band numbered according to a standard system. The Paris classification, which with minor modifications is in current use, is shown in Figure 2–3.

Symbols for Chromosome Nomenclature

With the accumulation of knowledge concerning numerical and structural aberrations of the chromosomes, it has become necessary to devise a set of sym-

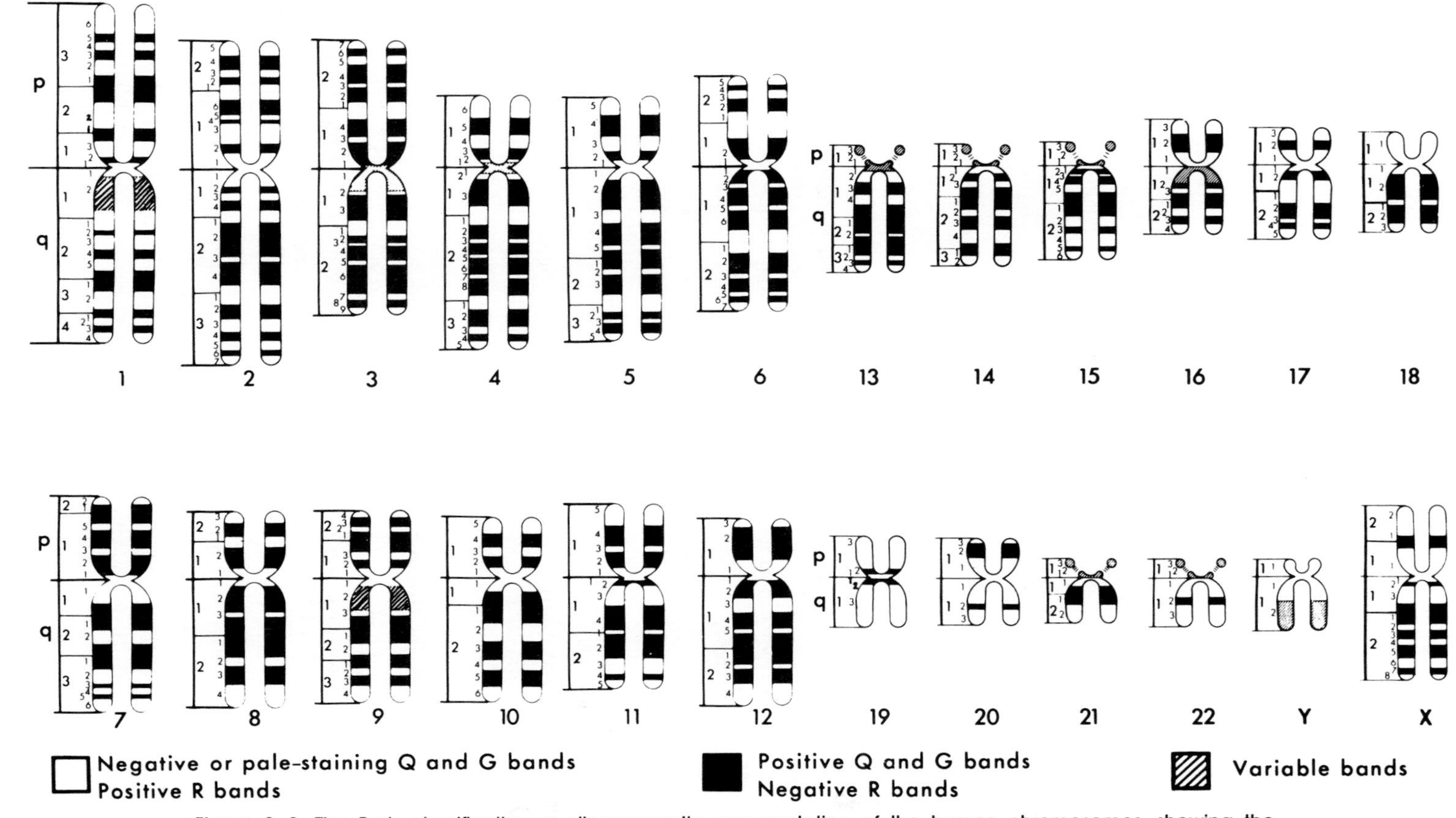

Figure 2–3. The Paris classification: a diagrammatic representation of the human chromosomes, showing the banding patterns and numbering scheme adopted at the Paris Conference. Reprinted from Paris Conference (1971). Standardization in human cytogenetics. Birth Defects: 1972;8(7).

Table 2–1. SYMBOLS FOR CHROMOSOME NOMENCLATURE (Partial List)

Symbol	Meaning
A–G	The chromosome groups
1–22	The autosome numbers
X,Y	The sex chromosomes
/	Diagonal line indicates mosaicism, e.g., 46/47 designates a mosaic with 46-chromosome and 47-chromosome cell lines
del	Deletion
der	Derivative of chromosome
dup	Duplication
i	Isochromosome
ins	Insertion
inv	Inversion
mar	Marker chromosome
p	Short arm of chromosome
q	Long arm of chromosome
r	Ring chromosome
s	Satellite
t	Translocation
	rcp Reciprocal translocation rob Robertsonian translocation
ter	Terminal (may also be written as pter or qter)
→	From → to
+ or −	Placed before the chromosome number, these symbols indicate addition (+) or loss (−) of a whole chromosome; e.g., +21 indicates an extra chromosome 21, as in Down syndrome.
	Placed after the chromosome number, these symbols indicate increase or decrease in the length of a chromosome part; e.g., 5p− indicates loss of part of the short arm of chromosome 5, as in cri du chat syndrome.

For further details, see ISCN 1978.

bols to designate certain features of the karyotype. Table 2–1 lists some of the more commonly used symbols, most of which will be used later to describe chromosome abnormalities of various types.

CHROMOSOME TECHNIQUES

Cells for chromosome analysis must be capable of growth and rapid division in culture. The most readily accessible cells that meet this requirement are the white cells of the blood. (Red cells have no nucleus and therefore no chromosomes.) To prepare a short-term culture, a sample of peripheral blood is obtained and mixed with heparin to prevent clotting. It is then centrifuged at a carefully regulated speed so that the white blood cells form a distinct layer. Cells of this layer are collected, placed in a suitable tissue culture medium and stimulated to divide by the addition of a mitogenic (mitosis-producing) agent, phytohemagglutinin, which is an extract of red bean. The culture is then incubated until the cells are dividing well, usually for about 72 hours. When the cultured cells are multiplying rapidly, a very dilute solution of colchicine is added to the medium. This interferes with the action of the spindle by binding specifically to the tubulin of the spindle microtubules, and also prevents the centromeres from dividing. Because colchicine stops mitosis at metaphase, cells in metaphase accumulate in the culture. A hypotonic solution is then added to swell the cells and to separate the chromatids while leaving the centromeres intact. The cells are fixed, spread on slides and stained by one of several techniques. They are then ready for microscopic examination, photography and karyotype preparation.

Chromosome cultures prepared from peripheral blood have the disadvantage of being short-lived. Long-term cultures can be derived from other tissues, such as skin. A skin biopsy is a minor surgical procedure. The sample, which must include dermis, grows in culture and produces fibroblasts, which are elongated, spindle-shaped cells capable of continuous multiplication in vitro for many cell generations. These cells can be used for a variety of biochemical and histochemical studies as well as for chromosome analysis. Fetal cells from amniotic fluid (amniocytes) obtained by the procedure of amniocentesis can be cultured by a similar technique.

When the chromosomes of a cell have been karyotyped, the karyotype can be examined for abnormalities of number or structure. Though numerical abnormalities are easy to identify, detection of structural defects requires excellent technique and careful observation. Even then, many structural abnormalities may remain beyond the limits of analysis.

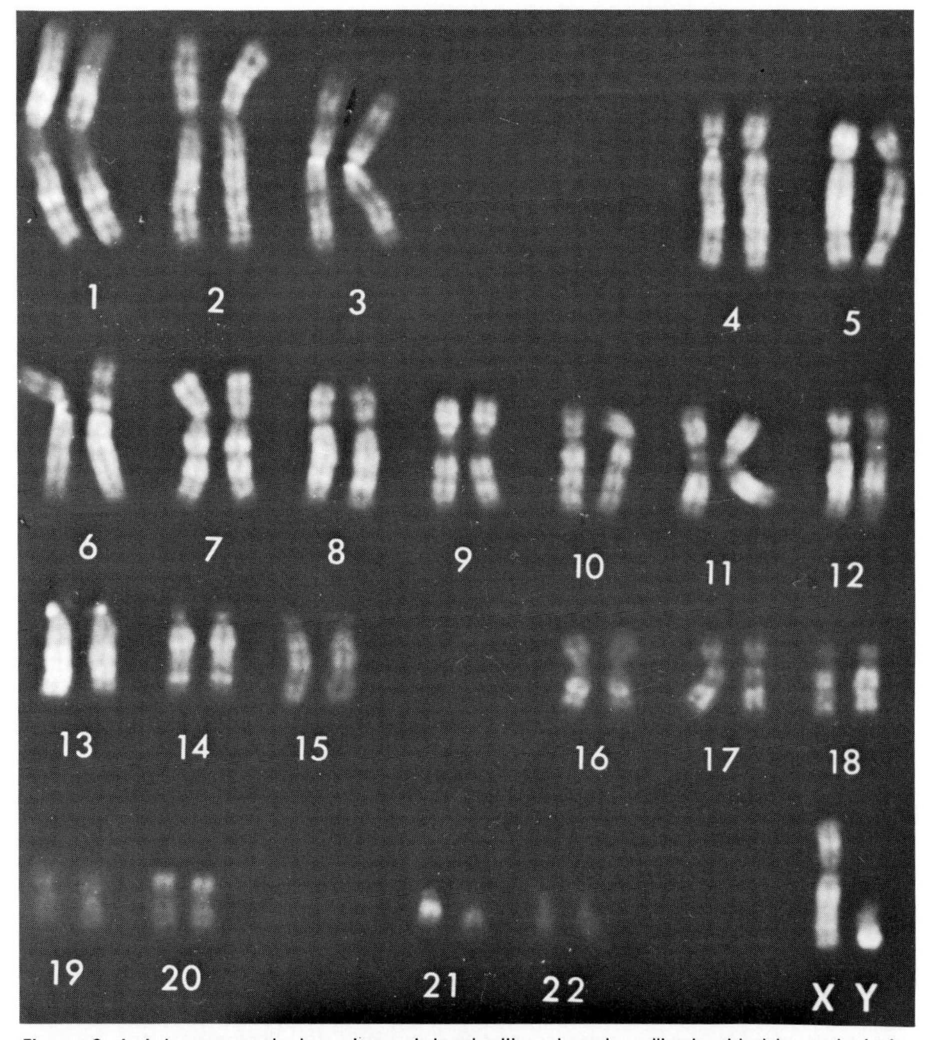

Figure 2–4. A human male karyotype stained with quinacrine dihydrochloride and photographed under fluorescent light to show Q bands. Photomicrograph courtesy of I. Uchida.

Staining Methods

Until 1970, solid staining (as shown in Fig. 2–2) was the only staining method available, but since that time several special techniques have been developed for staining chromosomes in banded patterns. Some but not all of the special staining methods are included in the following list.

Q Banding. Caspersson and his colleagues (1970) found that, when chromosomes are stained with quinacrine mustard or related compounds and examined by fluorescence microscopy, each pair stains in a specific pattern of bright and dim bands (Q bands), as shown in Figure 2–4. The Q bands were used as the reference bands for the standard classification.

G Banding. In this widely used technique chromosomes are treated with trypsin, which denatures chromosomal protein, then stained with Giemsa stain. The chromosomes take up stain in a pattern of dark and light staining bands (G bands), with the dark bands corresponding to the bright Q bands (Fig. 2–5).

R Banding. If the chromosomes receive a heat pretreatment, then Giemsa staining, the resulting dark and light stained bands (R bands) are the reverse of those produced by Q and G banding. R banding gives much the same information as Q or G binding but is less widely used.

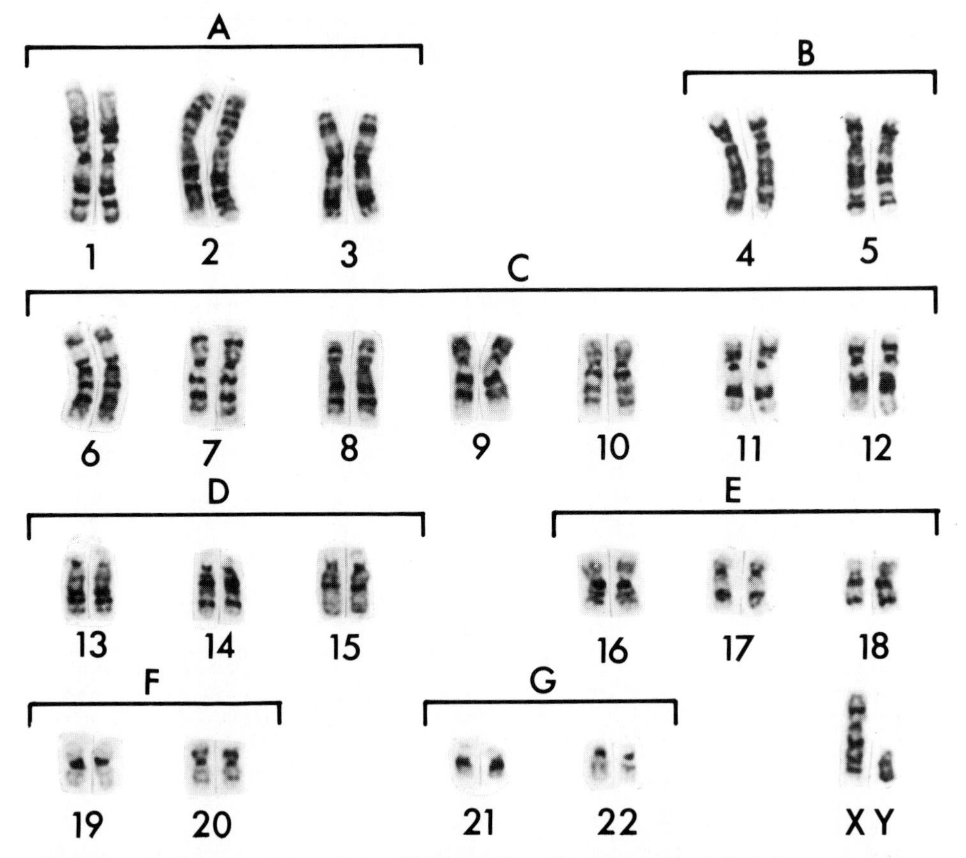

Figure 2–5. A human male karyotype with Giemsa banding (G banding). The chromosomes are individually labeled, and the seven groups A to G are indicated. Photomicrograph courtesy of R. G. Worton.

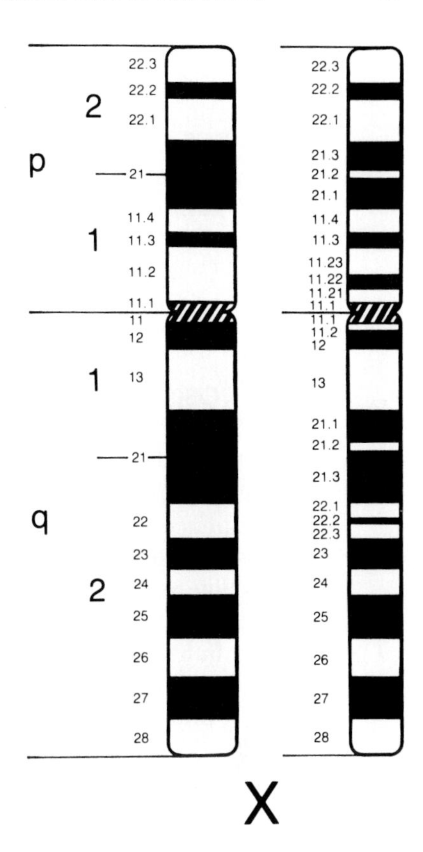

Figure 2–6. High-resolution banding. The G-banding pattern of the human X chromosome, as shown in the Paris classification (left) and at a less condensed stage (right). Reprinted from ISCN (1981); Birth Defects 17 (5). Also Cytogenet Cell Genet 1981; 31:1–23.

C Banding. This method specifically stains the centromere regions and other regions containing constitutive heterochromatin, that is, the secondary constrictions of chromosomes 1, 9 and 16 and the distal segment of the long arm of the Y chromosome. (Heterochromatin is chromatin that stains differently from the majority of the chromosomal material, which is euchromatin.)

NOR Staining. This method uses ammoniacal silver to stain the "nucleolar organizing regions," that is, the stalks of the satellited chromosomes, which contain the 28S and 18S ribosomal genes.

High-Resolution Banding

This technique, often called prophase banding, is becoming widely used in clinical genetics. Prophase and prometaphase chromosomes reveal a much larger number of bands (800 to 1400 in all) than can be seen in metaphase preparations (about 200) (Fig. 2–6). High-resolution banding can therefore be helpful in delineating precise breakpoints or demonstrating small alterations in chromosome structure that would not otherwise be observed. The technique requires blocking cultured cells in the S phase of the cell cycle, releasing the block and harvesting the culture at a time when the maximum number of cells are in late prophase or prometaphase.

Demonstration of the Fragile Site on the X Chromosome

"Fragile sites" are present on several human chromosomes, but the only one known at present to be clinically important is on the X chromosome, near the distal end of the long arm, in a proportion of cultured cells from males with a specific type of familial X-linked retardation and some carriers of the same genetic defect. Demonstration of the fragile site depends on culturing the cells under conditions of thymidine deprivation, either by using a medium low in thymidine and folic acid or by adding an inhibitor of thymidine synthetase to the culture.

The fragile site does not show up well when chromosomes are stained by the G-banding or Q-banding techniques used routinely in most chromosome laboratories. Instead, solid staining must be used.

Mitosis

In mitotic division, the cytoplasm of the cell apparently simply cleaves into two approximately equal halves, but the nucleus undergoes a complicated sequence of activities. Four stages of mitosis can be distinguished: prophase, metaphase, anaphase and telophase. These stages are shown diagrammatically in Figure 2–7. In this figure, each homologous chromosome pair has one member outlined and one in solid black to signify that one homologue is derived from the father, the other from the mother.

Interphase. A cell that is not actively dividing is said to be in interphase (Fig. 2–7A). Most of the metabolic activities of the cell, including DNA replica-

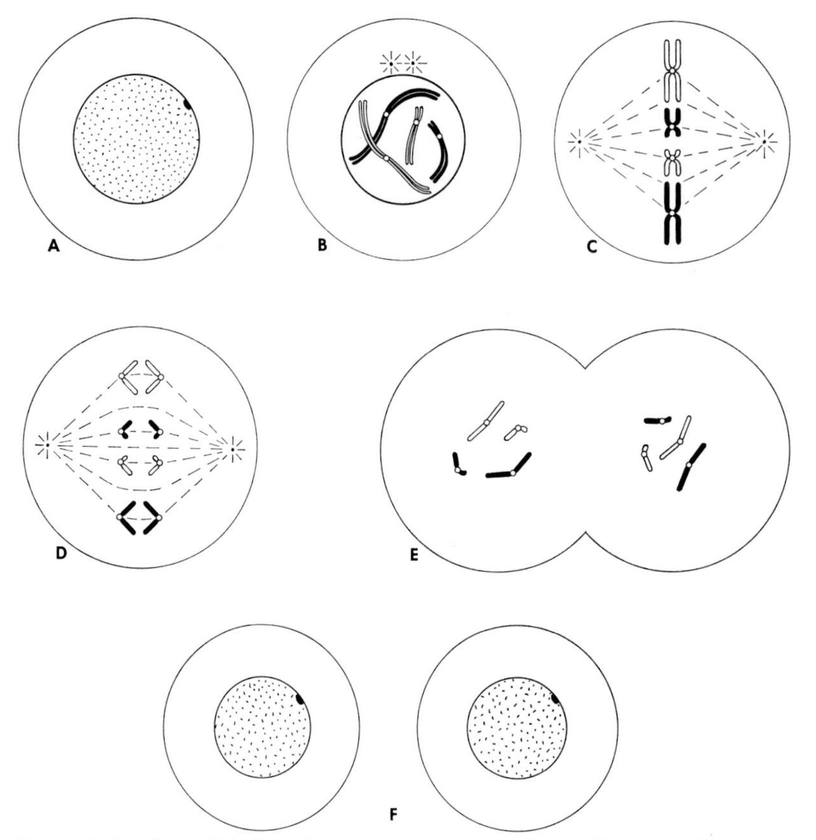

Figure 2–7. Mitosis. Only two chromosome pairs are shown. One parent's chromosomes are shown in outline, the other parent's in black. A, Interphase; B, prophase; C, metaphase; D, anaphase; E, telophase; F, interphase. For further details, see text.

tion, occur during this stage of the cell cycle. The total length of the extended interphase chromosomes has been estimated as 174 cm. In female cells the Barr body or sex chromatin (an inactive X chromosome) appears at this stage as a compacted mass of chromatin, though the other chromosomes are metabolically active and not individually distinguishable. As the cell prepares to divide, the chromosomes begin to condense by a complicated process of folding and coiling and thus become visible as deeply staining bodies. As soon as the appearance of the nucleus changes and the chromosomes begin to become visible, the cell has entered the first stage of cell division, prophase.

Prophase. When the chromosomes are discernible, but before there is any obvious pattern in their arrangement, the cell is in prophase (Fig. 2–7B). Each chromosome can now be seen to consist of a pair of long, thin parallel strands, or **chromatids** (sister chromatids), which are held together at one spot, the **centromere**. The nuclear membrane disappears, and the nucleus begins to lose its identity. Meanwhile the **centriole**, an organelle just outside the nuclear membrane, duplicates itself, and its two products migrate toward opposite poles of the cell.

Metaphase. When the chromosomes have reached their maximal contraction and maximal staining density, they move to the equatorial plane of the cell, which is now in metaphase (Fig. 2–7C). This is the stage at which chromosomes are most easily studied, because they are highly contracted, densely stained and arranged in a more or less two-dimensional **metaphase plate** along the equatorial plane. Meanwhile, the **spindle** has formed; this is a structure consisting of microtubules of protein (spindle fibers) that radiate from the centrioles at either pole of the cell up to the equatorial plane and from the **kinetochores**, attachment sites that are associated with the centromeres of the chromosomes.

Anaphase. The cell enters anaphase (Fig. 2–7D) when the centromeres divide and the paired chromatids of each original chromosome disjoin, becoming new **daughter chromosomes**. The spindle fibers contract and draw the daughter chromosomes, centromere first, to the poles of the cell. The molecular mechanism by which the spindle fibers draw the chromosomes apart is not fully understood. It is known that spindles contain actin, so actin-myosin interaction between the fibers may be involved.

Telophase. The arrival of the daughter chromosomes at the poles of the cell signifies the beginning of telophase (Fig. 2–7E), the final stage of the cell division. Concurrently with the onset of telophase, the division of the cytoplasm **(cytokinesis)** begins with the formation of a furrow in the area of the equatorial plane. Eventually a complete membrane is formed across the cell, which is thereby divided into two new cells with identical chromosome complements. Meanwhile, the chromosomes are unwinding and consequently stain less densely. Eventually they no longer stain as individual entities and again become enclosed by a nuclear membrane. Each daughter cell now appears as a typical interphase cell (Fig. 2–7F).

THE MITOTIC CYCLE

Mitotic division takes up only a small part of the life cycle of a cell. Three other stages of the mitotic cycle are recognized (Fig. 2–8). After division the new cell enters a postmitotic period during which there is no DNA synthesis. This

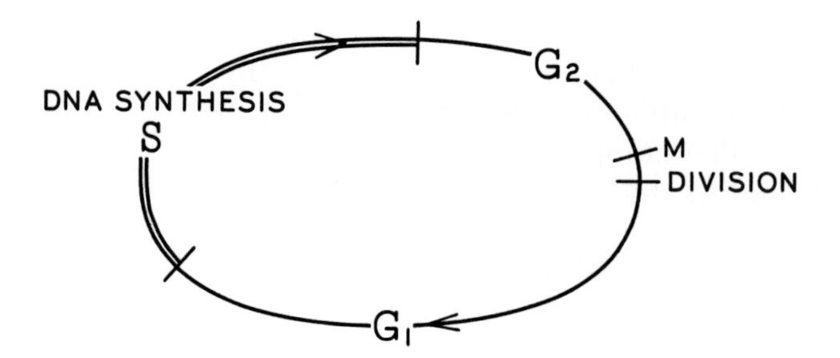

Figure 2–8. The mitotic cycle. For description, see text.

stage is called G_1 (Gap 1). The next stage is the S stage, the period of DNA synthesis, in which the DNA content of the cell doubles, each DNA molecule serving as a template to make a complementary copy of itself. There is then a premitotic non-synthetic period, G_2 (Gap 2), which is ended by the onset of mitosis. Typical studies of cultured human cells have shown that the complete cycle may last 12 to 24 hours, about one hour of which involves mitosis.

Information about the timing of DNA synthesis in cultured cells can be obtained by autoradiography or by the bromodeoxyuridine (BUdR) technique described later (see sister chromatid exchange). Autoradiography involves adding thymidine that has been labeled with the radioactive isotope of hydrogen, tritium (^3H), to the culture. Tritiated thymidine is taken up only by cells that are actively synthesizing DNA. Cells are cultured in the presence of ^3H thymidine, harvested and prepared for chromosome analysis. The chromosome slides are then covered

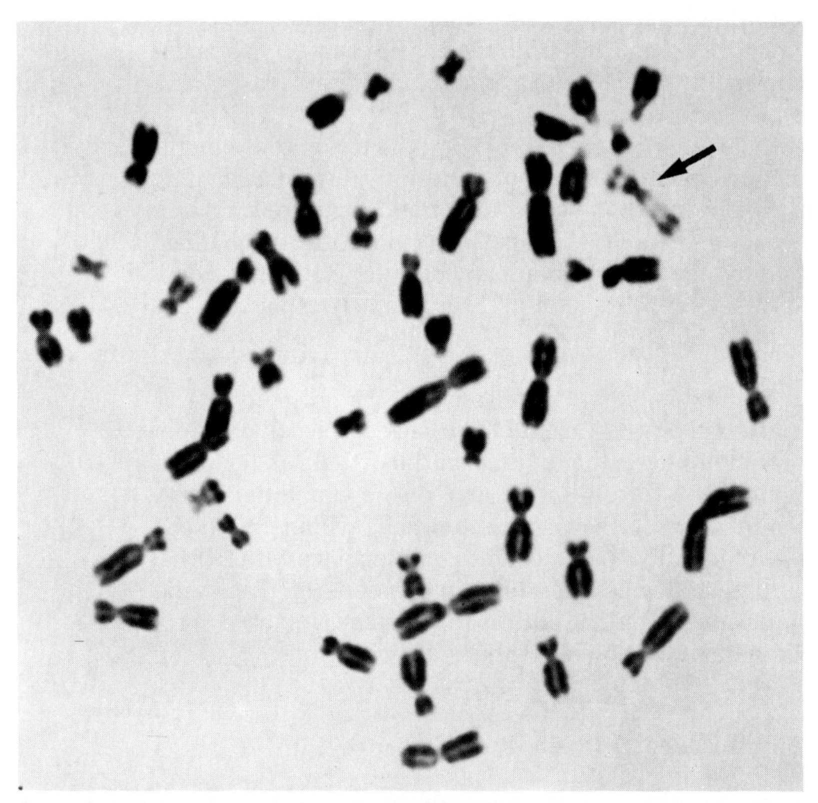

Figure 2–9. Metaphase plate from lymphocyte culture of human female after BUdR incorporation. The arrow indicates the late-replicating X chromosome, in which the light-stained area has incorporated BUdR into both chromatids. Photomicrograph courtesy of C. Verellen.

with a photographic emulsion and left in the dark for a time. When the slides are developed to make "autoradiographs," silver granules are seen only over chromosomes that have incorporated ^3H thymidine, that is, chromosomes that were actually synthesizing DNA while the radioactive material was present in the culture medium. By varying the time and duration of exposure of cell cultures to ^3H thymidine, the duration of the various stages of cell cycle has been determined. Not all chromosomes replicate simultaneously. In particular, as shown by BUdR incorporation in Figure 2–9, one of the two X's of the female is late-replicating. This X chromosome forms the sex chromatin body in interphase cells.

SOMATIC RECOMBINATION

Somatic recombination (crossing over between homologous chromosomes in mitosis) is much less common than meiotic recombination (described later) because homologous chromosomes are not usually closely associated except at meiosis. However, it is known from experimental work that somatic crossing over does occasionally take place. The consequences are shown in Figure 2–10. In an individual heterozygous for albinism, if nonsister but homologous chromatids that carry *A* and *a* undergo a crossover event between the centromere and the A locus, at anaphase the two chromatids with *A* alleles may pass to the

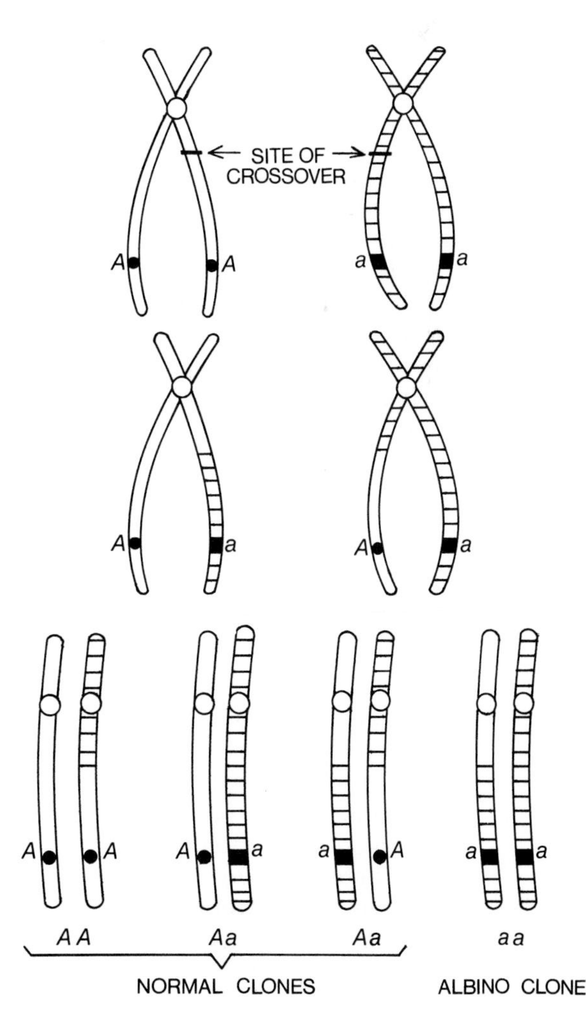

Figure 2–10. A possible consequence of somatic crossing over in a heterozygote for albinism, described in the text. The albino clone, if it included pigment cells, could produce a patch of albino tissue.

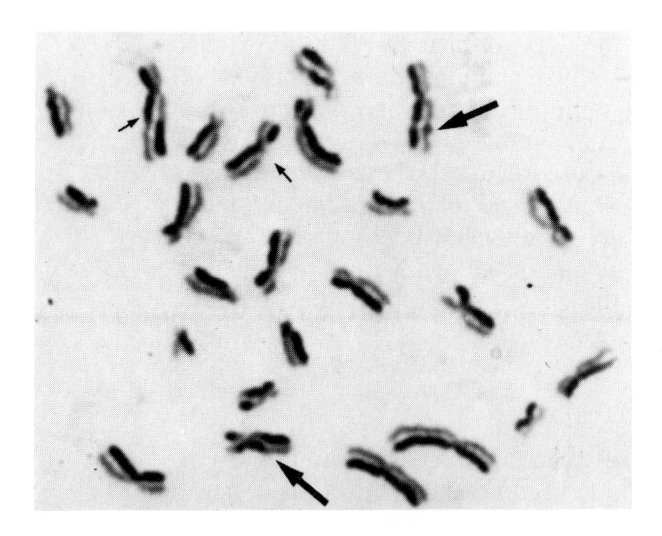

Figure 2–11. Sister chromatid exchange (SCE). These chromosomes are from a cultured lymphocyte prepared by growth in BUdR, as described in the text. Small arrows indicate two of the five chromosomes in which single exchanges have occurred, and larger arrows indicate chromosomes with two exchanges. Photomicrograph courtesy of R. G. Worton.

same daughter cell and the two with *a* alleles to the other daughter cell. (This also applies, of course, to any other locus between the crossover and the end of the chromosome.) The clone of cells descended from each daughter cell would all have the same genotype as the original recombinant.

SISTER CHROMATID EXCHANGE

Crossing over between the sister chromatids of a single chromosome at mitosis is a phenomenon that was first recognized by Taylor and associates in 1957 and has been widely studied since a special technique was developed by Latt in 1973. Cultured cells are allowed to replicate twice in bromodeoxyuridine (BUdR), allowing incorporation of BUdR into newly synthesized DNA in the place of thymine. BUdR modifies the staining properties of the chromatids; for example, the fluorescent stain Hoechst 33258 stains the chromatid in which only one strand is BUdR-substituted. If a sister chromatid exchange (SCE) has occurred, this is readily recognized by the bright and dim fluorescence patterns along the chromatids (Fig. 2–11). The frequency of SCE is greatly increased in a particular genetic disorder, Bloom syndrome. At present, the relationship of the curious increase in chromosome breakage and SCE frequency to the other features of the syndrome is not clear (see chromosome breakage syndromes, Chapter 6).

Meiosis

Meiosis is the special type of cell division by which gametes are formed. Each daughter cell formed by meiosis has the haploid chromosome number, with one representative of each chromosome pair. This is in contrast to mitosis, in which each daughter cell is identical in chromosome complement to the parent cell. Some of the stages distinguished in meiosis are illustrated diagrammatically in Figures 2–12 and 2–13.

There are two successive meiotic divisions. In meiosis I, the reduction division, homologous chromosomes pair during prophase and disjoin from one another during anaphase, each chromosome's centromere remaining intact. Meiosis II follows meiosis I without DNA replication but, as in ordinary mitosis,

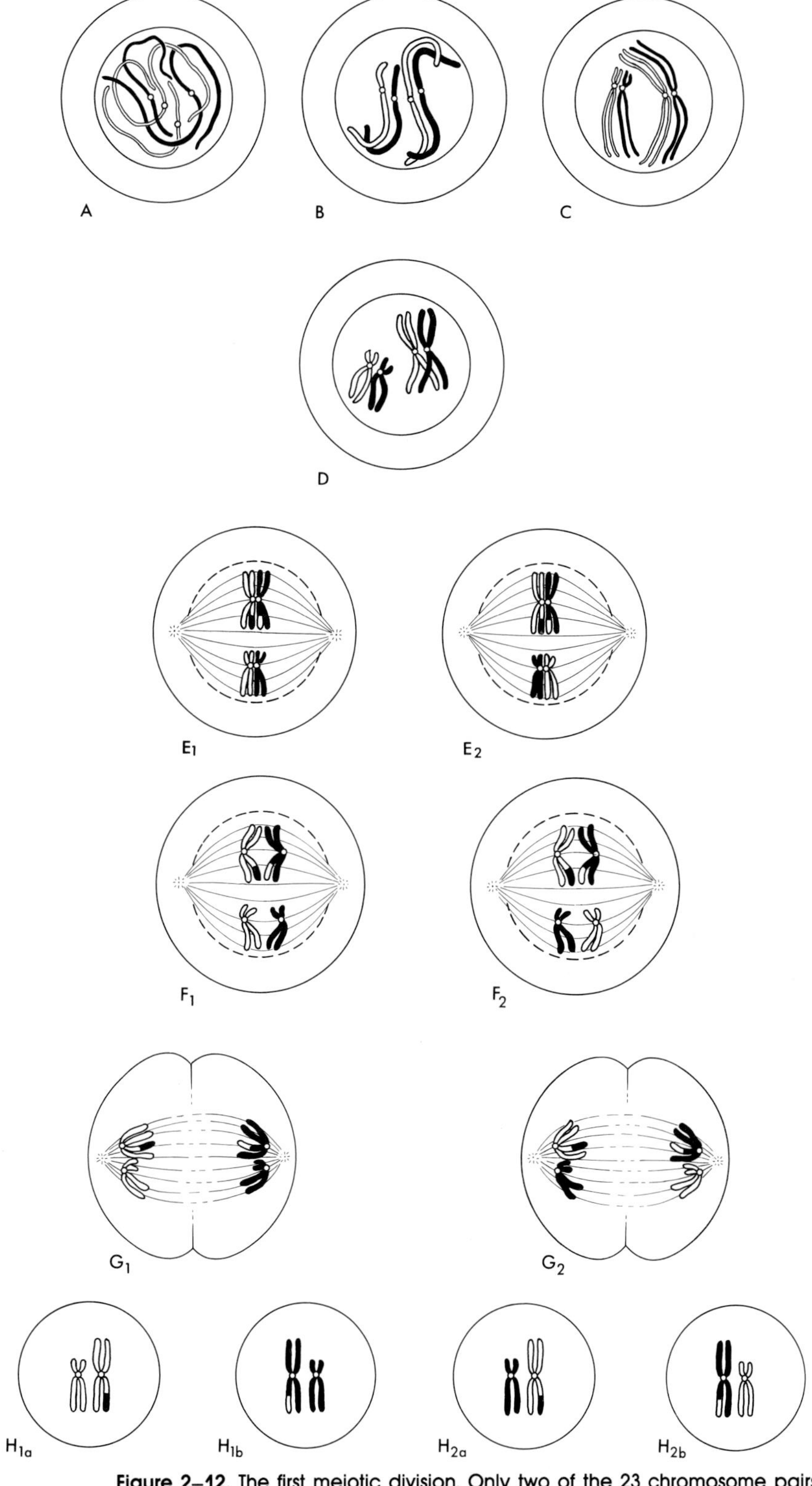

Figure 2–12. The first meiotic division. Only two of the 23 chromosome pairs are shown; chromosomes from one parent are shown in outline, from the other in black. A, Leptotene; B, zygotene; C, pachytene; D, diplotene; E_1 and E_2, metaphase; F_1 and F_2, early anaphase; G_1 and G_2, late anaphase; H_{1a}, H_{1b}, H_{2a}, H_{2b}, telophase. One possible distribution of the parental chromosome pairs is shown in illustrations E_1 to H_1, the alternative combination in illustrations E_2 to H_2.

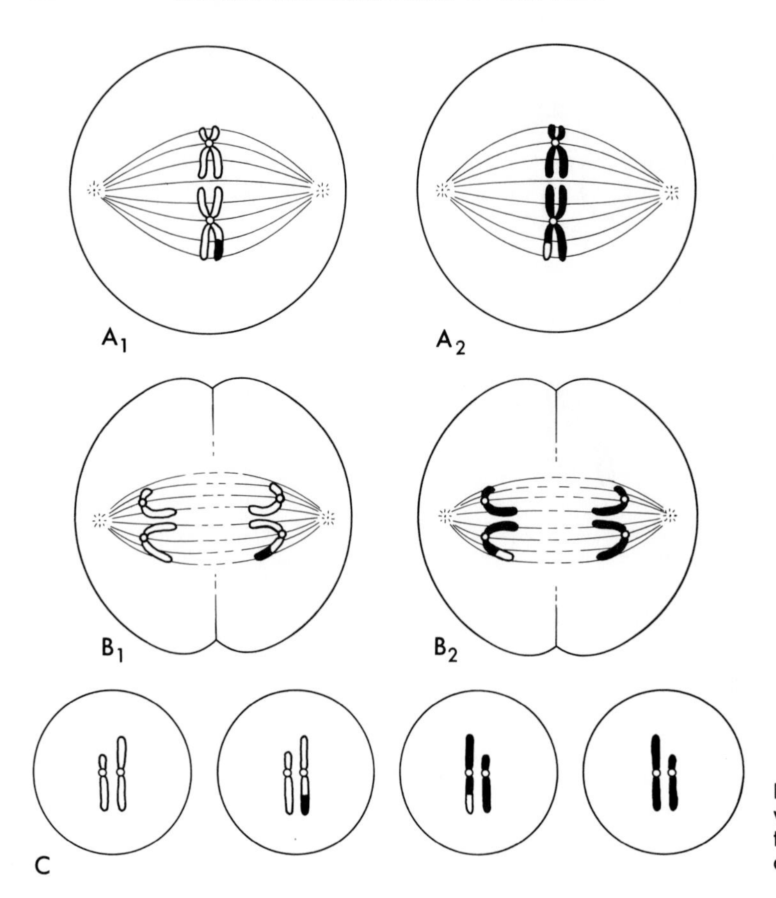

Figure 2–13. The second meiotic division. A, metaphase; B, anaphase; C, telophase. A_1 and A_2 represent H_{1a} and H_{1b} of Figure 2–12.

the centromere of each chromosome divides and the chromatids segregate from one another.

THE FIRST MEIOTIC DIVISION (MEIOSIS I)

Prophase of Meiosis I

The prophase of meiosis I is a complicated process, with a number of important differences from the prophase of a mitotic division. Several stages are shown in Figure 2–12.

Leptotene. Leptotene (Fig. 2–12A) is characterized by the first appearance of the chromosomes, seen as thin threads that are beginning to condense. Although the DNA has duplicated before this stage, the threads still appear to be single on microscopic examination. Unlike mitotic chromosomes, they are not smooth in outline but consist of alternating thicker and thinner regions; the thicker regions, which are known as **chromomeres**, have a characteristic pattern for each meiotic chromosome.

Zygotene. Zygotene (Fig. 2–12B), as the name implies, is the stage of pairing (**synapsis**) of homologous chromosomes. The two members of each homologous pair lie parallel to one another in intimate, point-for-point association to form **bivalents**. *Pairing of homologous chromosomes does not occur in mitosis.* Unlike the homologous pairs, the X and Y chromosomes are associated only at the tips of their short arms (see Fig. 2–1).

Pachytene. Pachytene (Fig. 2–12C), is the main stage of chromosomal thickening. The chromosomes coil more tightly and stain more deeply, and the chromomeres become more pronounced. The bivalents (paired chromosomes) are in close association, and each chromosome is now seen to consist of two chromatids; each bivalent is a **tetrad** of four strands. This is the stage at which crossing over (exchange of homologous segments between two of the four strands) occurs.

Diplotene. Diplotene (Fig. 2–12D) is recognizable by the longitudinal separation that begins to appear between the two components of each bivalent. Although the two chromosomes of each bivalent separate, the centromere of each remains intact, so the two chromatids of each chromosome remain together. During the longitudinal separation the two members of each bivalent are seen to be in contact in several places, called **chiasmata** (*chiasma*, cross), only one of which is shown. Chiasmata mark the locations of crossovers, where chromatids of homologous chromosomes have previously exchanged material (see later discussion in this chapter). In human spermatocytes, the average number of chiasmata seen is about 50 (Hulten, 1974). Eventually the chromosomes draw apart and the chiasmata begin to terminalize (draw to the ends of the chromosome arms).

Diakinesis. Diakinesis, the final stage of prophase, is marked by even tighter coiling and deeper staining of the chromosomes and by terminalization of some chiasmata.

Metaphase, Anaphase and Telophase of Meiosis I

Metaphase I (Fig. 2–12E) begins, as in mitosis, when the nuclear membrane disappears and the chromosomes move to the equatorial plane. At anaphase I (Fig. 2–12F and G) the two members of each bivalent disjoin, one member going to each pole. The bivalents assort themselves independently of one another, so that the chromosomes received originally as a paternal and a maternal set are now sorted into random combinations of paternal and maternal chromosomes, with one representative of each pair going to each pole. The disjunction of paired homologous chromosomes is the physical basis of segregation, and the random assortment of paternal and maternal chromosomes in the gametes is the basis of independent assortment; thus the behavior of the chromosomes at the first meiotic division provides the physical basis for Mendelian inheritance. The parallel between the behavior of chromosomes and the transmission of inherited traits was first noted independently by Sutton in 1903 and Boveri in 1904, soon after the rediscovery of Mendel's laws.

By the end of meiosis (Fig. 2–12H) each product has the haploid chromosome number; hence meiosis I is often referred to as **reduction division.**

Crossing Over

As mentioned above, a constant feature of meiosis I is the presence of chiasmata, which hold paired chromosomes together from the diplotene stage through metaphase and mark the positions of crossovers, sites where prior to metaphase chromatids of homologous chromosomes have exchanged segments by breakage and recombination. Though only two chromatids take part in any one crossover event, all four may be simultaneously involved in different crossovers. Chiasmata may play an important biological role in holding bivalents together and thus preventing premature disjunction.

Because crossing over causes genes to become reorganized into new combinations, it increases genetic variability. As noted previously, the paternal and maternal chromosomes assort independently at meiosis, allowing for 2^{23} (about 8 million) different chromosome combinations in the gametes of a single individual. Crossing over markedly increases the number of different combinations possible, thus increasing the likelihood of favorable new combinations; it also allows favorable genes to become separated from deleterious new mutations that may arise in the same chromosome.

Linkage and crossing over are discussed further in Chapter 11.

THE SECOND MEIOTIC DIVISION (MEIOSIS II)

The second meiotic division follows upon the first without DNA replication and without a normal interphase. It resembles ordinary mitosis in that in each cell formed by meiosis I, the centromeres now divide and the sister chromatids disjoin, passing to opposite poles to produce two daughter cells (Fig. 2–13). With the exception of areas in which crossovers occurred during meiosis I, the daughter cells have identical chromosomes.

Thus the end result of the two successive meiotic divisions is the production of four haploid daughter cells, formed by only one doubling of the chromosomal material.

Human Gametogenesis

According to the theory of "continuity of the germ plasm," germ cells are potentially immortal and are set aside for their special role at the very beginning of development. This is probably not strictly true; in the mouse, mutations at the *W* (dominant spotting) locus lead not only to reduced numbers of germ cells but also to reduced numbers of hematopoietic stem cells and defective functioning of pigment cells, indicating that these three cell types share a common ancestry. In humans the earliest embryonic history of the primordial germ cells is unknown, but by the fourth week of development they are known to lie outside the embryo proper, in the endoderm of the yolk sac. From there, they migrate to the genital ridges and associate with somatic cells to form the primitive gonads, which later (at about 46 days after fertilization) differentiate into testes or ovaries in accordance with their genetic constitution.

SPERMATOGENESIS

Spermatogenesis occurs in the seminiferous tubules of the testis of the male from the time of sexual maturity onward. The process is shown diagrammatically in Figure 2–14. At the periphery of the tubules are **spermatogonia** of several types, ranging from the self-renewing stem cells of the series to more specialized forms committed to the pathway of sperm formation; the latter are derived from the stem cells by a series of mitoses. The last stage in this developmental sequence is the **primary spermatocyte**, the cell that undergoes meiosis I. It divides to form two **secondary spermatocytes,** each with 23 chromosomes. These cells rapidly undergo meiosis II, each forming two **spermatids.** The spermatids mature without further division into **sperm** (spermatozoa) that are released into

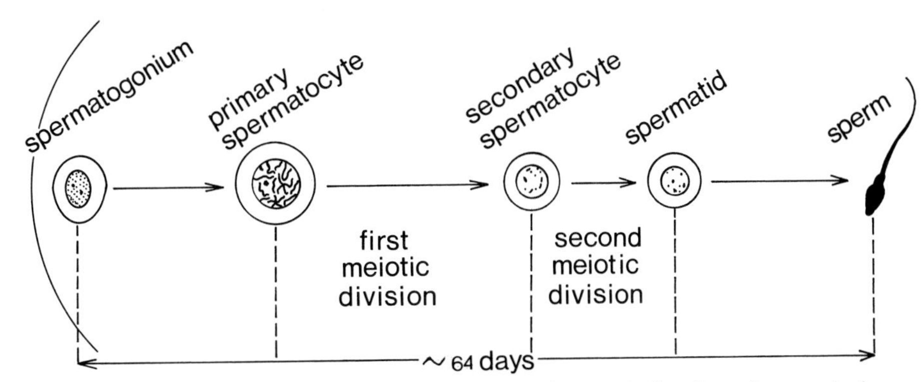

Figure 2–14. Diagram to illustrate human spermatogenesis. For discussion, see text.

the lumen of the tubule. The total time involved in all of the stages from the beginning of meiosis to the formation of the mature sperm is about 64 days.

Sperm are produced in enormous numbers, as many as 200 million per ejaculate. To provide such numbers over a long period of time, several hundred successive cell divisions are necessary. The older a man is when he becomes a father, the greater the number of DNA replications in the history of the germ cell he gives to the child, hence the greater the possibility that an error in DNA replication (a mutation) will occur.

OOGENESIS

In contrast to spermatogenesis, the process of oogenesis is largely complete at birth. The ova develop from **oogonia**, cells in the cortical tissue of the ovary that have originated from the primordial germ cells by a series of mitoses. Each oogonium is the central cell in a developing follicle. By about the third month of prenatal development the oogonia of the embryo have begun to develop into **primary oocytes**, some of which have already entered first meiotic prophase. The process is not synchronized, both early and later stages coexisting in the fetal ovary. The primary oocytes remain in "suspended prophase" (**dictyotene**) until sexual maturity is reached. Then, as each individual follicle begins to mature, the meiotic division of its oocyte resumes. Meiosis I is completed at about the time of ovulation, which may be more than 40 years after the beginning of the division.

As shown diagrammatically in Figure 2–15, the primary oocyte completes meiosis I in such a way that, while each daughter cell receives 23 chromosomes, one receives most of the cytoplasm and becomes the **secondary oocyte** and the

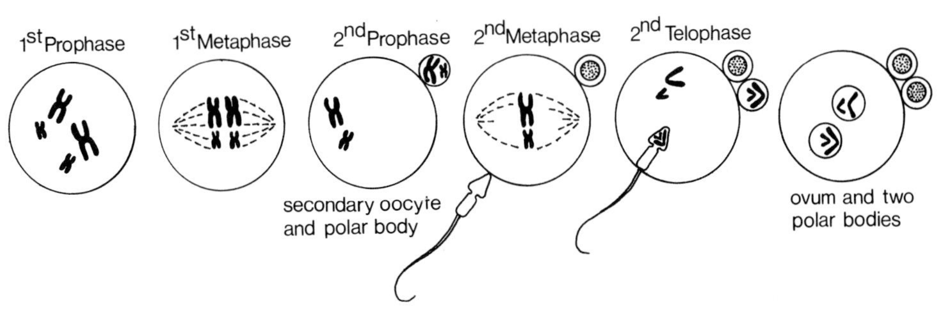

Figure 2–15. Diagram to illustrate human oogenesis and fertilization. For discussion, see text.

other becomes a **polar body.** The second meiotic division commences almost immediately and proceeds as the ovum passes into and down the uterine tube. It is not completed until after fertilization, which usually takes place (if it takes place at all) before the ovum reaches the uterus. The second meiotic division produces the mature **ovum** with virtually all the cytoplasm and a second polar body. The first polar body may also divide. The polar bodies are ordinarily incapable of forming embryos, though rare exceptions occur (see discussion of dispermic chimeras, Chapter 16).

The maximum number of germ cells present in the human female is about 7 million, found in the five-month fetus. By birth the number has dropped to 2 million and by puberty fewer than 200,000, only 3 percent of the original number, still remain. The total number of female germ cells could be formed by no more than 23 successive mitoses.

The differences between spermatogenesis and oogenesis have genetic significance. The long duration of meiotic prophase in females may be causally related to the increasing risk of meiotic nondisjunction (failure of paired chromosomes to disjoin) with increasing maternal age, whereas the opportunity for error in the numerous replications of genetic information that take place during spermatogenesis in mature males may account for the fact that the fathers of new mutants are often older than the average paternal age.

FERTILIZATION

The process of fertilization normally takes place in the uterine tube within a day or so of ovulation and requires about 24 hours. There is considerable speculation but little hard information about whether "aging" of the ovum or sperm in the uterine tube is harmful in any way to the development of the resulting child. Ordinarily, only a single ovum is released in any one menstrual cycle, whereas very large numbers of sperm may be present. Nevertheless, when one sperm penetrates the ovum a series of biochemical events is set into motion that normally prevents the entry of any other sperm.

After entering the ovum, the sperm head rounds up to form the **male pronucleus**. Meanwhile the second meiotic division of the ovum is being completed, producing the **female pronucleus** and the second polar body. The male and female pronuclei approach one another, lose their membranes and combine to form the **zygote**, with the restored diploid chromosome number. Though DNA synthesis has been entirely shut off in the ovum, soon after the entry of the sperm it recommences, each haploid chromosome set replicates and the combined 46 chromosomes divide as at any mitosis to form two 46-chromosome daughter cells. This is the first of the series of cleavage divisions that initiate the process of embryonic development.

Though development begins with the formation of the zygote (conception), in clinical medicine the duration of pregnancy is usually measured from the date of the mother's last menstrual period, which on the average is about 14 days earlier.

Medical Applications of Chromosome Analysis

Apart from their intrinsic genetic interest, chromosomes are important in medicine in a number of ways. Though the applications noted here are also discussed later, they are summarized here for convenience. Their significance may not be clear without additional background information, which is supplied in later chapters. These medical applications include the following:

1. Clinical Diagnosis. Chromosome studies are widely used in clinical diagnosis, especially in patients with congenital malformations involving several organ systems, mental retardation, failure to thrive or disorders of sexual development.

2. Linkage and Mapping. Chromosome studies are essential for the assignment of specific human genes to their linkage groups and chromosomal positions.

3. Polymorphisms. Minor heritable differences in banding patterns are not unusual, especially for chromosomes 1, 9 and 16 and the Y chromosome. The short arms and satellites of the acrocentric chromosomes also vary, both in fluorescence with Q banding and in size. These polymorphisms may be used to trace individual chromosomes through families. Thus they can serve as markers in family studies or for determination of the source of the abnormal gamete in chromosome abnormalities. As described later, a familial chromosome polymorphism (a secondary constriction) allowed Donahue and his colleagues (1968) to map the Duffy blood group locus to chromosome 1, thus making the first assignment of any gene to a specific autosome. In a triploid infant who survived for a few hours after birth, Uchida and Lin (1972) found that the Q banding pattern showned that the extra chromosome set had in all probability come from the father in a diploid sperm. Chromosome polymorphisms are being analyzed in trisomic children and in their parents to determine the relative frequency of nondisjunctional events in the first meiotic division as compared with the second, and in fathers as compared with mothers (Hassold et al., 1984).

4. Chromosomes and Neoplasia. One of the first applications of Q banding was the demonstration that the so-called Philadelphia chromosome (a G chromosome with deletion of part of the long arm) found in the bone marrow cells of most patients with chronic myelogenous leukemia is chromosome 22 (Nowell and Hungerford, 1960). Later Rowley (1973) showed that the Philadelphia chromosome is actually due to a translocation between the distal segment of chromosome 22 and the long arm of chromosome 9. It is now apparent that chromosomal defects are present in most neoplasias.

5. Reproductive Problems. The very high incidence of chromosome abnormalities found in spontaneous first-trimester abortions is described in a later section. Chromosome analysis is also helpful in determining the cause of infertility or repeated abortion, although only a small proportion of couples with such problems have a chromosome abnormality in one or the other parent that could account for their reproductive difficulties.

6. Prenatal Diagnosis. Because of the comparative ease and safety with which the karyotype of the fetus can be determined and the known association of chromosome abnormalities with late maternal age, many older pregnant women now undergo amniocentesis or sampling of chorionic villi to allow analysis of the chromosomes of the fetus. Familial chromosome abnormalities can also be detected prenatally.

GENERAL REFERENCES

Hamerton JL. Human cytogenetics. Vol. 1, Clinical cytogenetics. New York: Academic Press, 1971.
ISCN: An International System for Human Cytogenetic Nomenclature. Birth Defects 1978; 14(8). Also Cytogenet Cell Genet 1978; 21:309–404.
ISCN: An International System for Human Cytogenetic Nomenclature. High resolution banding. Birth Defects: 1981; 7(5). Also Cytogenet Cell Genet 1981; 31:1–23.
Moore KL. The developing human. Clinically oriented embryology, 3rd ed. Philadelphia and Toronto: W. B. Saunders, 1982.
Swanson CP, Merz T, Young WJ. Cytogenetics: The chromosome in division, inheritance and evolution, 2nd ed. Englewood Cliffs, NJ: Prentice-Hall, 1981.
Therman E. Human chromosomes: structure, behavior, effects. New York: Springer-Verlag, 1980.
Yunis JJ (ed). Molecular structure of human chromosomes. New York: Academic Press, 1977.
Yunis JJ. The chromosomal basis of human neoplasia. Science 1983; 221:227–236.

PROBLEMS

1. At a certain locus an individual is heterozygous, having the genotype *Aa*.
   a) What are the genotypes of his gametes?
   b) When do *A* and *a* segregate:
      1) If there is no crossing over between the A locus and the centromere of the chromosome?
      2) If there is a single crossover between this locus and the centromere?

2. How many different genotypes are possible in the ova of a woman who is:
   a) heterozygous at a single locus?
   b) heterozygous at 5 independent loci?
   c) heterozygous at n independent loci?

3. What is the proportion of normal human germ cells that contain chromosome sets in which there has been no assortment?

4. How do Mendel's laws reflect chromosome behavior?

5. It is estimated that there are 50,000–100,000 structural genes in the haploid human genome. On the average, approximately how many genes are present per band in standard metaphase preparations? How many genes per band in 800–band and 1400–band karyotypes prepared by high-resolution banding techniques?

6. In spermatogenesis, at what stage of cell division do the X and Y chromosomes disjoin? At which cell stage?

3

THE MOLECULAR STRUCTURE AND FUNCTION OF CHROMOSOMES AND GENES

In the last few years, remarkable progress has been made in our knowledge of the structure and function of genes at the molecular level. This has come about chiefly through the discovery and widespread application of recombinant DNA technology, which has provided a distinctive new approach to medical genetics and has already been usefully applied to many clinical problems. This chapter is not intended to be an extensive description of the experimental procedures of modern molecular biology and the wealth of new information that has already been revealed; it is not a replacement for a university level molecular biology course. It is intended as a review of the aspects of molecular genetics that are required for an understanding of the approach and potential of the new genetics.

Man is a **eukaryote**. This means that his cells, like those of many other organisms (including protozoa and fungi), have a genuine nucleus. In contrast, in prokaryotes such as *E. coli (Escherichia coli*, the intestinal bacterium widely used in molecular genetic research) the genetic information is not enclosed in a nucleus. Eukaryotic cells differ from prokaryotic cells in a number of other important ways. They are larger in size, have an extensive internal organization of membranes and organelles in addition to the nuclear membrane and may be capable of phagocytosis and independent movement. The biological distinction between eukaryotic and prokaryotic cells is important in molecular genetics, but the term eukaryote has been so little used in human and medical genetics that it does not even appear in some leading textbooks. Nevertheless, because of its wide use in the literature of molecular genetics it is now an essential part of the medical genetics vocabulary.

The Nucleic Acids: A Brief Review

The nucleic acids, DNA and RNA, are macromolecules (polymers) composed of three types of units: a five-carbon sugar (deoxyribose in DNA, ribose in RNA), a nitrogen-containing base and a phosphate. The bases are of two types, purines and pyrimidines. In DNA there are two purine bases, adenine (A) and guanine

(G), and two pyrimidines, thymine (T) and cytosine (C). In RNA, uracil (U) replaces thymine. There are a few rare exceptions, such as the presence of unusual pyrimidine bases in the type of RNA known as transfer RNA. Nucleotides, each composed of a base, a phosphate and a sugar moiety, polymerize into long polynucleotide chains by 5′–3′ phosphodiester bonds.

DNA

The DNA molecule, the "Watson-Crick double helix" (Fig. 3–1), carries in its biochemical structure the genetic information that allows the exact transmission of genetic information from generation to generation and simultaneously specifies the amino acid sequence of the polypeptide chains of proteins. DNA has special features that give it these properties. The double helix of the DNA molecule resembles a spiral staircase, with its two polynucleotide chains running in opposite directions, held together by hydrogen bonds between pairs of bases, A of one chain paired with T of the other, and G with C. As a consequence, a DNA molecule can replicate precisely by separation of the strands followed by synthesis of two new complementary strands (Fig. 3–2). The complementary structure also allows repair of a broken strand. In mitosis, a stage of DNA synthesis is part of each cell cycle (see Fig. 2–8). Numerous enzymes take part in DNA replication or repair.

The genetic information is stored in DNA by means of a code in which three adjacent bases (a triplet) constitute a **codon**, coding for a specific amino acid. Almost infinite variations are theoretically possible in the arrangement of the bases along a polynucleotide chain. In any one position there are four possibilities; thus there are 4^n possible combinations in a sequence of n bases. For three bases there are $4^3 = 64$ possible combinations. These 64 codons comprise the genetic code.

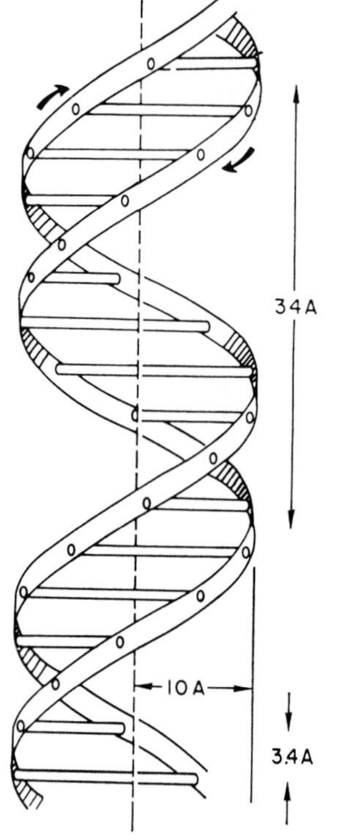

Figure 3–1. Structure of the DNA molecule, as proposed in 1953 by Watson and Crick in their original paper.

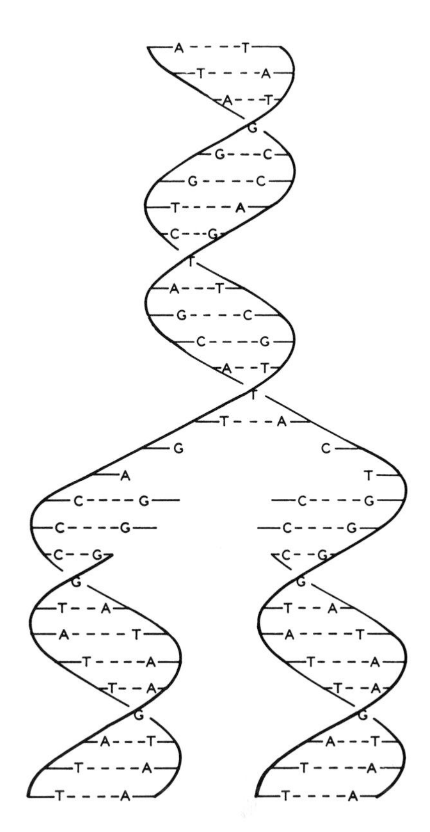

Figure 3–2. Replication of DNA, resulting in two identical daughter molecules, each composed of one parental strand and one newly synthesized strand.

THE GENETIC CODE

The genetic code (Table 3–1) was worked out through experiments using synthetic polynucleotides. The first synthetic polynucleotide used as a messenger RNA was polyU (polyuracil, a sequence of nucleotides in which all the bases are U). PolyU directs the synthesis of a polypeptide chain composed exclusively of phenylalanine, thus demonstrating that the code for phenylalanine is UUU. As might be expected, the sequence of codons in the coding sequences of the DNA molecule is colinear with the sequence of amino acids in the corresponding polypeptide. Since there are only 20 amino acids and 64 possible codons, most amino acids are specified by more than one codon; hence the code is said to be **degenerate.** For instance, the base in the third position of a triplet can often be either purine (A or G), or either pyrimidine (T or C), or in some cases any one of the four bases, without altering the coded message. Leucine and arginine are each specified by six different codons. Three of the 64 codons are "nonsense" codons designating termination of a message. With some exceptions, the code is **universal**; the same codons specify the same amino acids in all organisms, though recently an unexpected number of exceptions have been found in lower organisms.

RNA AND TRANSCRIPTION

Genetic information is contained in DNA in the chromosomes within the cell nucleus, but translation (polypeptide synthesis) takes place in the cytoplasm in association with cytoplasmic organelles known as **ribosomes**. The link between DNA and polypeptide is RNA. The rule that the flow of genetic information is

Table 3–1. THE GENETIC CODE*

First Base	Second Base: *U*		Second Base: *C*		Second Base: *A*		Second Base: *G*		Third Base
U	UUU	phe	UCU	ser	UAU	tyr	UGU	cys	U
U	UUC	phe	UCC	ser	UAC	tyr	UGC	cys	C
U	UUA	leu	UCA	ser	UAA	stop	UGA	stop	A
U	UUG	leu	UCG	ser	UAG	stop	UGG	try	G
C	CUU	leu	CCU	pro	CAU	his	CGU	arg	U
C	CUC	leu	CCC	pro	CAC	his	CGC	arg	C
C	CUA	leu	CCA	pro	CAA	gln	CGA	arg	A
C	CUG	leu	CCG	pro	CAG	gln	CGG	arg	G
A	AUU	ile	ACU	thr	AAU	asn	AGU	ser	U
A	AUC	ile	ACC	thr	AAC	asn	AGC	ser	C
A	AUA	ile	ACA	thr	AAA	lys	AGA	arg	A
A	AUG	met	ACG	thr	AAG	lys	AGG	arg	G
G	GUU	val	GCU	ala	GAU	asp	GGU	gly	U
G	GUC	val	GCC	ala	GAC	asp	GGC	gly	C
G	GUA	val	GCA	ala	GAA	glu	GGA	gly	A
G	GUG	val	GCG	ala	GAG	glu	GGG	gly	G

Abbreviations for amino acids:

ala	alanine	leu	leucine
arg	arginine	lys	lysine
asn	asparagine	met	methionine
asp	aspartic acid	phe	phenylalanine
cys	cysteine	pro	proline
gin	glutamine	ser	serine
glu	glutamic acid	thr	threonine
gly	glycine	try	tryptophan
his	histidine	tyr	tyrosine
ile	isoleucine	val	valine

Other abbreviation:
stop termination of a gene

*Codons are shown in terms of messenger RNA, which are complementary to the corresponding DNA codons.

from DNA to RNA to protein is often called the "central dogma" of molecular genetics. Though there are a few exceptions, the dogma generally holds true.

As already noted, the primary structure of RNA is similar to that of DNA, except that it contains a different sugar, ribose, and the base uracil in place of thymine. In its secondary structure it is single-stranded rather than double-stranded, though in special circumstances a single-stranded RNA molecule can form a double helix with part of its own structure. A number of specialized forms of RNA serve specific functions in transcription (see later).

Classes of DNA in the Genome

The chromosomal organization of DNA in man and other eukaryotes has not yet been explored fully, but is already known to be more complex than had been foreseen.

A gene that codes for any RNA or protein product other than a regulator is a **structural gene**. The structural genes, numbering an estimated 50,000 to 100,000 in man and ranging from 1000 to more than 200,000 base pairs (bp) in size, constitute only a small percentage of the 3×10^9 bp estimated to be in the haploid human genome. The structure of much of the remaining DNA is beginning to yield to analysis, but its function remains mysterious. Some of it may play a regulatory role or be involved in DNA replication and in chromosome pairing and recombination.

Three chief types of DNA are recognized:

1. **Unique Sequences.** These DNA sequences, which include but are not exclusively composed of the protein coding sequences, are present in single copies or only a few copies in the genome, and make up approximately 60 percent of the total DNA.

2. **Highly Repetitive Sequences.** Perhaps 10 percent of the genome or even more, depending on the species, consists of sequences that are not transcribed and are repeated hundreds of thousands or even millions of times. These sequences are probably all clustered in short (4–8 bp) tandem repeats at the heterochromatic regions of chromosomes 1, 9 and 16 and the long arm of the Y chromosome, or slightly larger tandem repeats at the centromeres of human chromosomes. Some sequences are called satellite DNA (not to be confused with chromosomal satellites, but named because it is usually readily separated from other DNA by centrifugation in certain types of cesium chloride gradient).

3. **Moderately Repetitive Sequences.** The moderately repetitive class of DNA is present in both dispersed and clustered forms, although the distinction is not always clear. The class contains a number of **gene families**, that is, functional genes that occur in several to many copies throughout the genome, such as the loci for ribosomal RNA, the histones and the immunoglobulins. Other moderately repetitive sequences are arranged as dispersed sequences that may be repeated several hundred to a hundred thousand times. Altogether some 30 percent of the genomic DNA is of this class.

Alu Family. In humans, the Alu family provides one example of a number of types of dispersed, moderately repetitive sequences, each of which may make up 3 percent or so of the genome. The Alu family includes at least 300,000 repeats of a 300 bp sequence. Other moderately repetitive sequences are substantially longer. Alu sequences are also present in other primates and in a shorter version (135 bp) in rodents. The name Alu indicates that the unit is cut by the bacterial restriction enzyme Alu. The function of Alu sequences and of the other families of moderately repetitive dispersed DNA is completely unknown. A puzzle also exists as to how the Alu sequences and other repetitive DNA sequences maintain their integrity both within an individual genome and within a species. The mechanisms that bring about divergence in unique-sequence DNA, which ultimately account for human variation and thus for single-gene defects, do not appear to operate for repetitive DNA. Instead, repetitive DNA seems to be highly conserved within a species.

Gene Structure

Until about 1977 the gene was visualized simply as a segment of a DNA molecule containing the code for the amino acid sequence of a polypeptide chain. This model is now known to be inadequate. In almost every structural gene of man and other vertebrates that has been analyzed, the coding sequences, often called **exons**, are interrupted at intervals by non-coding sequences known

Figure 3–3. Structure of a human globin gene. For details, see text.

as **introns** or intervening sequences, which are initially transcribed but are not represented in mature messenger RNA or in the protein product. The gene also includes extensive **flanking regions** of importance in regulation, and "start" and "stop" signals. Because the globin genes are among the best-known human genes, a human globin gene is used to illustrate gene structure (Fig. 3–3).

As shown, the globin gene contains three coding sequences (exons) split by two intervening sequences or introns. Upstream (5′) from the gene is an untranslated flanking region containing two specific sequences that play a significant role in regulation: CCAAT, the "CAT box," about 70 bp upstream from the beginning of the first coding sequence, and TATA, the "TATA box," between the CAT box and the site where transcription begins. Downstream (3′) is a second flanking region, containing a sequence AATAAA that appears to be a signal for the addition of the polyA tail to the end of the mRNA strand.

A number of other genes have also been analyzed in detail, and certain generalizations have emerged. In typical genes of vertebrates, the coding sequences are split by intervening sequences. Exceptions include the genes for histones and for interferons. (Interferons are a class of antiviral agents specified by a family of genes all located on chromosome 9.) In lower eukaryotes splitting is less common than in vertebrates, and in prokaryotes the DNA has no introns.

The exon-intron patterns of split genes appear to be strikingly conserved during evolution; for example, the α and β globin genes, which are believed to have arisen by duplication of a primitive precursor about 500 million years ago, each have two introns at precisely the same locations; the same pattern is also seen in the nine other α-like or β-like globin genes. Within the exons alterations in sequence occur slowly, by fixation of rare advantageous mutations through natural selection. The average rate is estimated to be in the range of 10^{-9} substitutions per codon per year. Within introns changes occur much more rapidly, and the inference is that intron sequences, except at their splice junctions (the sequences immediately surrounding the exon-intron boundaries), are under much less rigid selective control than are the coding sequences.

GENE FAMILIES

Even before the development of recombinant DNA technology, a number of "families" of closely related genes had been defined.

Figure 3–4 shows two such families (or a single extended family), the α-like and β-like globin gene clusters, on chromosomes 16 and 11 respectively. In each of these two linked clusters, the genes code for closely related proteins expressed at different developmental stages; the sequence of the genes along the chromosome is the same as the sequence of their activation during development. The individual genes within the clusters are separated by non-coding "spacer" sequences considerably larger than the coding sequences themselves. The function of the spacer DNA, if any, is unknown.

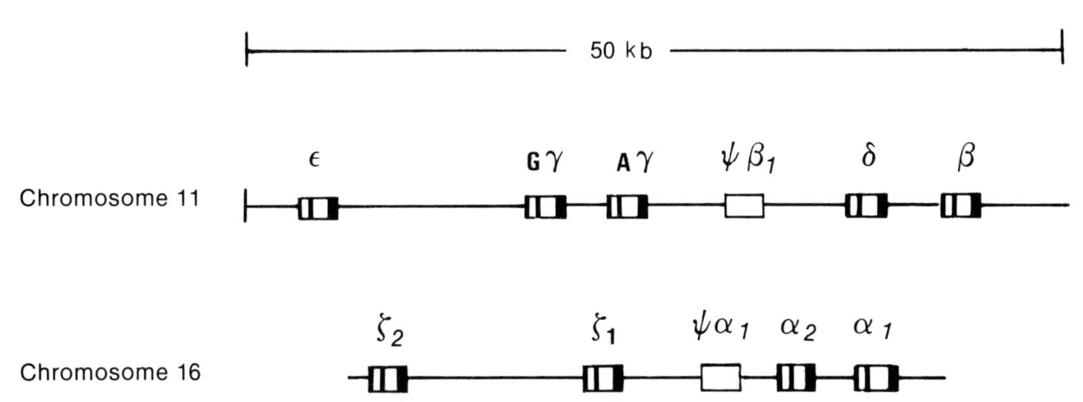

Figure 3–4. The α and β globin gene clusters. On chromosome 16, two genes coding for an early embryonic hemoglobin (ζ globin), a pseudo-α (ψα) globin gene and two identical α globin genes (α_1 and α_2) are linked. On chromosome 11, the cluster of globin genes contains a gene for embryonic hemoglobin (ε); two genes for fetal hemoglobin, differing in a single amino acid and designated Gγ and Aγ; a pseudo-β gene; and the δ and β genes. See also chapter 5.

Other gene families, such as the ribosomal RNA genes, are tandem arrays of closely similar or identical genes, without spacer sequences. Examples are also known of genes that are clustered on a chromosome and show considerable homology, but code for different proteins with distinct functions; for example, the albumin and α-fetoprotein genes on chromosome 4.

Some gene families, including both globin gene clusters, contain **pseudogenes**. Pseudogenes are sequences that show striking homology to functional genes but are not transcribed, probably because their regulatory regions have been altered by mutation. They are thought to represent genes that at one evolutionary stage were functional but are now vestigial. Some pseudogenes lack introns, and it is believed that this type of pseudogene, a so-called "processed pseudogene," may represent a **cDNA** (complementary or copy DNA) sequence synthesized on a messenger RNA template and reinserted into the genome at a site usually not linked to its progenitor.

Transcription

With the background provided above, we can now examine transcription and translation in further detail.

After transcription of a split gene, its introns are excised and the exons ligated to make a complete messenger RNA molecule. This process is known as RNA splicing. The splice junctions of introns are characterized by "consensus sequences" of bases which invariably include GT at the 5′ end and AG at the 3′ end. Splicing can be disrupted by mutations that change one of these sequences or alter a sequence elsewhere so that it closely resembles a splice junction. The mechanism that brings together the two ends of the intron and juxtaposes the two exons that are to be spliced is not yet clear; a type of RNA known as U_1RNA, so far poorly understood, seems to be involved.

Transcription of a β-globin gene and maturation of the corresponding mRNA is shown in Figure 3–5. The first step is the formation of a long primary transcript, which includes the important flanking regions at both ends of the gene. The primary transcript is processed by excision of the introns, splicing of the exons and addition of a CAT at the 5′ end and a polyA tail (sequence of adenylic acid residues) at the 3′ end. These steps take place within the cell nucleus. The mature messenger then passes to the cytoplasm, where translation takes place.

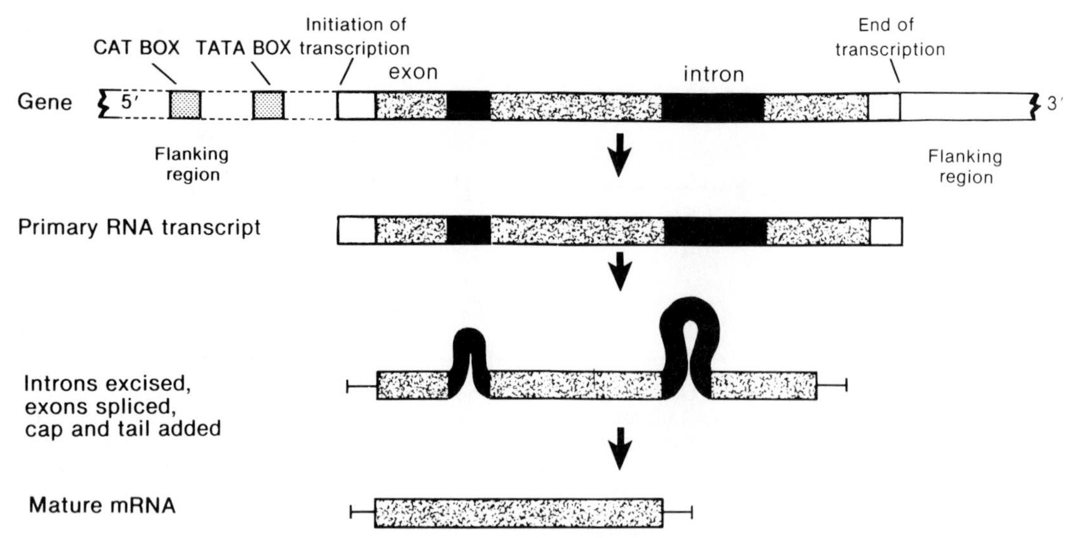

Figure 3–5. Transcription and processing of a β globin gene to form mature messenger RNA. See text for details.

Translation

In the cytoplasm, mRNA is translated into protein in cooperation with **transfer RNA** (tRNA), which has the role of transferring amino acids from the cytoplasm to their positions along the mRNA template (Fig. 3–6). A tRNA molecule is only 80 nucleotides long. Its three-dimensional structure is complex, with hairpin folds that allow base pairing within their length. One unpaired site on the molecule is an anticodon that is complementary to a specific codon on the mRNA chain and bonds to it, thus bringing its specific amino acid into position on the growing polypeptide chain. Amino acids are added sequentially from the 5′ to the 3′ end of the chain. The ribosomes, composed of ribosomal RNA (rRNA) with protein, participate in this process. Initiation of chain synthesis is followed by chain elongation, with peptide bonding between the successive amino acids, until the chain-termination codon is reached.

Many proteins undergo extensive modification after synthesis. The polypep-

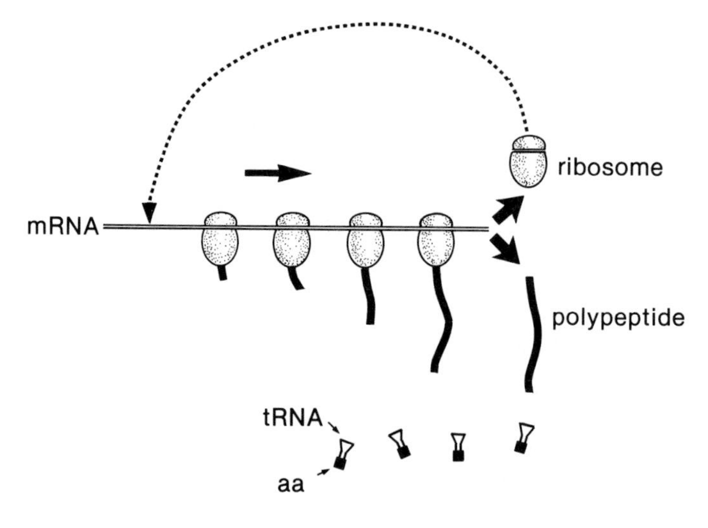

Figure 3–6. Diagram to represent translation of mRNA into a polypeptide chain. See text for details.

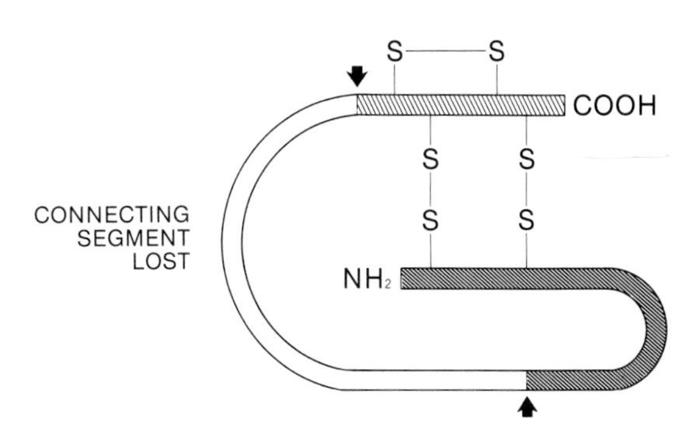

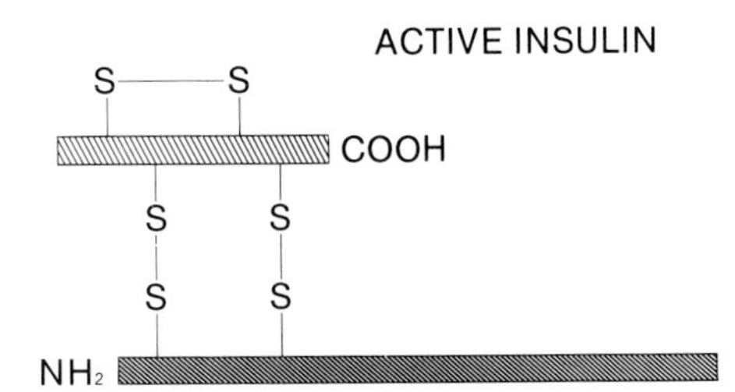

Figure 3–7. Synthesis of the insulin molecule. See text for details.

tide chain that is the primary gene product is folded and bonded in a very specific way that appears to be determined by the amino acid sequence itself. Two or more chains, alike or different, may combine to form a single protein. When different chains combine, the molecular size and amino acid sequence of the component polypeptides tend to suggest that the corresponding genes have arisen by duplication of a single gene in the distant past.

Proteins may also be modified by removal of a portion of the molecule. Insulin is a case in point. This protein is a hormone that is responsible for reducing the level of glucose in the blood; when insulin is insufficient, diabetes mellitus results. Insulin is made up of two chains, one with 21 and the other with 30 amino acids, held together by disulfide bonds. By analogy with other proteins it was originally suspected that each chain was coded by a separate gene, but this proved not to be the case. Instead, the primary gene product, proinsulin, is a single polypeptide 82 amino acids long which, after folding and bonding, loses a large interstitial section (Fig. 3–7). Though insulin production is defective in diabetes mellitus, the insulin molecule itself does not have abnormal structure; in other words, except for a very few rare instances, diabetes is not caused by a mutation in the insulin gene.

Recombinant DNA Technology

Though delineation of the nucleotide sequences of DNA was a long-term goal of molecular geneticists, little progress could be made in the earlier years because there was no way to cut DNA at specific sites to obtain specific sequences for analysis. The picture changed with the discovery in bacteria of specific **restriction nucleases**. A restriction nuclease is a nucleic acid-cleaving enzyme that acts upon DNA, cutting it into fragments at a specific sequence (restriction site). In bacteria, which lack the immunological protection eukaryotes possess

Table 3–2. EXAMPLES OF RESTRICTION ENZYMES AND THEIR RECOGNITION SEQUENCES

Name	Source	Recognition Sequence
BamHI	*Bacillus amyloliquefaciens* H	G GATCC CTTAG G
EcoRI	*Escherichia coli* RY 13	G AATTC CTTAA G
HaeIII	*Haemophilus aegyptius*	GG CC CC GG
HindII	*Haemophilus influenzae* Rd	GTPy PuAC* CAPu PyTG
HindIII	*Haemophilus influenzae* Rd	A AGCTT TTCGA A
HpaII	*Haemophilus parainfluenzae*	C CGG GGC C

*Py, any pyrimidine; Pu, any purine.

against invading foreign DNA, these enzymes serve the function of breaking down infecting DNA molecules. Over 200 such enzymes are now known. Examples are given in Table 3–2 and, as shown, the names are abbreviations of the microorganisms from which the enzymes were isolated. In some cases the enzyme recognizes a sequence of four bases; by chance the identical sequence could occur on average once in 4^4 bases. Others (the majority) recognize groups of six bases; because on the average matching sequences would be 4^6 bases apart, the enzymes that recognize 6-base sequences cut DNA into fewer and longer segments than those that recognize 4-base sequences. Some enzymes cut both strands at the same site, leaving blunt ends, and others make offset cuts, leaving "sticky" ends. Usually, as in the examples shown, the sequences are palindromes; that is, they read the same forward or backward.

DNA SEQUENCING

Methods for sequencing the DNA fragments produced by the use of restriction enzymes were independently developed by Walter Gilbert and Frederick Sanger, who won Nobel prizes in 1980 for their work. This was Sanger's second Nobel prize; he was also awarded one in 1958 for being the first to determine the amino acid sequence of a protein. The sequence of virtually any DNA segment can now be determined; from this the amino acid sequence of the protein for which it codes can be deduced. Thus, in time, a catalog of the entire human genome can be expected.

MAKING RECOMBINANT DNA

"Recombinant" DNA molecules contain DNA sequences from different organisms. They are constructed by insertion of foreign DNA into vectors, which may be either bacterial plasmids or phages. Plasmids are circular chromosomal elements in bacteria that replicate independently of the main bacterial chromosome. They vary in size; one plasmid of *E. coli* that has been specially constructed for recombinant DNA technology, known as pBR322, is 4362 bp long. This

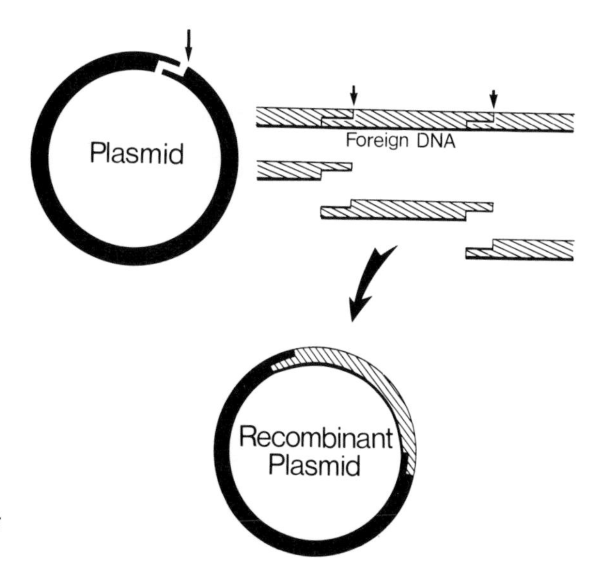

Figure 3–8. Construction of recombinant DNA. See text for details.

plasmid contains several restriction sites at which the circular DNA can be opened up by specific restriction nucleases to allow a sequence of foreign DNA (DNA from a different organism), cut with the same enzyme, to be spliced in or "recombined" (Fig. 3–8), forming recombinant plasmids.

The recombinant plasmids are then reintroduced into bacteria, which are then allowed to multiply under selective pressure. Thus, clones of bacteria carrying a specific restriction fragment of foreign (e.g., human) DNA can be generated.

Gene "libraries," representing essentially all the genes of an organism or some part of the genome such as the X chromosome, are prepared by fragmenting the DNA with a particular restriction enzyme and making recombinants between the individual fragments and a vector (plasmid or phage) cut by the same enzyme. The recombinants, which may be of a million or more different kinds, are then cloned in bacteria. Libraries can also be prepared from cDNA under special conditions.

Since the process of making recombinant DNA begins with attempts to insert a large number of more or less random restriction fragments into plasmids, it is obviously crucial to be able to recognize a particular gene. This is feasible if messenger RNA of the desired type is available. The mRNA can then be purified and used to generate cDNA, by means of the enzyme reverse transcriptase (isolated from certain RNA tumor viruses), through a complicated procedure that will not be described here. The end product is a radioactive cDNA molecule called a **probe** that will hybridize to the gene of interest (or to its messenger RNA) under appropriate conditions.

Phages (bacteriophages) are bacterial viruses that can infect bacteria, multiply within the bacteria and cause lysis of the bacterial cell with release of the phage progeny, which in turn can infect other bacteria. In this process, the chromosome of the virus sometimes becomes integrated into the bacterial chromosome and replicates with it thereafter. Phages have an advantage over plasmids in some recombinant DNA work because much larger DNA fragments, up to 15 kb in length, can be cloned in them. Still larger fragments of DNA, 35 to 45 kb in size, can be cloned in cosmids, which are ingeniously constructed combinations of a phage and a plasmid.

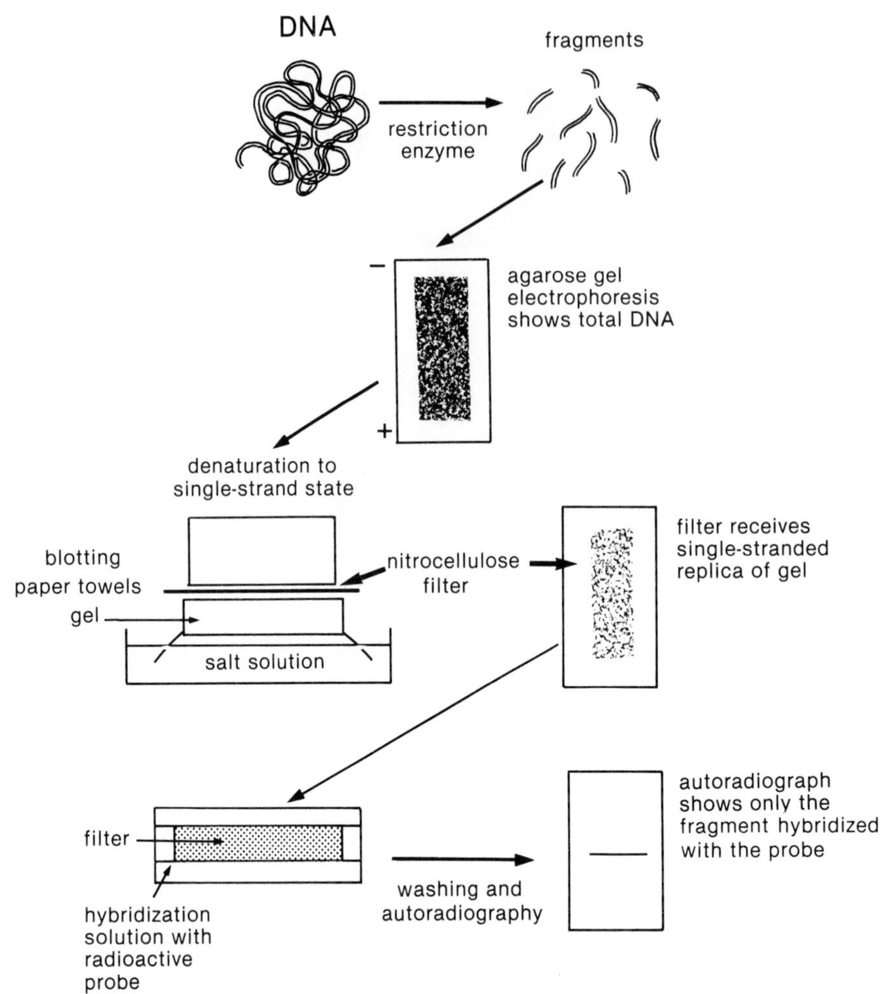

Figure 3–9. The Southern blotting technique for analysis of DNA cleaved by restriction enzymes. See text for details.

SOUTHERN BLOTTING

The Southern blotting technique, devised by Edwin Southern in Edinburgh in 1975, is the standard way to analyze the structure of DNA cleaved by restriction enzymes (Fig. 3–9). The DNA fragments are separated by agarose gel electrophoresis according to their size, small fragments moving much more rapidly than larger ones. After DNA denaturation (to separate the two DNA strands), the gel is overlaid with a nitrocellulose filter, and the DNA fragments are transferred to the filter by blotting. To identify the fragment of interest among the numerous pieces of DNA on the filter, a specific radioactively labeled probe is used. The probe is allowed to anneal to the complementary DNA sequence on the filter, which is then autoradiographed to show the position of the fragment of interest.

RESTRICTION FRAGMENT LENGTH POLYMORPHISMS (RFLPs)

When restriction endonuclease technology was applied to DNA analysis, it was soon discovered that the hybridization pattern showed variation; at a specific site a particular enzyme might cut the DNA of some individuals but not of others. Since cleavage with restriction nuclease is site-specific, obviously sites were being recognized at which the DNA sequence had been altered so that cleavage was no longer possible. These sites have been named restriction fragment length polymorphisms (RFLPs) or, more broadly, DNA markers. Note, however, that this use of the term polymorphism is different from its classic use to define a locus at which there are at least two alleles, of which the rarer one has an appreciable frequency (see Glossary for definitions). For this reason, the sites are now sometimes referred to as restriction fragment length variants (RFLVs).

RFLPs have great potential for human gene mapping, and in fact they have already been applied to the analysis of certain genetic diseases with considerable success. (RFLPs are known in many other organisms, but this discussion will be restricted to humans.) They are inherited as simple Mendelian codominant markers. Though at a single site only the presence or absence of cutting by a specific restriction enzyme can be identified, additional variation can be shown by the use of several different enzymes that cleave the DNA at different but closely linked sites, thus generating **haplotypes**. (By definition a haplotype is a set of closely linked alleles usually inherited as a unit, but in molecular genetics the term has come to indicate a set of closely linked restriction sites, each site being either cut or left unchanged by a given enzyme.) RFLPs can be used as markers of a gene even when the gene itself has not been isolated or even mapped, and without any knowledge of its biochemical product. It has been estimated that 150 RFLPs spaced out throughout the genome would, in principle, be enough to allow any human gene to be mapped (Solomon and Bodmer 1979, Botstein et al. 1980). The implications for prenatal diagnosis and preclinical identification are major and are already beginning to be realized in, for example, the thalassemias, the hemoglobinopathies, Huntington disease and the X-linked muscular dystrophies; an application of RFLP technology to each of these conditions is described later (see Chapter 11). In cytogenetics, Y-specific DNA is already detectable and in the future the identification of small structural abnormalities below the limit of microscopic analysis should become a possibility. Because of the potential of RFLP analysis for clinical medicine, it is likely that service-oriented molecular genetics laboratories will be established in major hospitals within the next few years, and that clinicians will need to be familiar with the concepts, potential and limitations of the application of RFLP analysis to clinical practice.

Mutation

Any change in the sequence of genomic DNA constitutes a mutation. Normally DNA replication is highly accurate, but if an error in the DNA sequence occurs it is copied at subsequent replications. The phenotypic effects of a mutation depend on the nature of the alteration, if any, in the corresponding gene product. There are three different mechanisms of mutation: substitution, deletion or insertion (Fig. 3–10).

Single Base Substitutions. A single base change in a DNA sequence (point mutation) can alter the code in a triplet of bases and cause the replacement of one amino acid by another in the gene product. Since the code is degenerate, not all base substitutions lead to actual changes in amino acid sequence. Point mutations that alter the active site of the resulting protein have particularly severe effects.

Deletions and Insertions. Either deletion or insertion of a base (or any number of bases that is not a multiple of three bp) alters the whole reading frame

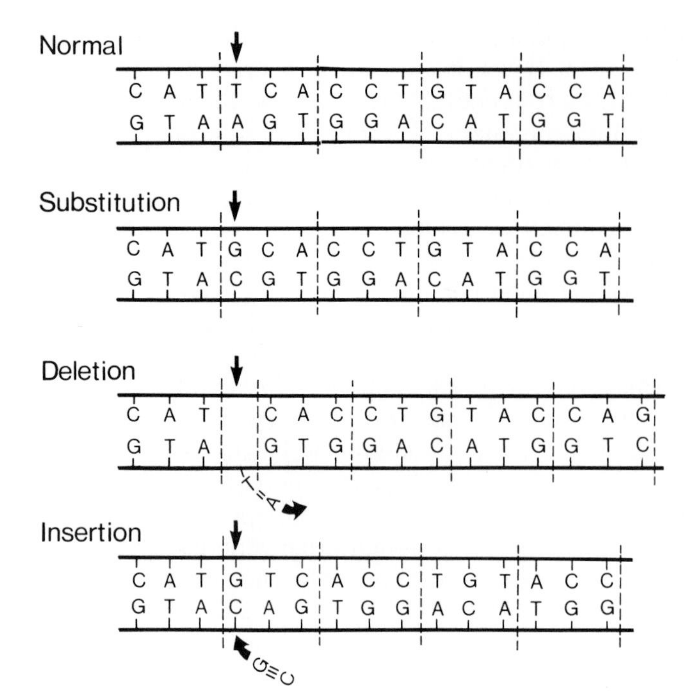

Figure 3–10. Mechanisms of mutation. There may be substitution, deletion or insertion of a base pair. See text for details.

of the DNA strand and thus of the amino acid sequence. These are "frameshift" mutations. Such massive changes cannot be expected to result in active proteins.

Certain types of mutation either within or outside coding sequences of a gene have extensive effects on the gene product or may even bring about suppression of transcription of the gene. These types include:

Chain Termination Mutations. Normally, transcription of the DNA ceases when a termination codon is reached. A mutation that creates a termination codon can cause premature cessation of transcription; a mutation that destroys a termination codon allows transcription to continue until the next termination codon is reached. (A coding sequence without any termination codons is referred to as an open reading frame; such a sequence is potentially translatable into protein.)

Splice Mutations. Examples are known of mutations that affect the normal mechanism by which introns are excised and exons spliced together during the synthesis of mature messenger RNA. Such changes can lead to complete failure of synthesis of the gene product.

Mutations in Regulatory Sequences. Changes in DNA sequence in the CAT box or TATA box region upstream of the structural gene can lead to reduced transcription of the gene.

The Fine Structure of Chromosomes

The description of cell division in Chapter 2 passed over the question of how the total extended length of DNA in the chromosomes of interphase cell nuclei, estimated as 174 cm, becomes condensed about 10,000-fold into the metaphase chromosomes, which nevertheless maintain their regional organization with characteristic and consistent banding patterns. So far, we do not have a complete, unified view of the process, but its broad outlines are becoming clear.

To make the following description of chromosome condensation more concrete, consider the size of the longest chromosome, no. 1, and the shortest, no. 21. Each chromosome 1 contains about 2.5×10^8 bp of DNA, and would be about 15 cm long as a naked double helix. Chromosome 21 contains only 0.5×10^8 bp of DNA, and would be about 3 cm long if it were fully extended.

The DNA molecule of the chromosome is complexed with histones (small basic proteins) and non-histone proteins in roughly equal amounts to form chromatin. The DNA of a single chromosome is a single molecule which forms a single fiber (unineme) of chromatin, continuous through the centromere. Because the amino acid sequences of histones have been conserved through a wide range of organisms, from yeast to man, it is believed that histones must play an essential biological role. In contrast, the non-histone proteins of chromatin are extremely heterogeneous.

Figure 3–11 is a model of chromosome structure and condensation as

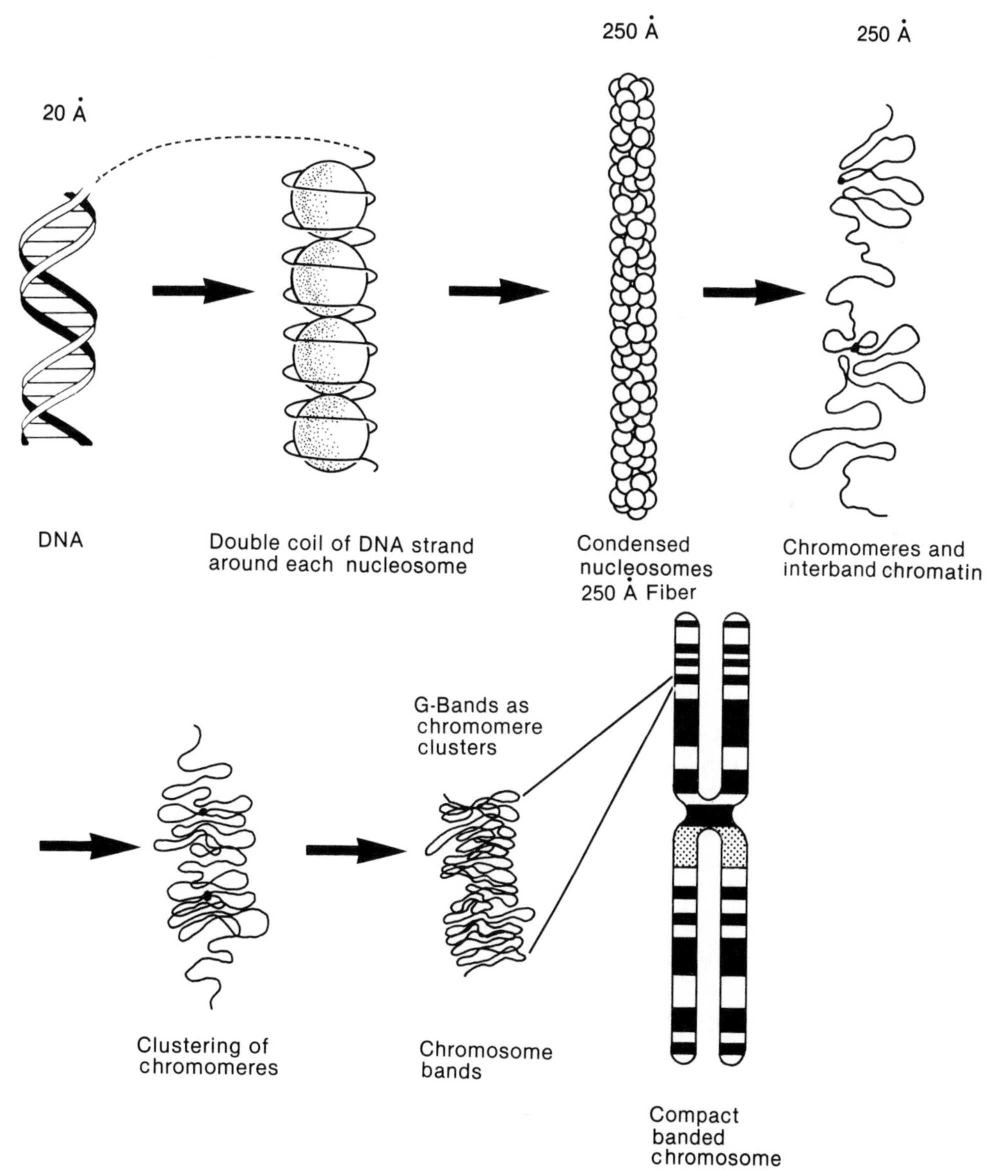

Figure 3–11. A model of chromosome structure and condensation. For description, see text. Modified from Comings, 1978, p. 39.

presently conceptualized, though it should be kept in mind that the first and second stages of condensation are not part of a cycle in nature but simply represent a view of how the DNA fiber might be arranged within the nucleus. At the first level of condensation, 146 bp DNA sequences are wound twice around close-packed histone cores to form units known as **nucleosomes**. Successive nucleosomes are separated by linkers, giving a beads-on-a-string structure containing about 200 bp per unit and visible in electron micrographs. Each histone core consists of two units each of four major histones. The fifth histone does not form part of the core, but is attached at the ends of the nucleosome, where the DNA enters and leaves. The "packing ratio" at this stage is about 10; that is, chromosome 1 DNA is now condensed to 1.5 cm and chromosome 21 DNA to 0.3 cm in length.

At the second level of condensation, the nucleosomes are further compacted to form a "thick" helical fiber termed a **solenoid**, with six or seven nucleosomes per turn. Formation of the solenoid may be controlled by cross-linkage of histones attached to adjacent nucleosomes. At this stage the packing ratio is increased to about 50. Chromosome 1 DNA is now condensed to 0.3 cm and chromosome 21 to 0.06 cm.

The solenoids are next packed into loops or **domains** surrounding a protein matrix or **scaffold**. The loops are the beginning of the **chromomeres**, the bead-like thickenings observable under the microscope in early prophase chromosomes. It appears that in these domains the genes being expressed are primarily in the outer parts of the loops, while the suppressed genes are buried on the inside. In other words, one level of the control of gene expression depends in some way on how genes are packed into chromosomes, and their association with chromosomal proteins in the packing process. Unlike the chromosomes seen in stained preparations under the microscope or in photographs, the chromosomes of living cells are flexible, dynamic structures intimately involved in the selective expression of genes.

As the chromosomes condense further, adjacent chromomeres merge into larger ones. In stained prophase chromosomes 800 or more bands can be recognized, whereas in metaphase chromosomes the number is about 200. The clusters of chromomeres eventually become the darkly-stained bands of G-banded chromosomes. At metaphase, chromosome 1 has contracted to 10-15 μm and chromosome 21 to 2-3 μm in length.

Three major types of chromatin have been defined. The regions of a chromosome that stain lightly and represent the interchromomeric regions of prophase chromosomes are the regions that are transcribed into RNA. They tend to replicate early in the DNA synthesis stage of the cell cycle. The term euchromatin is used to define this type of chromatin. In contrast, chromosome regions that are more condensed and deeply staining are made up of heterochromatin. Centromeric heterochromatin, which remains condensed and inactive and shows up particularly well with C-banding techniques, is usually composed of satellite DNA, that is, highly repetitive sequences. The third major type, termed intercalary heterochromatin, is present in chromomeric regions and G-bands and is late-replicating. Further analysis of DNA structure is needed before the relationship of chromosome packaging to the structural genes and their regulation will be fully understood.

GENERAL REFERENCES

Emery AEH. An introduction to recombinant DNA. Chichester: John Wiley & Sons, 1984.

Igo-Kemenes T, Horz W, Zackau HG. Chromatin. Ann Rev Biochem 1982; 51:89–121.

Watson JD, Tooze J, Kurtz DT. Recombinant DNA. A short course. Scientific American Books, distributed by W. H. Freeman and Company, New York, 1983.

Weatherall DJ. The new genetics and clinical practice. London: Nuffield Provincial Hospitals Trust, 1982.

PROBLEMS

1. What kind of mutation is illustrated by each of the following amino acid sequences?
   Wild type--lys arg his his tyr leu---
   Mutant I--lys arg his his cys leu---
   Mutant II--lys arg ile ile ile-----
   Mutant III--lys glu thr ser leu ser---

2. Match the following:

A	B
–DNA molecule	1) cleaves DNA at specific sites
–eukaryote	2) unit at early stage of DNA condensation
–repetitive DNA	3) conserved in evolution
–nucleosome	4) removed during RNA maturation
–satellite of acrocentric chromosome	5) chromosome
–unique sequence DNA	6) contains genes for ribosomal RNA
–upstream flanking sequence of gene	7) plays regulatory role in transcription
–probe	8) includes structural genes
–solenoid	9) organism with nucleated cells
–intron	10) radioactive cDNA molecule
–restriction nuclease	

4

PATTERNS OF SINGLE-GENE INHERITANCE

In Chapter 1 the three main types of genetic disorders were named and briefly defined. This chapter is concerned with a more complete description of the patterns of transmission shown by traits determined by genes at a single locus, and some of the ways these patterns may be modified. Single-gene phenotypes are sometimes said to be "Mendelian" or "Mendelizing" traits because they segregate sharply within families and, on the average, occur in fixed proportions as did the characteristics studied by Mendel in garden peas. As noted earlier, the pedigree patterns shown by such traits depend on two factors: (1) whether the gene responsible is on an autosome or on the X chromosome and (2) whether it is **dominant**, that is, expressed even when present on only one chromosome of a pair, or **recessive,** expressed only when present on both chromosomes. Thus there are only four basic patterns:

Autosomal { dominant / recessive   X-linked { dominant / recessive

An individual pedigree is also determined by the chance distribution of genes from parents to children through the gametes. Especially with the small family size typical of most developed countries, the patient may be the only affected family member. Complications arising from lack of information, difficulties in diagnosis, genetic heterogeneity, variation in clinical expression and environmental effects may make a pedigree difficult to interpret.

Though some pedigrees show the pattern of transmission so clearly that it can hardly be misinterpreted, a single pedigree is usually not enough to establish the genetics of a disorder. Fortunately, by now clinical geneticists have made considerable progress in the description and cataloging of single-gene defects. If the diagnosis is clear (and it must be stressed that a correct diagnosis is the essential basis of accurate genetic counseling), the pedigree can be examined to see if it is consistent with the expected pattern. If the diagnosis is not clear, the family history can sometimes lead to a very good guess of what it might be, and the guess can be followed up by specific diagnostic tests.

At this point several terms with special connotations in genetics must be introduced and defined. The family member who first brings a family to the attention of an investigator is the **proband** (index case, propositus). **Sibs** (or siblings) are brothers or sisters, of unspecified sex. The parent generation is designated the P_1, and the first generation offspring of two parents the F_1, but these terms are used more frequently in experimental genetics with inbred lines of plants or animals than in human genetics.

Recall that genes at the same locus on a pair of homologous chromosomes are **alleles.** In more general terms, alleles are alternative forms of a gene. When an individual has a pair of identical alleles, he or she is **homozygous** (a homozygote); when the alleles are different, he or she is **heterozygous** (a heterozygote or carrier). The term **compound**, or compound heterozygote, is used to describe a genotype in which two different abnormal alleles are present, rather than one normal and one abnormal. These terms (homozygous, heterozygous and compound) can be applied either to an individual or to a genotype. The term **mutation** is used in two senses, sometimes to indicate a new genetic change that has not previously been known in a kindred, but at other times merely to indicate an abnormal allele.

An allele that is expressed, whether homozygous or heterozygous, is **dominant;** an allele that is expressed only when homozygous is **recessive.** Strictly speaking, it is the trait (phenotypic expression of a gene) rather than the gene itself that is dominant or recessive, but the terms dominant gene and recessive gene are in common use.

The distinction between dominants and recessives is not absolute. In heterozygotes for autosomal conditions, as a general rule each member of the gene pair is transcribed to form a gene product, whether or not the heterozygous phenotype is distinguishable from the homozygous forms. The gene product may be nonfunctional, or at least defective, but the point is that both alleles are expressed. By definition a recessive has no detectable expression in heterozygotes, but some genes are defined as recessive simply because when heterozygous they are not phenotypically evident *under the conditions of analysis,* though when examined in a different way there may be a detectable expression. For example, if the anticipated expression is a manifest disease, a phenotype may be regarded as recessive even when on the biochemical level the "recessive" gene is identifiable in heterozygotes. Later in this chapter and in Chapter 5 we will refer to a number of examples of so-called recessives that are biochemically detectable in heterozygotes.

In dominant disorders, the gene product is of a type that can produce clinical abnormality when the quantity is 50 percent of normal. The abnormal genes responsible for these disorders are thought to encode two main classes of proteins: (1) nonenzymatic or structural proteins, such as hemoglobin or collagen, and (2) proteins that regulate complex metabolic pathways, such as membrane receptors (as in familial hypercholesterolemia) or rate-limiting enzymes in biosynthetic pathways under feedback control (as in acute intermittent porphyria in which an enzyme that catalyzes a rate-limiting step in heme biosynthesis is abnormal). However, in many autosomal dominants the basic mechanism remains unknown.

In contrast, in typical recessive disorders, which are by definition expressed only in homozygotes, the gene products are enzymes. The margin of safety in an enzyme system allows normal function even if only one of a pair of alleles is actively producing normal enzyme.

Family data can be summarized in a **pedigree**, which is merely a shorthand method of classifying the data for ready reference. Symbols used in drawing up a pedigree are shown in Figure 4–1. Variants of these symbols may be invented to demonstrate special situations. By convention, gene symbols are always in italics.

The human gene nomenclature and symbols have recently been standardized in order to facilitate computerization (Shows et al., 1983). Most loci are now designated by capital letters or a combination of capital letters and Arabic numerals not more than four characters in length, with no superscripts or subscripts. An asterisk separates the gene symbol from the allele symbol, which

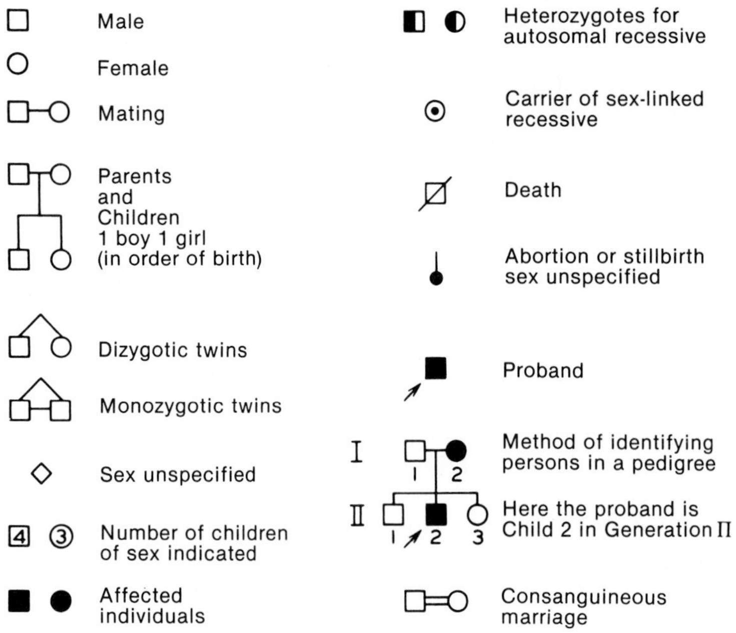

Male

Female

Mating

Parents and Children 1 boy 1 girl (in order of birth)

Dizygotic twins

Monozygotic twins

Sex unspecified

Number of children of sex indicated

Affected individuals

Heterozygotes for autosomal recessive

Carrier of sex-linked recessive

Death

Abortion or stillbirth sex unspecified

Proband

I 1 2
II 1 2 3

Method of identifying persons in a pedigree

Here the proband is Child 2 in Generation II

Consanguineous marriage

Figure 4–1. Symbols commonly used in pedigree charts.

is also capitalized. In the following pages standard gene symbols are sometimes used, but in most instances abbreviations are employed. A genotype is conventionally shown with a slash (to symbolize the chromosome pair) between the symbols for a pair of alleles; thus, *T/t*. In common usage the slash is often omitted, and it has been omitted throughout this book except when sets of linked genes, rather than single gene pairs, are discussed. As an abbreviation, a plus sign (+) may indicate any normal allele, which is often called the "wild type."

Many traits are determined by genes at a single locus, in either homozygous or heterozygous state. Over 3000 phenotypes, most of which are abnormalities rather than normal variants, are catalogued in a most useful reference: *Mendelian Inheritance in Man* by V.A. McKusick (6th edition, 1982). This is an appreciable proportion of the estimated 50,000 to 100,000 human structural genes. Of the total, 1827 are autosomal dominant, 1298 autosomal recessive and 243 X-linked.

Autosomal Inheritance

GENES, GENOTYPES AND MATING TYPES

A simple example of a common hereditary trait governed mainly by a single pair of autosomal alleles is the ability to taste phenylthiocarbamide (PTC). To most people PTC has a bitter taste, but others cannot taste it at all. Ability to taste PTC is dominant to inability to taste it. Formally the alleles are designated *PTC*1* and *PTC*2*, but here we will use *T* and *t* for brevity. The phenotype of both the *TT* (homozygous dominant) and *Tt* (heterozygous) genotypes is taster, and that of the *tt* (homozygous recessive) genotype is nontaster. In summary:

Genes	Genotypes	Phenotypes
T and *t*	*TT*	taster
	Tt	taster
	tt	nontaster

Table 4–1. AUTOSOMAL INHERITANCE: MATING TYPES AND EXPECTED PROPORTIONS OF PROGENY FOR A PAIR OF AUTOSOMAL ALLELES *T* AND *t*

Mating Types		Progeny	
Genotypes	*Phenotypes*	*Genotypes*	*Phenotypes*
TT × *TT*	taster × taster	All *TT*	All taster
TT × *Tt*	taster × taster	½ *TT* ½ *Tt*	All taster
TT × *tt*	taster × nontaster	All *Tt*	All taster
Tt × *Tt*[1]	taster × taster	¼ *TT* ½ *Tt* ¼ *tt*	¾ taster ¼ nontaster
Tt × *tt*[2]	taster × nontaster	½ *Tt* ½ *tt*	½ taster ½ nontaster
tt × *tt*	nontaster × nontaster	All *tt*	All nontaster

[1]This is the usual pattern of inheritance for a rare autosomal recessive trait.
[2]This is the usual pattern of inheritance for a rare autosomal dominant trait.

Since there are three possible genotypes and since either male or female may have any of the three, there are six possible mating types, as shown in Table 4–1. Each gamete will have just one allele of the pair, and the possible gametes of each genotype are shown below.

Genotype	TT	Tt	tt
Gametes	T, T	T, t	t, t

To illustrate how the information contained in Table 4–1 is derived, consider the progeny of a heterozygous (*Tt*) male and female. A "checkerboard" is sketched (below left) to show the gametes of each parent (in the margins) and their possible progeny (in the closed squares). A checkerboard is sometimes called a Punnett square because it was invented by R.C. Punnett, who is better known as the codiscoverer, with William Bateson, of genetic linkage in 1906.

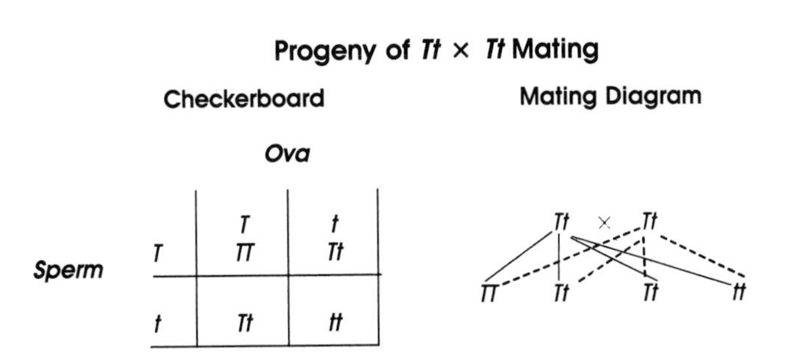

Each of the offspring has a 1/4 chance of being a dominant homozygote (*TT*), and a 1/2 chance of being a heterozygote; either of these genotypes causes the taster phenotype. There is a remaining 1/4 chance that the child will be a recessive homozygote, with the nontaster phenotype. (See Table 4–1, fourth line.)

Another way of working out the possible combinations of gametes in the progeny of a cross is by a "mating diagram" (above right).

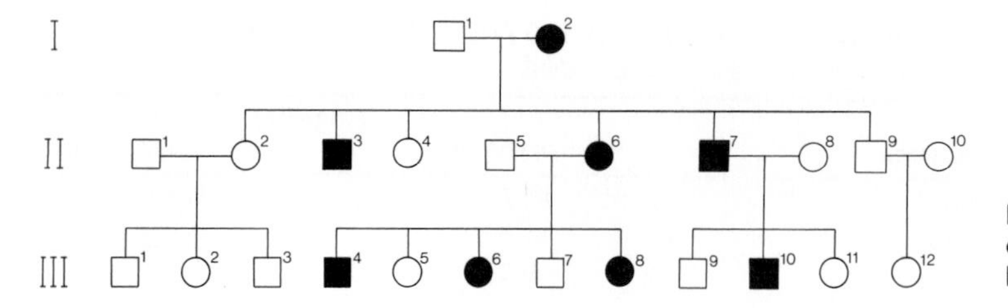

Figure 4–2. Stereotype pedigree of autosomal dominant inheritance. Note male-to-male transmission.

All the possible mating types involving a single pair of autosomal alleles are set out in Table 4–1, but two of these are of particular importance in medical genetics: the $Tt \times Tt$ mating, the common one for autosomal inheritance of a rare recessive gene t, and the $Tt \times tt$ mating, typical of autosomal inheritance of a rare dominant gene T.

AUTOSOMAL DOMINANT INHERITANCE

Table 4–1 shows five patterns of inheritance involving autosomal dominant genes in both parent and child but, if the gene is rare, only the $Tt \times tt$ mating is frequent enough to be of practical significance. In this mating, one parent is heterozygous for a rare autosomal dominant gene and the other parent is homozygous for the normal allele. Each child of the affected parent has a 1/2 chance of receiving the abnormal allele T and thus being affected, and a 1/2 chance of receiving the normal allele t and thus being normal. (The normal parent will give a normal allele to each child.) Though on average half the children will have the trait, the formation of each zygote is statistically an independent event, so that in any single family the ratio of affected to normal may be quite different from 1:1. The distribution of the trait in the family is not affected by sex, that is, both males and females can manifest the condition, and both can transmit it to children of either sex. Figure 4–2 is a stereotype pedigree illustrating these characteristics.

Dentinogenesis imperfecta is an example of an autosomal dominant phenotype. It is a dental defect with an incidence of about 1 in 8000. The teeth have a peculiar opalescent brown color and their crowns wear down readily; the abnormality is so obvious that the condition is quite easy to ascertain in members of a kindred in which the gene is segregating. Figure 4–3 is a pedigree of dentinogenesis imperfecta in four generations of a family. The condition of the teeth and gums of one of the affected children is shown in Figure 4–4.

If D symbolizes the dominant gene for dentinogenesis imperfecta and d its normal allele, an affected person has the genotype Dd and a normal person has the genotype dd. In the general population, half the progeny of $Dd \times dd$ matings

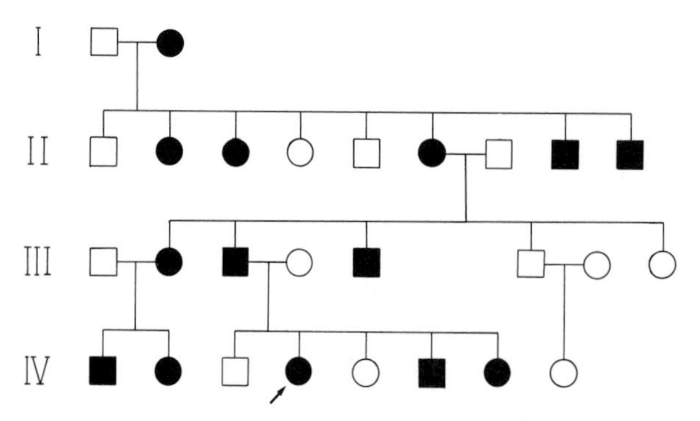

Figure 4–3. Pedigree of dentinogenesis imperfecta, an autosomal dominant disorder of dentin formation.

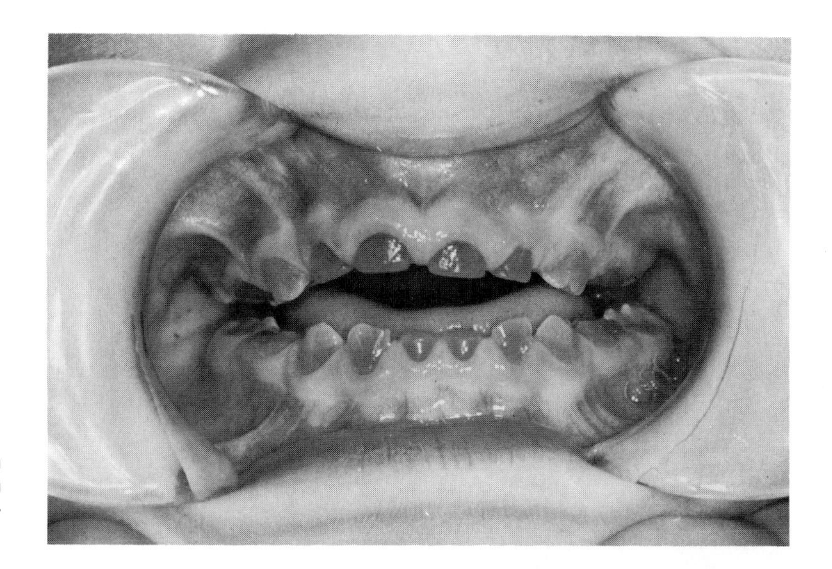

Figure 4–4. Dentinogenesis in the proband of the family shown in Figure 4–3. Courtesy of N. Levine.

are *Dd* and half *dd*. In individual families, each child has a 1/2 chance of being affected and a 1/2 chance of being normal. The *Dd* parent forms *D* and *d* gametes in equal numbers. The *dd* parent forms *d* gametes only:

Progeny of *Dd* × *dd* Mating

		Normal Parent *d*	Normal Parent *d*
Affected Parent	*D*	*Dd* Affected	*Dd* Affected
	d	dd Normal	dd Normal

The Homozygous Dominant Genotype

A dominant gene, by formal definition, has the same expression when heterozygous as when homozygous, but in medical practice homozygotes for rare dominants are seldom encountered. This is because homozygotes are produced only by the mating of two heterozygotes, and a mating of two persons both with the same rare dominant trait is statistically improbable. Any child of two heterozygous parents has a 1/4 risk of being homozygous affected.

Progeny of *Dd* × *Dd* Mating

	D	*d*
D	*DD* Homozygous affected	*Dd* Heterozygous affected
d	*Dd* Heterozygous affected	*dd* Homozygous normal

Proof that an individual is homozygous for a dominant mutation is that, regardless of the genotype of the spouse, all the offspring are affected (since each must receive one copy of the abnormal gene). In experimental animals, with

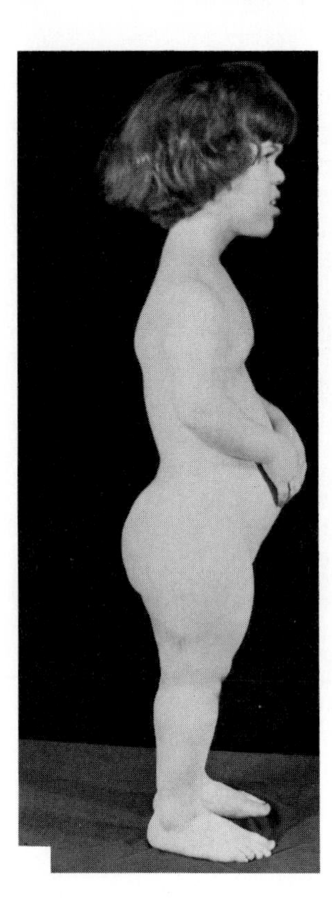

Figure 4–5. Achondroplasia, an autosomal dominant disorder that often occurs as a new mutation. Note small stature with short limbs, large head, low nasal bridge, prominent forehead, lumbar lordosis. From Tachdjian MO. Pediatric orthopedics, Vol. 1. Philadelphia: W. B. Saunders Company, 1972, p. 284.

large numbers of progeny, this test is practical. However, in human families, by chance a *Dd* person with a *dd* spouse might have only *Dd* children even though the chance of a *dd* child is 1/2 in each pregnancy. Thus it may be difficult to identify a homozygote positively, even among the offspring of two heterozygotes.

As it happens, many rare dominants are more severe in homozygotes and so can be identified by phenotype (or at least an informal judgment can be made that a child with very severe expression of a disorder that is present in milder form in both parents may well be a homozygote). Obviously this is not in agreement with the formal definition of a dominant given above. In medical genetics intermediate expression seems to be the rule in heterozygotes for dominant mutations.

Achondroplasia is an example of an autosomal dominant phenotype in which the homozygous form is lethal in early infancy. This is a skeletal disorder of short-limbed dwarfism and large head size, with bulging forehead and "scooped out" nose bridge (Fig. 4–5). True achondroplasia can be distinguished radiologically from the many other forms of short-limbed dwarfism because the interpeduncular distance of the vertebrae narrows, rather than widens, toward the caudal end of the vertebral column. Most achondroplastics have normal intelligence and wish to lead normal lives within their physical capabilities. Understandably, marriages between two achondroplastics are not uncommon, and in these marriages achondroplastic offspring may even be preferred to those of normal height. The homozygous achondroplastic offspring have a severe skeletal disorder and do not survive.

Another example is **familial hypercholesterolemia,** in which the heterozygotes have high plasma LDL cholesterol and develop premature coronary artery disease. The population incidence is high (about 1 in 500), so it is not surprising that occasionally two affected persons marry and produce homozygous offspring, who develop clinically severe coronary artery disease in early childhood.

Since a heterozygote has one normal allele and a homozygote has none, it is not really surprising to see a more severely abnormal phenotype in homozygotes. The observation demonstrates that classification as dominant and recessive is a convenient label rather than an essential distinction between dominance and recessivity.

New Mutation in Autosomal Dominant Disorders

When a disorder impairs the ability of the affected individual to reproduce ("reproductive fitness"), an appreciable proportion of patients with the disorder may be new mutants who have received the defective gene as a fresh mutation in a germ cell from a genetically normal parent. New mutation is not an unusual observation in autosomal dominant or X-linked pedigrees of severe disorders. In achondroplasia, at least 80 percent of the patients are new mutants. New mutation is much less common in autosomal recessives, where a phenotypically normal parent is much more likely to be a heterozygote than a new mutant.

There is an inverse relationship between the fitness of a condition and the proportion of all cases of the disorder that are caused by new mutation. Some disorders have a fitness of zero, or in other words they are invariably lethal before the reproductive age is attained. Others, including some very serious but late-onset diseases such as Huntington disease, which typically is not expressed until after the reproductive years, have high reproductive fitness and rarely arise through fresh mutation.

The significance of new mutation for genetic counseling is that when a particular patient can confidently be classified as a new mutant, the risk that the parents (or other relatives) will produce another such child is not greater than the general population risk. The risk for the offspring of the patient, however, is the usual Mendelian risk (50 percent for each child of an achondroplastic, for example).

There have been rare reports of more than one achondroplastic child in a sibship with normal parents (Reiser et al., 1984; Opitz 1984). Though chance alone would explain a few such cases, gonadal mosaicism is another possibility to be considered (see delayed mutation, Chapter 15). In general, exceptions to the rule are so infrequent that, practically speaking, they do not raise the theoretical recurrence risk above the general population risk.

Criteria for Autosomal Dominant Inheritance

In summary, the criteria for autosomal dominant inheritance are as follows:

1. The trait appears in every generation, with no skipping.

Although this statement is theoretically true, in clinical genetics there are apparent exceptions, when the proband is a new mutant or because of failure of penetrance or variability of expression (described later in this chapter).

2. Any child of an affected person has a 50 percent risk of inheriting the trait. Because statistically each child is an "independent event," in an individual family wide variation from the expected 1:1 ratio may be present.

3. Unaffected family members do not transmit the trait to their children.

As in point 1 above, failure of penetrance of the condition can lead to exceptions to this rule.

4. The occurrence and transmission of the trait are not influenced by sex;

that is, males and females are equally likely to have the trait and equally likely to transmit to children of either sex.

Codominance and Intermediate Inheritance

If both alleles of a pair are fully expressed in heterozygotes, they are said to be **codominant.** Many examples of codominance are provided by the various blood group and enzyme systems. For example, a person of blood group AB has both A and B antigens on his red cells. The allelic genes *A* and *B* are therefore codominant. DNA markers (RFLPs) are always codominant.

If the heterozygote is different from both homozygotes, the genes concerned are said to show **intermediate inheritance.** When red and white snapdragons are crossed, the pink progeny are clearly intermediate. In humans, as noted above, many heterozygous dominant phenotypes such as achondroplasia may well be intermediate between the normal and the homozygous abnormal forms, but this is speculative because of the difficulty of proving the genotype is homozygous. In sickle cell anemia, heterozygotes for the abnormal allele do not have the severe sickle cell anemia found in homozygotes; though they have hemoglobin S as well as normal hemoglobin A, a proportion of their red cells show the sickling phenomenon, and they have mild anemia. In other words, heterozygotes are intermediate in expression between normal homozygotes and sickle cell homozygotes and are said to have sickle cell trait. (The use of the term trait to indicate a heterozygous phenotype, as in sickle cell trait, is a specialized clinical usage. In human genetics, trait is used interchangeably with phenotype to indicate the visible or detectable expression of a gene.)

The difference between codominance and intermediate inheritance is sometimes indistinct. As noted above, heterozygotes for the sickle cell anemia gene do not have an "intermediate" type of hemoglobin but have both normal hemoglobin A and sickle cell hemoglobin (hemoglobin S). In other words, these genes are codominant as far as their hemoglobin production is concerned.

Multiple Alleles

All the examples used so far in this chapter have involved only a single pair of alleles, usually one "normal" and the other "abnormal." This simple situation is not the only possibility, for at many loci more than two different alleles are known—for some human loci, far more. When at a single locus more than two alternative alleles exist in the population, they are called multiple alleles.

The classic example of multiple allelism is provided by the series of alleles that determines the ABO blood groups. The best known alleles of the series are designated *O*, A^1, A^2 and *B*. For simplicity we will disregard A^2 and consider only the three main alleles. On this basis, the relationships of the genes, genotypes and phenotypes are as follows:

Genes	Genotypes	Phenotypes (Blood Groups)
O, A, B	*OO*	O
	AA *AO*	A
	BB *BO*	B
	AB	AB

Note that gene *O* is recessive to genes *A* and *B*, which are codominant. If *A* and *B* are present, both corresponding antigens are formed. *O* is an amorph, that is, a gene that has no effect, leaving the substrate (H antigen) unaltered.

It was originally believed that the ABO blood groups were determined by genes at two independent loci, that is, by two pairs of nonallelic genes. Alleles can be distinguished from nonalleles by analysis of family data, since alleles segregate in the progeny, whereas nonalleles assort independently. The progeny of an AB × O mating are always A or B, never AB or O. In other words, gene *A* and gene *B* are never both transmitted from the parent to the child but always segregate; hence they must be allelic.

It is not always easy in human genetics to prove whether two rare genes are alleles or whether they are at independent loci. One reason is that families suitable for analysis are not often seen. Furthermore, even if two genes do not assort independently, they may be linked rather than allelic. The critical test for distinguishing linkage from allelism is that recombination can occur between the loci of linked genes (by crossing over in meiotic prophase), but not within a locus, that is, not between alleles. However, at the molecular level there are exceptions even to this rule (see hemoglobin Lepore and haptoglobin in Chapters 5 and 9).

AUTOSOMAL RECESSIVE INHERITANCE

An autosomal recessive trait is expressed only in homozygotes, who have received the recessive gene from both parents. The trait may or may not appear among the sibs of the proband but typically does not occur in relatives other than sibs. Exceptionally, other relatives may also be affected, especially in large inbred kindreds. Table 4–1 shows three mating types that can produce recessively affected offspring, but only one of these (heterozygote × heterozygote) is at all common.

The most frequent autosomal recessive disorder in white ("Caucasian") children is **cystic fibrosis,** a condition in which there are abnormalities of several exocrine secretions, including pancreatic and duodenal enzymes, sweat chlorides and bronchial secretions. The thick, viscid mucus produced by the bronchi is particularly serious, since it makes the affected children highly susceptible to pneumonia. Males are infertile as a secondary consequence of abnormal mucous secretions in the vas deferens. The loss of chlorides in the sweat may be severe enough to cause heat prostration in warm weather.

In whites, cystic fibrosis affects perhaps 1 child in 2000 births. About 1 person in 22 is a heterozygous carrier. Both homozygotes and heterozygotes are much less common in Orientals. (The mathematical relationship between gene frequency and genotype frequency is discussed in Chapter 15.)

In a hypothetical family of four children produced by carrier parents (as shown in Figure 4–6), one child will be homozygous affected, two will be heterozygous and phenotypically normal, and one will be homozygous for the normal allele and also phenotypically normal. These proportions are rarely observed in actual families but are averages for the general population.

Because children with recessive traits usually have phenotypically normal (but heterozygous) parents, usually the only families that can be recognized and studied are those in which there is already at least one affected child. Families in which no child is affected merge with the general population and are not ascertained. Bias of ascertainment creates a statistical problem in that the proportion of affected children in the sibships that can be ascertained will be

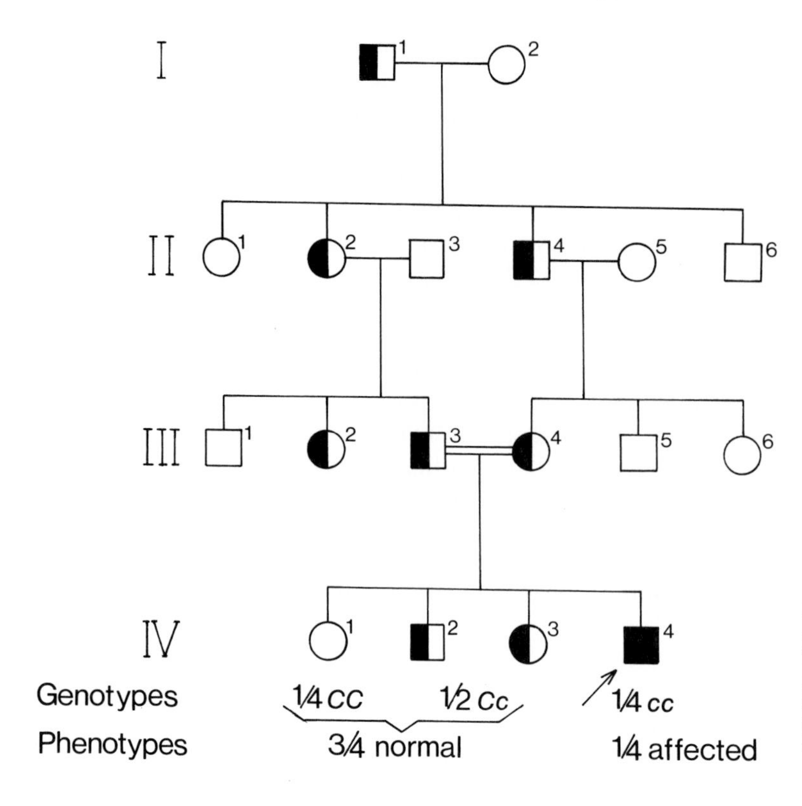

Figure 4–6. Stereotype pedigree of autosomal recessive inheritance, showing parental consanguinity. A gene from a common ancestor I-1 has been transmitted through two lines of descent to "meet itself" in IV-4.

well above the theoretical one fourth, unless the sibship size is very large (about 12). We will return to this problem and ways of dealing with it in Chapter 14.

Consanguinity and Recessive Inheritance

A carrier of an autosomal recessive gene can have affected children only if the other parent is also a carrier (or less frequently, a recessive homozygote). The risk that a carrier of cystic fibrosis will mate with another carrier is one in 22 (the incidence of carriers in the population). However, since rare recessive genes are passed down in families and so are concentrated in family groups, the risk that one carrier will mate with another is usually higher by at least one order of magnitude if he mates with a near relative than if he mates at random. (In genetics, random mating means marriage without regard to the genotype of the spouse.) For example, consider the chance that a carrier of cystic fibrosis

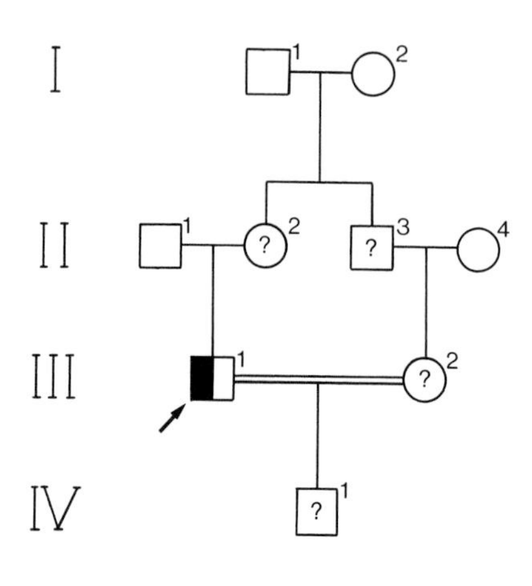

Figure 4–7. The probability that III-1, a carrier of cystic fibrosis (CF) who marries a first cousin, will have an affected child. The probability that II-2 is a carrier is 1/2. If II-2 is a carrier, the probability that II-3 is a carrier is 1/2. If II-3 is a carrier, the probability that III-2 is a carrier is 1/2. Thus the chance III-2 is a carrier is $(1/2)^3 = 1/8$, and the chance IV-1 will be homozygous for CF is $1/8 \times 1/4 = 1/32$. For simplicity, this discussion ignores the additional chance that III-2 will inherit the CF gene from another line of descent. This probability is equal to the population frequency of the CF gene, which is about 1/44 in white North Americans.

Table 4–2. EFFECT OF COUSIN MARRIAGE ON THE INCIDENCE OF A RECESSIVE CONDITION

	Affected Individuals		
Frequency of Recessive Gene	*In General Population*	*Among Children of Cousin Marriages*	*Ratio*
0.2	0.04	0.05	1.25
0.02	0.0004	0.0016	4.00
0.002	0.000004	0.00013	32.25

Data from Li CC. Amer J Med 1963:34:702.

who marries a first cousin will have children homozygous for the gene. Figure 4–7 sets out the various probabilities involved. In brief, the risk that a first cousin of a carrier is also a carrier because of inheritance of the gene from a common ancestor is one in eight. The mathematical aspects of consanguinity are treated in additional detail in Chapter 15.

The effect of inbreeding in producing homozygous recessives is the biological basis of the prohibition of cousin marriage in many societies. For a condition as common as cystic fibrosis, the chance that a carrier who marries a cousin will have an affected child is only a little greater (about three times as high, on the basis of the incidence figure used here) than the general population risk. For less common conditions, the ratio is higher. Some examples are given in Table 4–2, which compares the incidence of recessive traits of different frequencies in children of cousin matings with the incidence in the general population.

In clinical medicine, parental consanguinity is a strong clue, though not proof, that a disorder is autosomal recessive. Even if the parents consider themselves unrelated they may have common ancestry within the last few generations, especially if they are of closely similar ethnic or geographic origin. Thus in taking a family history it is important to ask about consanguinity and background.

RARE RECESSIVES IN GENETIC ISOLATES

There are many small groups in which the frequency of certain rare recessive genes is quite different from that in the general population. Such groups, genetic **isolates,** may have become separated from their neighbors by geographic, religious or linguistic barriers.

In Ashkenazi Jews in North America, the gene for **Tay-Sachs disease** (G_{M2} gangliosidosis) is very common. Tay-Sachs disease is a neurological degenerative disorder that develops at about six months of age. Affected children become blind and regress mentally and physically. A "cherry-red spot" in the fundus of the eye is a striking diagnostic sign. The disease is usually fatal in early childhood. The frequency is 100 times as high in Ashkenazi Jews (one in 3600) as in other populations (one in 360,000). A specific lysosomal enzyme, hexosaminidase A (hex A), is virtually absent in affected children.

When a recessive trait has a high frequency in a population, cousin marriage is not a striking feature of pedigrees of the trait. This is because the gene frequency is so high that a carrier who marries another member of the same group is almost as likely to marry another carrier as if he had married a close relative. Consequently, among Ashkenazi Jews the parents of affected children are usually not closely consanguineous, whereas in other populations the consanguinity rate in the parents of Tay-Sachs patients is high.

CARRIER IDENTIFICATION

As noted earlier, though by definition recessive genes are not expressed in carriers, many recessives that are not clinically significant in carriers nevertheless are detectable by appropriate tests. In Tay-Sachs disease the carrier test involves demonstration of intermediate levels of hex A. It has been applied for mass screening of the Ashkenazi Jewish population in many centers.

There are many other examples of rare recessives in genetic isolates. **Tyrosinemia,** a very rare and lethal hepatic disease of early infancy, has been recognized in many French-Canadian children in the isolated Lac St. Jean-Chicoutimi region of Quebec in recent years, but is virtually unknown elsewhere. Laberge (1969) traced both parents of each affected child to a couple who were in Quebec City by 1644. Probably one member of this couple was a carrier of the tyrosinemia gene, which has been passed down 10 or more generations to "meet itself" in double dose in some of the remote descendants. The carrier frequency in the isolate is of the order of one in 30, and, as expected, the parents of the affected children are not closely consanguineous.

GENETIC COUNSELING IN CONSANGUINEOUS MARRIAGES

Are cousin marriages unwise? There is considerable disagreement among geneticists as to whether the extra risk of defective offspring is significant. A commonly quoted statistic from Japan is that the proportion of abnormal progeny is about 2 percent for random marriages and about 3 percent for consanguineous marriages. From the population standpoint this may be an inconsequential increase, and life insurance companies do not consider the offspring of cousins to have higher-than-normal insurance risks. Nevertheless, from comparative studies on inbred and outbred human populations, as well as studies of experimental organisms, certain general conclusions can be drawn:

1. Inbred children are at some increased risk of early mortality, congenital abnormalities and cognitive disability.
2. The disorders seen in inbred children are not different in nature from those for which any child is at risk, but their frequency may be higher.
3. The more closely related the parents, the higher the risk of defective offspring.
4. The risk is higher when the parents are from a normally outbred population than when they are from a relatively inbred group.
5. With minor exceptions, the increase in risk applies only to the first-generation offspring of related parents, not to subsequent generations.

If cousin parents have a child who suffers from a recessively inherited disease, they are thereby proved to be carriers, with a 1/4 chance of having a similar affected child at any later pregnancy. Furthermore, first cousins may share more than one deleterious recessive gene; in fact, they have 1/8 of their genes in common, and their progeny are, on the average, homozygous at 1/16 of their gene loci.

Consanguineous marriages more distant than those of first cousins have a correspondingly lower risk of producing affected offspring. For example, second cousins have only 1/32 of their genes in common and third cousins only 1/128. For couples less closely related than second cousins, the risk of defective offspring is near enough to the general population risk to be of little or no genetic consequence. Nevertheless, if a child has a rare or undiagnosed defect and his parents are even remotely consanguineous, it is a reasonable hypothesis that his defect has autosomal recessive inheritance, with a one-in-four recurrence risk for later-born sibs.

GENETIC CONSEQUENCES OF INCEST

Matings closer than first-cousin marriages are illegal in many jurisdictions (with the exception of marriages of double first cousins, which are not proscribed although the relationship of double first cousins is as close as the uncle-niece relationship). However, parent-child and brother-sister matings do take place and do produce children. The children of such closely related parents are homozygous at one-fourth of their gene loci and have a correspondingly high risk of being homozygous for some deleterious gene.

In view of the social stigma attached to incest, it is not surprising that there is little genetic information about its consequences. The few studies available agree that children of incest are at high empiric risk of abnormality, severe mental retardation and early death. A report from British Columbia (Baird and McGillivray, 1982) describes 29 children born of incestuous unions, among whom most had severe abnormalities, low birth weight, developmental delay or other medical problems. Four of the 29 had specific autosomal recessive conditions.

The Offspring of Homozygous Recessives

To indicate the types of offspring a recessive homozygote can produce, we return to our original example, the nontaster trait. Table 4–1 shows three matings involving at least one nontaster parent: a *tt* person may have a *TT*, *Tt* or *tt* spouse. In brief:

Parents	Progeny	
	Genotypes	*Phenotypes*
tt × *TT*	All *Tt*	All taster
tt × *Tt*	1/2 *Tt*, 1/2 *tt*	1/2 taster, 1/2 nontaster
tt × *tt*	All *tt*	All nontaster

Note that the *tt* × *Tt* mating, in which an affected person has an apparently normal spouse but half the children are affected, mimics the pattern of inheritance of a rare dominant trait. This pattern is termed **quasidominant inheritance.** It cannot always be firmly distinguished from the ordinary autosomal dominant pattern, but a pedigree of such a trait over more than two generations will probably reveal that in each generation the characteristics of recessive inheritance are present; in particular, the trait will usually appear only in the parent sibship and the offspring sibship, not elsewhere in the kindred, and consanguinity may be present in the parents and/or in one set of grandparents. A typical quasidominant pedigree is shown in Figure 4–8.

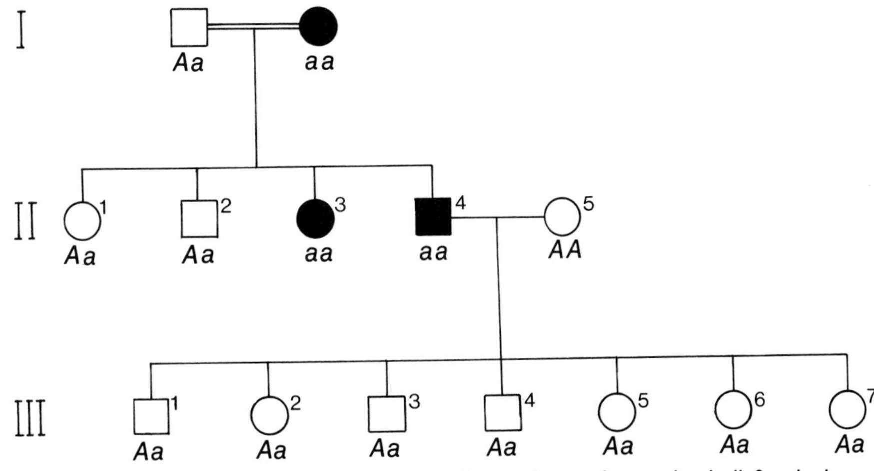

Figure 4–8. Quasidominant inheritance of an autosomal recessive trait. See text.

Criteria for Autosomal Recessive Inheritance

1. The trait characteristically appears only in sibs, not in their parents, offspring or other relatives.
2. On the average, one fourth of the sibs of the proband are affected; in other words, the recurrence risk is one in four for each birth.
3. The parents of the affected child may be consanguineous.
4. Males and females are equally likely to be affected.

X-Linked Inheritance

Genes on the sex chromosomes are distributed unequally to males and females within kindreds. The inequality produces characteristic and readily recognizable patterns of inheritance, and has led to the identification of many "sex-linked" conditions in man.

"Sex-linked" genes may be X-linked or Y-linked, but for all practical purposes only X linkage has any clinical significance. Apart from genes essential for male sex determination, little is known about the genetic constitution of the Y chromosome. Genes on the Y would show **holandric inheritance;** that is, they would be passed down rather like the family surname, in the male line exclusively, by an affected man to all his sons and to none of his daughters. From time to time pedigrees purporting to show holandric inheritance have been reported, but they have not withstood searching examination. The terms sex linkage and X linkage may be used synonymously, but most medical geneticists prefer to use X linkage because it is more specific.

The distribution of X-linked traits in families follows the course of the X chromosome carrying the abnormal gene. Since females have a pair of X chromosomes but males have only one, there are three possible genotypes in females but only two in males. One shorthand way to indicate that a gene is X-linked is this: let X_H represent a dominant gene H on the X, and X_h its recessive allele h. Expressed in symbols, the following are the possible combinations in males and females:

Males	Females
X_HY	X_HY_H
X_hY	X_HX_h
	X_hX_h

A male is said to be **hemizygous** with respect to X-linked genes. A female may be either homozygous or heterozygous (a carrier).

An important difference between autosomal and X-linked inheritance is that, whereas for autosomal alleles both members of a pair are genetically active, in females only one member of the pair of X's is active; the second X remains inactivated, appearing in interphase cells as the sex chromatin (Barr body). Thus in females as in males there is only one functional X. In heterozygous females it is a chance matter whether the paternal or maternal X is the functional one in a given cell. Consequently, though we distinguish X-linked "dominant" and "recessive" patterns of inheritance, it is important to keep in mind that a female heterozygous for either a dominant or a recessive X-linked mutant gene has the mutant as the only functional allele at that locus in about half her body cells. Meanwhile, the inherited allele is fully expressed in the hemizygous male whether it is dominant or recessive. Accordingly for X-linked traits the distinction between "dominant" and "recessive" is blurred.

If a female is heterozygous for an X-linked mutant gene, on the average approximately half her cells have the normal and half the abnormal allele as the functional member. However, there are exceptional carriers in whom far more cells have the normal allele as the functional one, and others in whom the mutant is active in the majority of cells. Thus X-linked traits are unusually variable in their expression in carriers. For further discussion, see X inactivation, Chapter 7.

X-LINKED RECESSIVE INHERITANCE

The inheritance of recessive genes on the X chromosome follows a well-defined pattern. An X-linked recessive trait is expressed by all males who carry the gene, but by females only if they are homozygous. Consequently, X-linked recessive diseases are practically restricted to males and are rarely if ever seen in females.

Hemophilia (classical hemophilia, hemophilia A) is an X-linked recessive disease in which the blood fails to clot normally because of a deficiency of antihemophilic globulin (Factor VIII). The clinical features, which include severe arthritis as a consequence of internal hemorrhages into the joints, are secondary to the clotting defect. The incidence is about 1 in 10,000 male births. The hereditary nature of hemophilia was recognized in ancient times; it has since achieved notoriety by its occurrence among descendants of Queen Victoria, who was a carrier.

We have used the symbol X_h to represent a recessive gene *h* on the X chromosome, and X_H to represent its dominant allele. To demonstrate the pedigree patterns of X-linked recessive inheritances, these symbols will now be used to denote the hemophilia gene and its normal counterpart.

The following checkerboard shows the offspring of an affected male and normal female:

$X_hY \times X_HX_H$

	X_H	X_H
X_h	X_HX_h	X_HX_h
Y	X_HY	X_HY

Daughters: all heterozygotes
Sons: all normal

Note that an affected male does not transmit the gene to any of his sons, but he gives it to all of his daughters, who are therefore carriers. If a carrier daughter marries a normal male, four genotypes are possible in the offspring, with equal probabilities.

$X_HY \times X_HX_h$

	X_H	X_h
X_H	X_HX_H	X_HX_h
Y	X_HY	X_hY

Daughters: 1/2 normal, 1/2 carriers
Sons: 1/2 normal, 1/2 affected

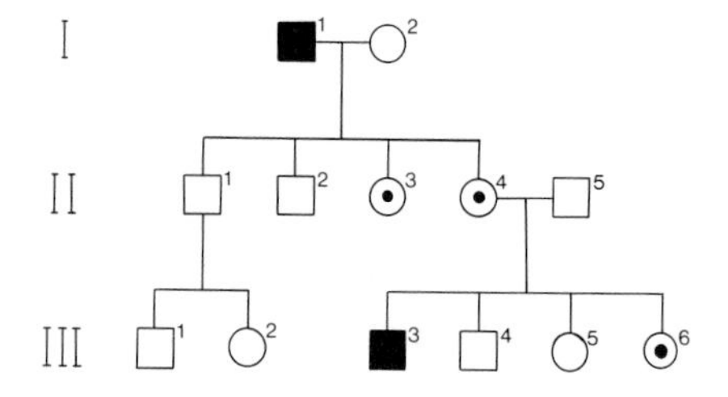

Figure 4–9. Stereotype pedigree of X-linked recessive inheritance.

Note that the X-linked recessive trait of the maternal grandfather, which did not appear in any of his own progeny, now reappears among his grandsons. Figure 4–9 is a stereotype pedigree to show the characteristics of X-linked recessive inheritance.

Note also that the daughter of a carrier has a 1/2 chance of being a carrier. By chance, an X-linked trait may be transmitted through a series of carrier women before it makes its appearance in an affected male.

For reference, the mating types, the gametes produced and the progeny expected on the basis of X-linked inheritance are shown in Table 4–3.

Duchenne muscular dystrophy (pseudohypertrophic muscular dystrophy) is an X-linked disease of muscle which affects young boys (Fig. 4–10). It is usually apparent by the time the child begins to walk and progresses inexorably, so that the child is confined to a wheelchair by about the age of 10 and is unlikely to survive his teens. This disorder is a genetic lethal, in that its nature prevents its transmission by affected males, but it may be transmitted by carrier females, who themselves rarely show any clinical manifestation of muscular dystrophy.

Table 4–3. X-LINKED INHERITANCE: MATING TYPES, GAMETES AND EXPECTED PROPORTIONS OF PROGENY FOR A PAIR OF X-LINKED ALLELES *H* AND *h*

	Gametes		Progeny	
Mating Types	*Ova*	*Sperm*	*Genotypes*	*Phenotypes*
$X_H X_H \times X_H Y$	X_H	X_H Y	$X_H X_H$ $X_H Y$	all normal
$X_H X_h \times X_H Y$	X_H X_h	X_H Y	$X_H X_H$	daughters — 1/2 normal
			$X_H X_h$	daughters — 1/2 carriers
			$X_H Y$	sons — 1/2 normal
			$X_h Y$	sons — 1/2 affected
$X_h X_h \times X_H Y$	X_h	X_H Y	$X_H X_h$ $X_h Y$	daughters all carriers sons all affected
$X_H X_H \times X_h Y$	X_H	X_h Y	$X_H X_h$ $X_H Y$	daughters all carriers sons all normal
$X_H X_h \times X_h Y$	X_H X_h	X_h Y	$X_H X_h$	daughters — 1/2 carriers
			$X_h X_h$	daughters — 1/2 affected
			$X_H Y$	sons — 1/2 normal
			$X_h Y$	sons — 1/2 affected
$X_h X_h \times X_h Y$	X_h	X_h Y	$X_h X_h$ $X_h Y$	all affected

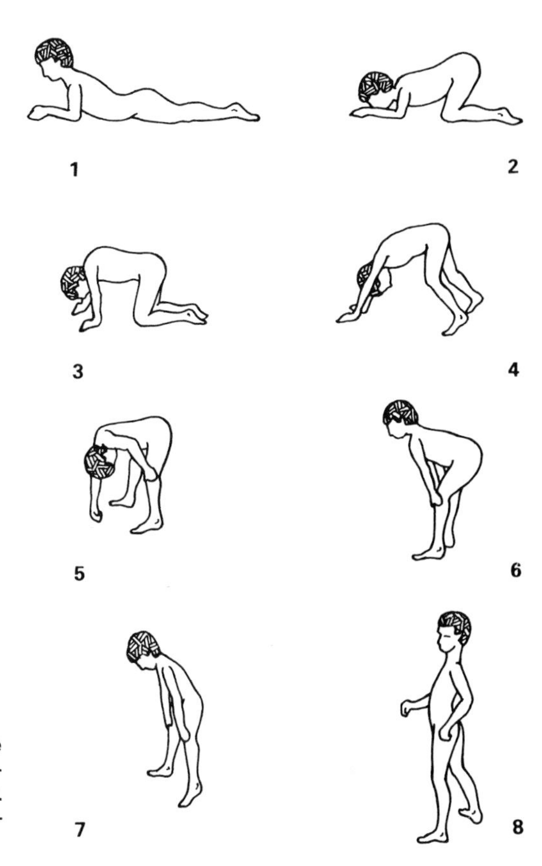

Figure 4–10. Duchenne muscular dystrophy, illustrating the Gowers sign, the characteristic "climbing up himself" maneuver by which the child rises from the prone position. From Dubowitz V. Muscle disorders in childhood. Philadelphia: W. B. Saunders Company, 1978, p. 26.

A representative pedigree of Duchenne muscular dystrophy is shown in Figure 4–11. The genetics of this disorder is discussed further in Chapter 14.

Criteria for X-Linked Recessive Inheritance

1. The incidence of the trait is much higher in males than in females.
2. The trait is passed from an affected man through all his daughters to, on the average, half their sons.
3. The trait is never transmitted directly from father to son.
4. The trait may be transmitted through a series of carrier females; if so, the affected males in a kindred are related to one another through females.
5. Carriers show variable expression of the trait.

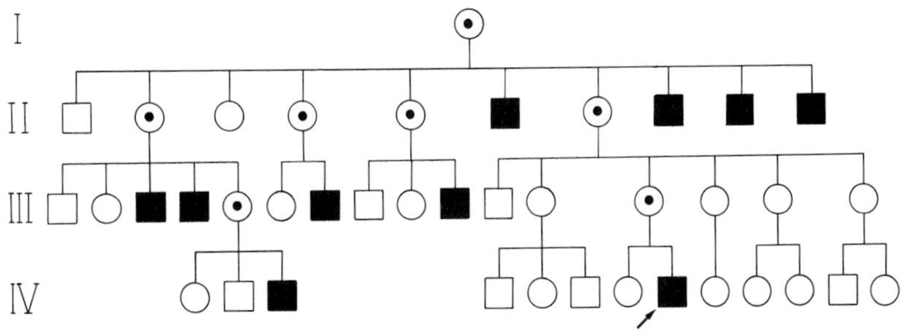

Figure 4–11. A pedigree of Duchenne muscular dystrophy (DMD), an X-linked disorder in which affected males almost never reproduce. See further discussion of the inheritance of DMD in Chapter 14.

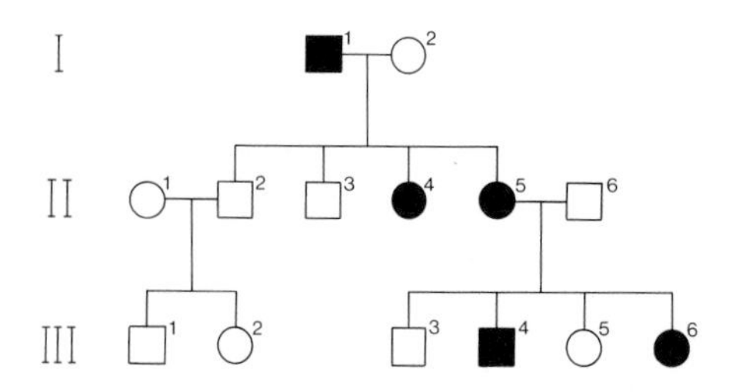

Figure 4–12. Stereotype pedigree of X-linked dominant inheritance. Affected males have no affected sons and no normal daughters.

X-LINKED DOMINANT INHERITANCE

Whereas X-linked recessive traits typically occur in males only, rare X-linked dominants are approximately twice as common in females as in males. The chief characteristic of X-linked dominant inheritance is that an affected male transmits the gene (and the trait) to all his daughters and to none of his sons (Fig. 4–12). If any daughter is normal or any son affected, the gene concerned must be autosomal, not X-linked. Because females have a pair of X chromosomes just as they have pairs of autosomes, it is not possible to distinguish between autosomal and X-linked inheritance of a dominant trait by observing its transmission to the progeny of affected females; whether a trait is autosomal or X-linked, affected females transmit it to half their progeny of either sex.

The X-linked blood group system Xg is an example of X-linked dominant inheritance. The Xg system is governed by a pair of alleles, Xg^a and Xg, which produce two phenotypes, Xg(a+) and Xg(a−). (Blood group terminology and symbols are described in Chapter 9.) The possible genotypes and phenotypes in the two sexes are as follows:

Males		Females	
"Genotypes"*	**Phenotypes**	**Genotypes**	**Phenotypes**
Xg^aY	Xg(a+)	*Xg^aXg^a* *Xg^aXg*	Xg(a+)
XgY	Xg(a−)	*XgXg*	Xg(a−)

*This is not really a genotype, because the Y chromosome is shown as well as the X-linked gene, but it is easier to visualize the pattern of transmission when both members of the chromosome pair are shown.

The transmission of Xg blood groups in families demonstrates the general pattern of X-linked dominant inheritance.

Mating: Xg(a+) Male × Xg(a−) Female (*Xg^aY × XgXg*)

	Xg	*Xg*
Xg^a	*Xg^aXg*	*Xg^aXg*
Y	*XgY*	*XgY*

	Genotypes	**Phenotypes**
Daughters:	*Xg^aXg*	Xg(a+) (like father)
Sons:	*XgY*	Xg(a−) (like mother)

Mating: Xg(a−) Male × Heterozygous Xg(a+) Female (*XgY × Xg^aXg*)

	Xg^a	*Xg*
Xg	*Xg^aXg*	*XgXg*
Y	*Xg^aY*	*XgY*

	Genotypes	**Phenotypes**
Daughters:	*Xg^aGXg, XgXg*	1/2 Xg(a+), 1/2 Xg(a−)
Sons:	*Xg^aY, XgY*	1/2 Xg(a+), 1/2 Xg(a−)

Males and females each have a 1/2 chance of receiving the dominant allele.

Only a few genetic disorders exhibit the X-linked dominant pattern. One example is hypophosphatemia, also called vitamin D–resistant rickets. In rare X-linked dominants, affected females (heterozygotes) are twice as common as affected males (hemizygotes) but usually have a milder form of the defect. Hypophosphatemia fits this criterion in that, although both sexes are affected, the serum phosphorus is less depressed and the rickets less severe in heterozygous females than in males.

X-Linked Traits with Expression Limited to Females

A few rare genetic defects expressed exclusively or almost exclusively in females appear to be determined by X-linked dominant genes that are usually lethal in males before birth. Typical pedigrees of these conditions, all of which are rare, show transmission by affected females, who produce affected daughters, normal daughters and normal sons in equal proportions, with no affected sons. The pedigrees may also show a large number of miscarriages, some of which may represent losses of affected males.

Criteria for X-Linked Dominant Inheritance

1. Affected males have no normal daughters and no affected sons.
2. Affected heterozygous females transmit the condition to half their children of either sex. Affected homozygous females transmit it to all their children. Transmission by females follows the same pattern as an autosomal dominant. In other words, X-linked dominant inheritance cannot be distinguished from autosomal dominant inheritance by the progeny of affected females, but only by the progeny of affected males.
3. Affected females are more common than affected males, but as they are almost always heterozygotes they usually have milder (but variable) expression.

THE FRAGILE X SYNDROME

It is well known that there is an excess of males in the mentally retarded population, but recognition that the male excess is chiefly the consequence of X-linked genes is quite recent. A common form of X-linked mental retardation has now been defined in which there is a cytogenetic marker on the X chromosome. The incidence of this condition, usually called fragile X syndrome, may be as high as 1 in 1000 male births. The marker, or fragile site, is a constriction near the distal end of Xq (Fig. 4–13). The combination of a defect caused by a

Figure 4–13. The characteristic fragile site on the long arm of the X chromosome, seen in a proportion of cells from males with fragile X-linked mental retardation.

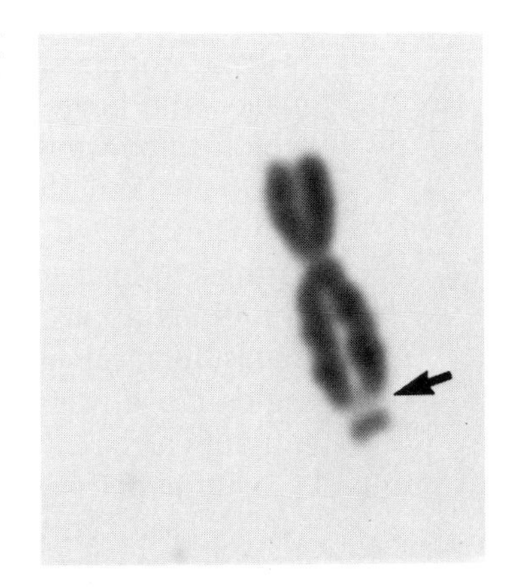

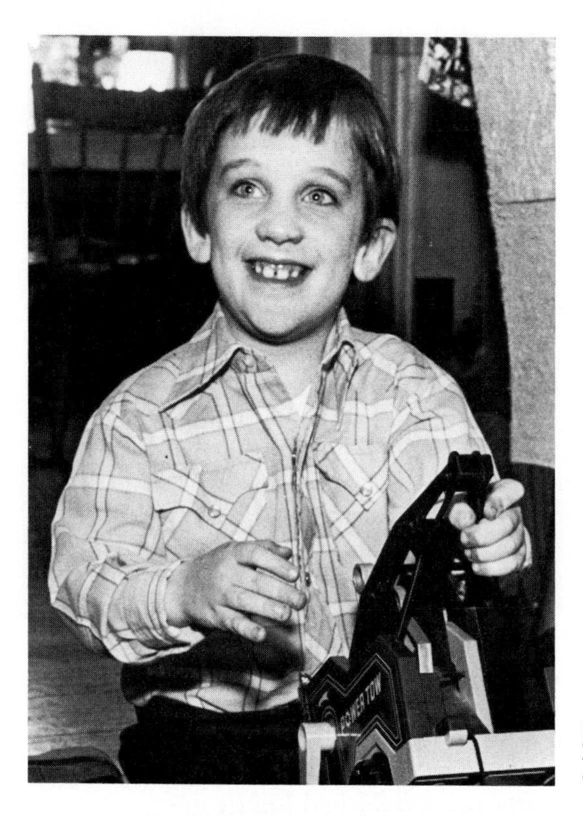

Figure 4–14. Characteristic facies of a patient with fragile X-linked mental retardation. Courtesy of M. W. Partington and the Kingston, Ontario, Whig-Standard.

mutant gene but associated with a specific cytogenetic abnormality is unique in medical genetics. Though the syndrome is undoubtedly X-linked, it cannot unequivocally be assigned to either the dominant or the recessive category.

The clinical features of the fragile X syndrome include large testes especially after puberty, large protuberant ears and a prominent chin (Fig. 4–14). All these features represent mild degrees of overgrowth. The average intelligence is in the moderately retarded range, but speech development is disproportionately delayed. Carrier females may be completely normal though one-third or more have learning difficulties and some are moderately retarded.

The fragile X is seen in only a proportion (35 percent or so) of cells of affected males. Many carrier females do not show the fragile site at all, and in those who do show it the frequency is much lower than in affected males. The expression depends on special culture conditions, especially the use of culture medium that has a relatively low concentration of folic acid and thymidine.

Genetic analysis of fragile X families has shown some unexpected findings (Turner and Jacobs, 1983; Sherman et al, 1984). It appears that 20 percent of males who have the gene are not retarded. There are several large pedigrees in which unaffected males have transmitted the condition to all their daughters, who have had affected sons, thereby proving their carrier status. The mutation rate is estimated to be 7.2×10^{-4}, an extremely high figure for a mutation rate in man. (Most mutation rates measured in man are of the order of 10^{-6}.) The mutations appear never to occur in eggs, but only in sperm; that is, affected males themselves are never new mutants. However, the mothers of affected males are always carriers, and over half of the mothers are themselves new mutant carriers.

For genetic counseling in fragile X families, the mother of an affected male can be assumed to be a carrier. Tentative risk figures indicate that any normal brother has a 17 percent chance of having the gene (20 percent of 50 percent of the brothers have the gene but do not express it; 50 percent do not have the gene; thus 10 percent + 50 percent = 60 percent of the brothers are not mentally retarded and 10/60 = 17 percent of the nonretarded brothers have the gene). This is a different situation from the usual one in genetic counseling for X-linked diseases, in which normal brothers do not have the mutant gene and thus cannot transmit it. A nonretarded sister of a patient can be confirmed as a carrier if she has the fragile X, but if she does not have it there still may be about a 30 percent chance that she is a carrier.

There is still much to be learned about fragile X syndrome, its expression and its genetics. A search for closely linked restriction fragment length polymorphisms that could be used as tags for the gene in families is being actively pursued. Success in the search could help to clarify many of the puzzling characteristics of the syndrome. Fragile X syndrome is such a common condition that it should be considered in the differential diagnosis of mental retardation in either males or females.

Variation in Gene Expression

Some genetic conditions segregate sharply; that is, the normal and abnormal phenotypes can be distinguished clearly. In ordinary experience, however, the clinical expression of a disorder may be extremely variable, the onset age may be late or variable, or the expression may be modified by other genes or by environmental factors. These problems are particularly characteristic of autosomal dominant phenotypes and can lead to difficulties in diagnosis and confusion in pedigree interpretation.

PENETRANCE AND EXPRESSIVITY

A mutant gene may not always be phenotypically expressed or, if it is expressed, the expression may vary widely in different individuals. Penetrance applies to a gene's likelihood of being expressed at all; expressivity refers to the degree of expression, i.e., whether clinically the condition is expressed in a mild, moderate or severe form.

When the frequency of expression of a trait is below 100 percent, that is, when some individuals who have the appropriate genotype fail to express it, the trait is said to exhibit **reduced penetrance.** For example, in Figure 4–15, the mutant gene must be carried by II–4, since she transmits it, though she does not express it; in her case, it is therefore said to be nonpenetrant. Penetrance is an all-or-none concept. In mathematical terms, it is the percentage of individuals who have the gene for a condition who actually show the trait.

If, on the other hand, a trait takes somewhat different forms in different members of a kindred, it is said to exhibit **variable expressivity**. Members of the same kindred may express the same gene in different ways and with different severity. Though expressivity may be roughly equated with clinical severity,

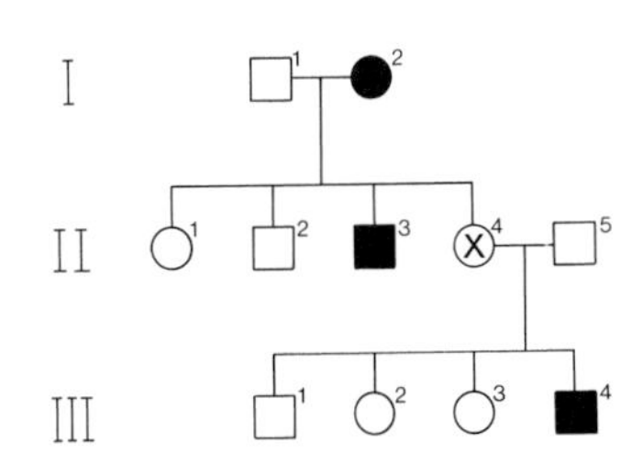

Figure 4–15. Reduced penetrance of an autosomal dominant trait. See text. For further discussion, see Chapter 14.

some single-gene syndromes have a variety of phenotypic effects, so expressivity must also take into account the range of abnormalities present.

Variability of expression is sometimes more obvious between members of different family groups than among members of a single kindred. Stated differently, some disorders that appear to have variable expression tend to "breed true" in families. It is not really clear why this should be so, especially since other disorders can be highly variable even within a sibship. A probable explanation for conditions that show less variation in clinical expression within than between families is that different mutations are responsible; in other words, genetic heterogeneity rather than variable expressivity is present. Recently, molecular methodology has revealed unexpected heterogeneity in the types of mutations that can occur in a single protein, such as collagen (Prockop 1984; Byers and Bonadio 1985). A second possibility is that modifying "background" genes elsewhere in the genome, which would vary more between than within families, might be involved.

It is a common clinical observation that failure of penetrance and variability of expression are much more common in autosomal dominant than in other pedigrees. The reason for this is not well understood. Female carriers of X-linked traits also show considerable phenotypic variability; this can readily be explained on the basis of the random nature of X inactivation, which is described more fully in Chapter 7.

PLEIOTROPY: ONE GENE, MORE THAN ONE EFFECT

Each gene has only one primary effect in that it directs the synthesis of a polypeptide chain. From this primary effect, however, many different consequences may arise. Multiple phenotypic effects produced by a single mutant gene or gene pair are examples of the principle of pleiotropy or pleiotropism.

In any sequence of events, interference with one early step may have ramifying effects. Thus a single defect occurring early in development can lead to various abnormalities in fully differentiated structures. In some cases a primary gene product might participate in a number of unrelated biosynthetic pathways, possibly at different times.

Clinical syndromes offer many examples of pleiotropy. The literal meaning of the word syndrome is "running together," and the term is used to describe the set of signs and symptoms characteristic of a specific disorder. In some syndromes, there is a clear or at least a plausible mechanism by which the primary effect of the gene could produce the diverse features of the syndrome. For example, in **phenylketonuria**, an autosomal recessive metabolic disease, the enzyme phenylalanine hydroxylase is lacking. The primary effect in homozygotes is the specific enzyme deficiency. There are multiple secondary effects, notably severe mental retardation, excretion of phenylketones in the urine and dilution of pigmentation. Though the pathways by which these secondary effects are produced are not known in full detail, the steps by which they are related to the primary defect can at least be conjectured. Similarly, in galactosemia lack of the enzyme galactose-1-phosphate uridyl transferase is the primary effect of homozygosity for the recessive gene concerned. The array of secondary effects, which include cirrhosis of the liver, cataracts, galactosuria and mental retardation, can be produced in experimental animals if they are fed a diet high in galactose. Furthermore, the secondary effects can be prevented in many genetically susceptible infants if the condition is recognized and the child is placed on a galactose-free diet at birth. Thus, the characteristic features of the disease are clearly secondary to the enzyme deficiency.

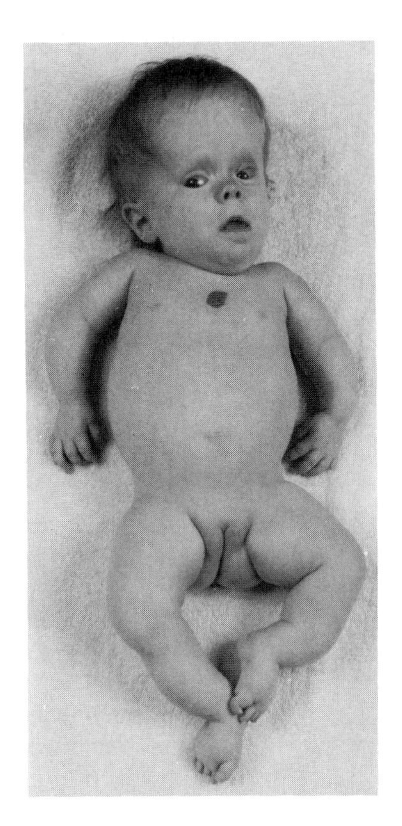

Figure 4–16. Autosomal dominant osteogenesis imperfecta in an eight-month-old female. Note triangular face, pointed nose, broad forehead, short chest and bowing of extremities secondary to fracture.

In some syndromes the various clinical signs and symptoms are obviously related to a single underlying structural defect. For example, in osteogenesis imperfecta type 1 (Fig. 4–16) the brittle bones, blue sclerae, otosclerosis and other manifestations are now known to stem from a basic defect in collagen synthesis (see later in this chapter). In the Marfan syndrome (Fig. 4–17), the

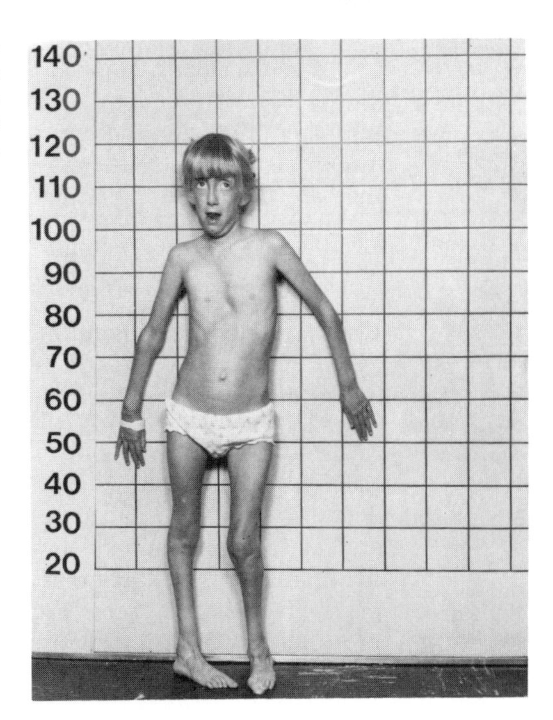

Figure 4–17. Marfan syndrome in a girl aged 5.5 years. Her height is above the 97th percentile for her age. Note limb length with long extremities and arachnodactyly, genu valgum, long facies and pectus excavatum. The child also has a high narrow palate, bilateral ectopia lentis and myopia. Photograph courtesy of V. A. McKusick.

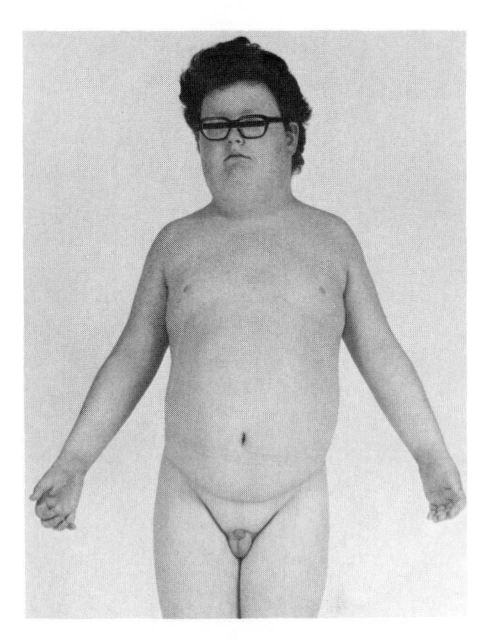

Figure 4–18. Laurence-Moon-Bardet-Biedl syndrome, an autosomal recessive disorder with variable pleiotropic effects including obesity, mental deficiency, polydactyly, genital hypoplasia, hypogonadism and retinitis pigmentosa.

pleiotropic effects (skeletal, ocular and cardiovascular anomalies) may have as a common basis a defect in the elastic fibers of connective tissue. By contrast, in the rare syndrome of hypogonadism, polydactyly, deafness, obesity, retinitis pigmentosa and mental retardation known as the Laurence-Moon-Bardet-Biedl syndrome (Fig. 4–18), so far there is certainly no one obvious common basis for the assortment of abnormalities.

Pleiotropy is not to be confused with linkage, in which two or more traits may be transmitted together for a number of generations simply because they are determined by linked genes.

GENETIC HETEROGENEITY: SEVERAL GENES, ONE EFFECT

When a genetic disorder that at first glance appears to be uniform is carefully analyzed, not infrequently it is found to comprise a number of separate conditions that are only superficially alike. If mutations at different loci or different mutations at the same locus can independently produce the same trait, or traits that are difficult to distinguish clinically, that trait is said to be genetically heterogeneous. Recognition of genetic heterogeneity is an important aspect of clinical diagnosis, genetic analysis and counseling for genetic defects.

Genetic Heterogeneity in Deafness

Profound childhood deafness shows a remarkable degree of genetic heterogeneity. In almost any community that has a high degree of consanguinity, congenital deafness is one of the autosomal recessive disorders found. Fraser (1976) estimates that there are about 16 to 18 types of autosomal recessive deafness alone and suggests that 10 percent or more of the general population are heterozygous carriers of one or more of these forms. There are also autosomal dominant types, such as Waardenburg syndrome, which combines deafness with pigmentary disturbances (Fig. 4–19); X-linked forms; malformation syndromes, probably not genetic, involving faulty embryological development of the first and second branchial arches; acquired deafness, the types resulting from prenatal

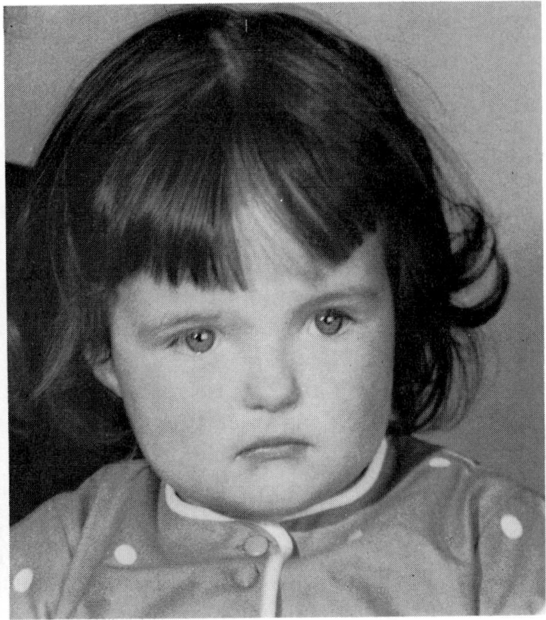

Figure 4–19. Waardenburg syndrome. Both mother and child have a white forelock and pale irides, as well as deafness. From Partington, MW. Arch Dis Child 1959; 34:154–157.

rubella or postnatal otitis media for example; and many cases of unknown etiology.

Profound childhood deafness produces difficulties in communication that may be so severe as to isolate affected persons from the rest of society. An understandable social consequence is that the deaf usually marry within the deaf community. Three such marriages are shown in Figure 4–20. In this pedigree, the marriage of I–3 and I–4 produces only deaf children, as does the marriage of II–10 and II–11. However, the marriage of III–7 and III–9 produces only hearing children. The probable explanation is that in the first two of these marriages the parents were homozygous for the same autosomal recessive gene, but in the third marriage the parents' deafness was caused by different recessive genes at different loci, each parent having normal alleles at the locus for which the partner had only abnormal alleles. Thus if *d* and *e* represent the two recessive genes concerned, the father could be *ddEE* (deaf because he is homozygous *dd*) and the mother *DDee* (deaf because she is homozygous *ee*). All the children would

Figure 4–20. Genetic heterogeneity in a pedigree of congenital deafness. For discussion, see text. Adapted from Stevenson AC, Cheeseman EA. Hereditary deaf mutism, with particular reference to Northern Ireland. Ann Hum Genet 1956; 20:177–231, Figure HD–2, p. 219, with the permission of Cambridge University Press.

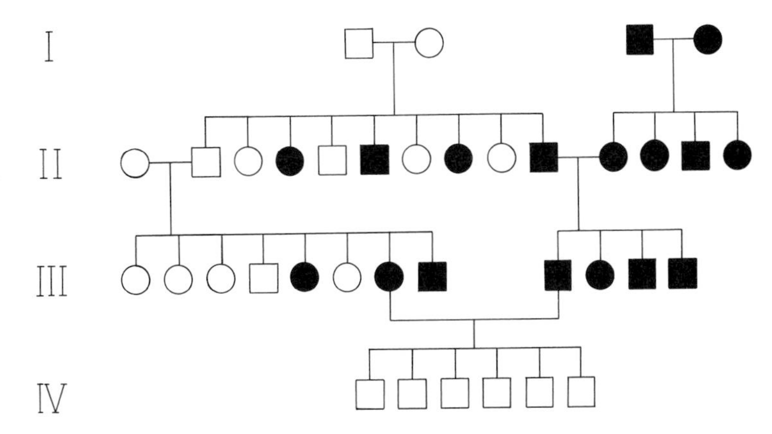

then be *DdEe*, and since there is a dominant normal allele at each locus, all have normal hearing.

Though genetic methods can be helpful in analyzing genetic heterogeneity, as in this example, often it can be recognized by clinical and biochemical observations.

Genetic Heterogeneity in the Muscular Dystrophies

The general term muscular dystrophy is used to describe a number of different primary disorders of muscle. The three main types differ both in clinical features and in genetics. The Duchenne type, mentioned earlier, is X-linked, expressed in early childhood and marked by involvement of the muscles of the pelvic girdle followed by those of the shoulder girdle. Hypertrophy of the calf muscles is an early and striking sign. The course is severe, with inability to walk by about age 10 and death usually by about age 20. Limb-girdle muscular dystrophy is autosomal recessive, typically has its onset in the second or third decade and, though milder than the Duchenne type, usually leads to severe disability by middle life. Facioscapulohumeral muscular dystrophy is autosomal dominant. The expression is irregular, and formes frustes (extremely mild forms) may occur. Onset is usually in the teens but may be much later. The incapacity usually progresses slowly and does not shorten life.

There are many other forms of muscular dystrophy, some well-defined and others still poorly characterized. In particular, a second X-linked type, the Becker type, distinguished from the Duchenne type by its later onset and generally milder course, may be allelic to the Duchenne type.

MYOTONIC DYSTROPHY

Myotonic dystrophy is a form of muscular dystrophy of particular genetic and clinical interest. Myotonia is the continued active contraction of a muscle after cessation of voluntary effort or stimulation, which is easily seen in the difficulty myotonic dystrophy patients have in releasing the grip. Other features of the disorder include cataracts, gonadal atrophy and frontal balding especially in males. Inheritance is autosomal dominant with considerable variability in severity and onset age.

Anticipation. The term anticipation is used for the apparent worsening and earlier onset age of a disease in successive generations. This pattern is often observed in myotonic dystrophy families, but it can be explained by bias in ascertainment of the families rather than by any biological mechanism. Families with early-onset, more severe cases are more likely to be ascertained, but patients with mild, late-onset disease are more likely to leave offspring. Examination of a family at a single point in time thus favors finding children with severe disease whose affected parents and grandparents have a milder form.

Maternal Transmission. Exclusively maternal transmission of the severe childhood and congenital cases of the disease is a phenomenon that has been found only in myotonic dystrophy so far. Virtually all the children with the severe infantile form are born to affected mothers, who themselves usually have a mild form of the disease. The children, if they survive, are unlikely to reproduce. So far we do not know what factor of the prenatal environment in the mothers causes their children who inherit the myotonic dystrophy gene to have the disease in such a severe, early-onset form while the children who are genetically normal seem not to be impaired by the mother's condition.

Genetic Heterogeneity in Osteogenesis Imperfecta

Osteogenesis imperfecta (OI) is a heritable disorder, or more specifically a group of disorders, that have in common a generalized connective tissue abnormality that leads to easy fracturing of bones even with little trauma. Other defects seen in many patients with OI are blue sclerae, conductive hearing loss relatively early in life, joint laxity and defective dentition. OI shows great variability in manifestation and severity, which was formerly ascribed to failure of penetrance and variability of expression of an autosomal dominant gene. A much more complex story has emerged as clinical and biochemical analysis has begun to clarify the underlying molecular defects in the different clinical types.

CLINICAL TYPES OF OSTEOGENESIS IMPERFECTA

Four major types of OI have been defined (Sillence et al, 1979; Sillence et al, 1984):

Type 1: The classic, mild autosomal dominant form, with brittle bones, blue sclerae and, in many affected members, presenile deafness.

Type 2: Perinatal lethal form. Severe congenital or prenatal fractures, poor calcification of the skull, broad crumpled femora, beaded ribs. At least three clinical subtypes are known.

Type 3: Fractures present at birth in many patients, progressive deformity, marked postnatal growth retardation, wormian bones in skull, sclerae less blue than in types 1 and 2.

Type 4: Propensity to fractures but normal sclerae.

MOLECULAR ABNORMALITIES OF COLLAGEN IN OSTEOGENESIS IMPERFECTA

Abnormalities of synthesis or structure of type I collagen have now been found in each form of OI, and the clinical heterogeneity has been shown to reflect even greater heterogeneity at the molecular level (Prockop, 1984; Byers and Bonadio, 1985).

Type I collagen is the major structural protein of bone and other fibrous tissues. A model of the collagen molecule is shown in Figure 4–21. It is a triple helix composed of two identical α1(I) chains coded by a structural gene on chromosome 17, and one non-identical but similar α2(I) chain coded by a gene on chromosome 7. The triple helix can form only with glycine at every third position in the amino acid sequence of the chains. The structure is important to the stability of the molecule. Each gene is translated into a proα chain, and there is extensive post-translational modification both before and after assembly of the three chains into the triple helix procollagen molecule. Eventually, in the extracellular space, mature collagen fibrils are assembled. There is a pair of alleles for each type of chain. If one proα1(I) chain is abnormal, the triple helix could contain zero, one or two abnormal proα1(I) chains; if one proα2(I) chain is abnormal, the triple helix could have a normal or an abnormal proα2(I) chain.

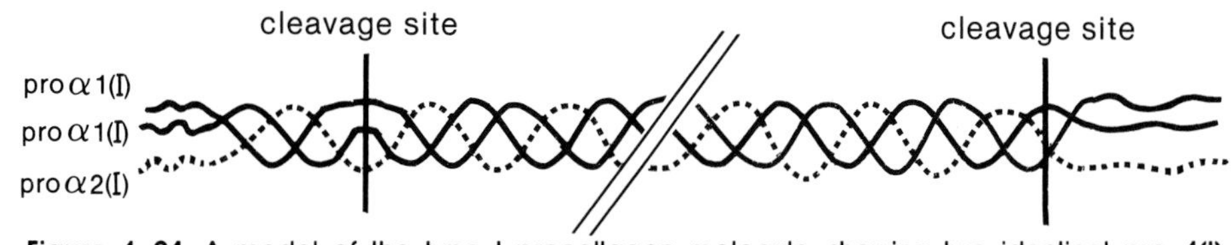

Figure 4–21. A model of the type I procollagen molecule, showing two identical proα1(I) chains and one proα2(I) chain, wound into a triple helix with globular extensions at each end. After secretion from the cell, the extensions are cleaved and the core regions self-assemble to form collagen fibrils.

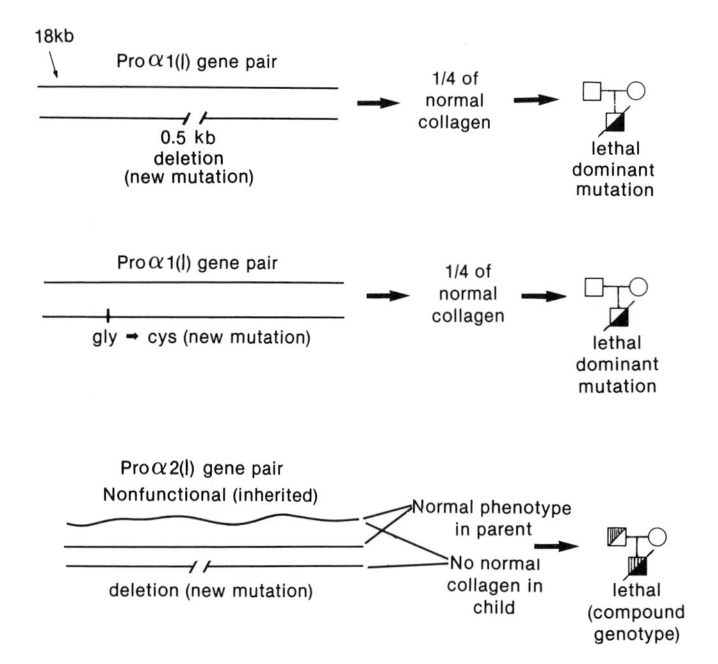

Figure 4–22. Three different structural modifications of procollagen. See text for details.

The molecular alterations that underlie the different clinical forms have been studied through analysis of cultured fibroblasts. In osteogenesis imperfecta type 1 there is decreased synthesis of type I procollagen, and usually it is only synthesis of the proα1(I) chain that is affected. There are probably a number of different mechanisms involved in different family groups.

In osteogenesis imperfecta type 2 not only is there less type I procollagen than normal, but much of what is produced is structurally abnormal. Various different structural modifications have been found in different patients. Figure 4–22 shows three examples:

1. Deletion of a 50-amino-acid length of one proα1(I) chain, reflecting a corresponding deletion in one of the patient's pair of α1(I) alleles. Genetically this is a new autosomal dominant lethal mutation.

2. Probable substitution of cysteine for glycine at one location in one proα1(I) chain. Genetically a new dominant lethal mutation would explain the presence of the abnormal chain in the child but not in either parent. Phenotypically the effect of the amino acid substitution is to destabilize the triple helix.

3. Two abnormal types of proα2(I), explained as one inherited nonfunctional α2(I) allele with no phenotypic effect in heterozygotes and a new mutation resulting in structurally abnormal proα2(I) in the other allele. Thus the patient was a compound heterozygote with two different α2(I) mutations. Since proα2(I) is a component of all type I chains, all the collagen is abnormal.

The different structural lesions at the molecular level lead to generally similar clinical phenotypes. This may be because whatever the nature of the molecular change, its effect is to produce a portion of the molecule that lacks the typical triple-helix structure. Genetic alterations could alter the amino acid sequence of the chain and thus destroy its structure.

The kind of in-depth knowledge of the different types of osteogenesis imperfecta that is currently emerging has useful applications to prognosis. Clinically, if a patient's molecular defect can be spelled out, eventually we should be able to predict the natural history of his disease as well as to approach the question of therapy. Genetically, demonstration of whether a defect is inherited from an affected parent (autosomal dominant), inherited in double dose from two unaffected but heterozygous parents (autosomal recessive) or a new mutation present in neither parent allows accurate recurrence risks to be provided. Prenatal diagnosis in type 2 OI, the perinatal lethal form, may be performed by examination of skull and limb length by ultrasonography the second trimester. The current molecular advances give hope of eventual first-trimester diagnosis by means of molecular analysis of chorionic villi.

In summary, osteogenesis imperfecta shows clinical heterogeneity, reflecting even greater biochemical heterogeneity that results from heterogeneity in the types of alteration present in the genes for the proα1(I) and proα2(I) procollagen chains.

SEX-LIMITED AND SEX-INFLUENCED TRAITS

If a trait is determined by a gene on the X chromosome, the sex ratio of the affected individuals is not 1:1. An abnormal sex ratio may provide the first hint that a trait is caused by an X-linked gene. However, there are many traits in which the sex ratio is abnormal even though the determining gene is autosomal. This should not be surprising, because the milieu in which any gene acts is determined in part by the sexual constitution of the individual; moreover, differential prenatal and postnatal survival can affect the observed prevalence of a condition in males and females.

Sex-Limited Traits

A trait that is autosomally transmitted but expressed in only one sex is said to be sex-limited. An example is **precocious puberty**, in which heterozygous

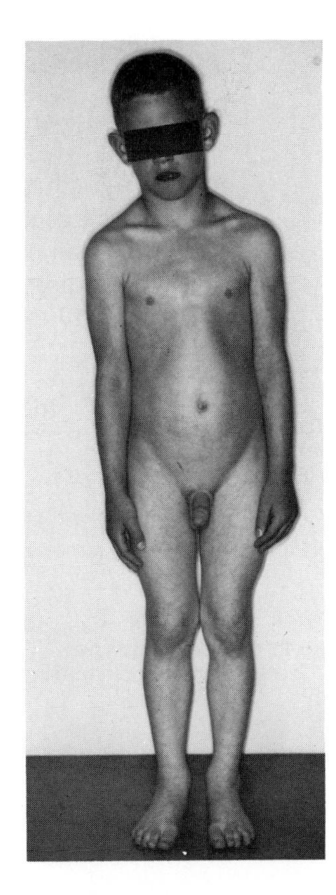

Figure 4–23. Precocious puberty. At age 4.75 years, the patient's height is 120 cm, far above the normal range.

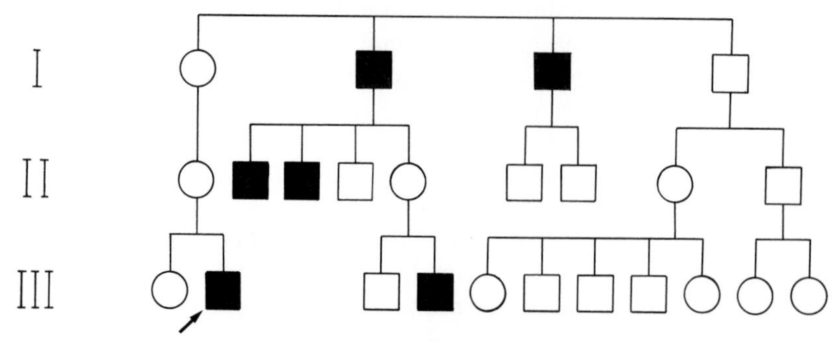

Figure 4–24. Pedigree of precocious puberty in the family of the child shown in Figure 4–23. Though the trait is transmitted from males to males through unaffected females, male-to-male transmission shows that the inheritance is autosomal.

males develop secondary sexual characteristics and undergo an adolescent growth spurt at about four years of age, sometimes even younger (Fig. 4–23). There is no expression in heterozygous females. Affected boys are much taller than their contemporaries at first, but because of early fusion of the epiphyses of the long bones they soon stop growing and end up as short men. Figure 4–24 is part of a large pedigree of precocious puberty, showing male-to-male transmission; the gene is therefore obviously autosomal dominant, not X-linked.

For traits in which males do not reproduce, sex-limited autosomal dominant inheritance cannot always be readily distinguished from X-linkage, because the most critical evidence of X-linkage, namely absence of male-to-male transmission, cannot be provided. Duchenne muscular dystrophy (described above) and testicular feminization (described in Chapter 7) are examples. For these two conditions and others in which the affected males are infertile, pedigree analysis alone cannot prove the pattern of inheritance and other lines of evidence are required.

Sex-Influenced Traits

Traits are said to be sex-influenced when they are expressed in both sexes, but with widely different frequencies. Table 4–4 lists the sex ratios in a number of different autosomal disorders, to illustrate the effect of sex on gene expression. Many multifactorial traits are also sex-influenced (see Chapter 12). Chromosomal aberrations involving the autosomes may also be more common in one sex than in the other; for example, in trisomy 21 there is male preponderance and in trisomy 18 four-fifths of the patients are female.

Table 4–4. SEX-INFLUENCED AUTOSOMAL PHENOTYPES

Sex-Limited Phenotypes	
Expression limited to males	Precocious puberty (see text) Hypogonadism (autosomal recessive male forms)
Expression limited to females	Hydrometrocolpos (transverse vaginal septum and other abnormalities)
Sex-Influenced Phenotypes	
Expression more common in males	Baldness (autosomal dominant in males, recessive in females?) Hemochromatosis (loss of blood in menses may protect females)
Expression more common in females	Congenital adrenal hyperplasia (more often recognized in females)

ONSET AGE

Many genetic disorders are not present at birth but become manifest later in life, some at a characteristic age and others at variable ages throughout the life span. Genetic disorders are, of course, not necessarily congenital, nor are congenital disorders necessarily genetic. To classify a disorder as genetic means that genes are plainly implicated in its etiology; to say that a disorder is congenital means only that it is manifest at birth.

Some genetic disorders have a prenatal onset. Dysmorphic conditions of many kinds originate during embryological development, and are recognized postnatally as "birth defects." Some disorders are not expressed until the infant begins its independent life, for example, phenylketonuria and galactosemia. Later, other defects of genetic origin may make their appearance, some at characteristic ages, others at variable times, in which case the phenomenon of anticipation (defined earlier in this chapter) may be observed.

That a genetic disorder need not be present at birth should cause no surprise, since many genes are known to be expressed at specific times during development. Actually, remarkably little is known about the action of many genes that govern normal development. Why does puberty occur when it does? What causes the menopause? Why do people become senile? These questions are just as puzzling as the problems of why the gene for Huntington disease remains dormant for many years before the disease becomes manifest and runs its progressive and eventually fatal course.

Huntington Disease

Huntington disease, often called Huntington's chorea, was first described by Huntington in 1872 in an American kindred of English descent. A recent monograph (Hayden, 1981) notes that the gene for Huntington disease seems to have spread from northwestern Europe throughout the world. Very few patients are new mutants; in other words, nearly all patients can be shown to have an affected parent. In fact, the presence of a family history is used as one of the diagnostic criteria for Huntington disease. Estimates of the heterozygote frequency in different populations range from about 1 in 5000 to 1 in 15,000, with the exception of a relatively isolated region in Venezuela where it is much more common. Among all heterozygotes, at any one time fewer than 50 percent have the clinical disease; the remainder are below the onset age.

The molecular genetic aspects of Huntington disease are described later (Chapter 11). It is one of the first disorders in which close linkage of an abnormal gene to a restriction fragment length polymorphism has been shown. In suitable families preclinical detection of heterozygotes is now feasible, though of course there are major ethical problems in applying such a test.

Natural selection against the Huntington gene is not strong since its late onset age means that it does not greatly impair the biological fitness of heterozygotes. In fact, some studies have actually shown that Huntington patients have larger families on average than their unaffected sibs. If this is generally true, over the course of many generations the theoretical consequence is that the Huntington gene gradually replaces the normal allele and becomes the new normal allele at the locus.

The onset age distribution of Huntington disease is shown in Figure 4–25, and the way to allow for the variability of onset age in genetic counseling appears in Figure 4–26. Of 100 babies each with an *Hh* parent, 50 (on average) will be *Hh* and 50 *hh*. By age 40, half the *Hh* persons (25) will already manifest the disease. Of the remaining 75, 25 (one-third) are still at risk. Thus the chance that

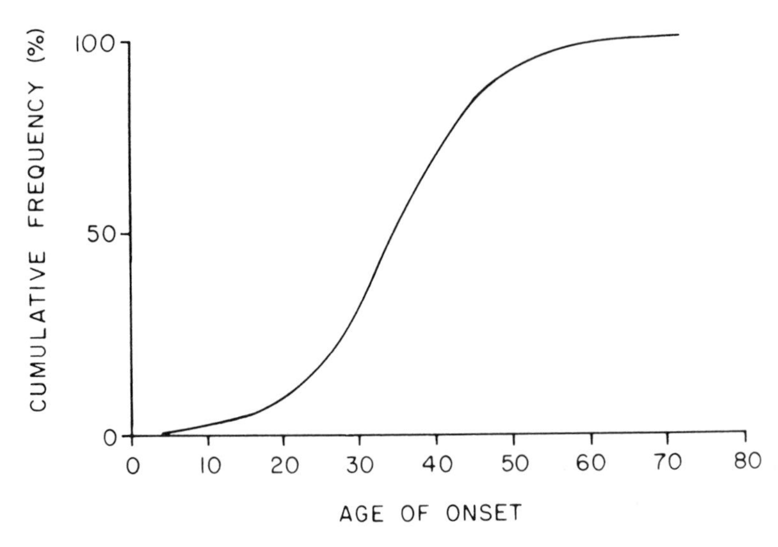

Figure 4–25. Onset age of Huntington disease in a sample of 999 individuals. From Conneally PM, Wallace MR, Gusella JF, Wexler NS. Huntington disease: estimation of heterozygote status using linked genetic markers. Genet Epidemiol 1984; 1:81-88.

any child of a parent with Huntington disease who is still unaffected at age 40 will eventually develop the disease has dropped from 1 in 2 to 1 in 3.

GENE INTERACTION

Although so far in this chapter the various patterns of genetic transmission have been discussed as though genes at any one locus were expressed independently of the remainder of the genome, this is an oversimplification. The expression of a gene may be affected by its own allele, by the presence of specific genes at other loci involved in the same protein product or pathway, or by the remainder of the genome, the "genetic background." Environmental effects, both prenatal and postnatal, can also modify gene expression.

Interaction of Allelic Genes. The expression of a gene may be affected by its allele. The most obvious examples are the different phenotypes resulting from homozygous, heterozygous and compound genotypic combinations, mentioned earlier. It also appears that isoallelism (the presence of more than one kind of normal allele at a locus) can be a significant cause of variation in expression of mutant genes. Renwick (1956) noted that in nail-patella syndrome (a rare autosomal dominant malformation characterized by variable degrees of dystrophy of the nails, absence or reduction of the patella, other bone dysplasias and, occasionally, nephropathy), the severity of the defect was closely correlated in sibs but not in parent-child pairs. The presence of isoalleles could explain this observation. Each of the affected sibs has inherited the affected parent's deleterious allele and not its normal counterpart; instead, each has a normal allele

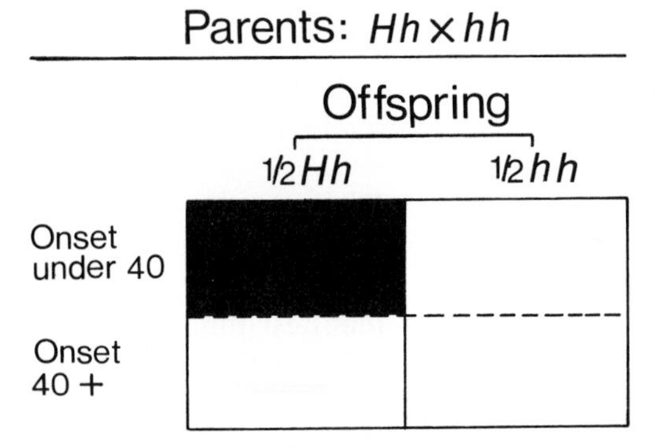

Figure 4–26. Risk of Huntington disease after age 40 in offspring of patients. For discussion, see text. See also Chapter 14.

from the unaffected parent. Thus the sibs have a 50 percent chance of having identical genotypes at the locus concerned, whereas if there are several possible normal alleles it is unlikely that parent and child would match exactly. The possible significance of isoalleles in the expression of other disorders has hardly been investigated.

Nonallelic Interaction. There are many examples of genes at two or more different loci acting together to produce a phenotype. Many such instances involve biosynthetic pathways, in which the product of one reaction acts as the substrate for the next. An example is the formation of the A, B and H blood group substances by the sequential action of genes at the H locus and the ABO locus (Chapter 9). In prokaryotes genes concerned with the same pathway have often been found to be genetically as well as functionally linked, but in higher organisms including man there is little evidence of genetic linkage of functionally related genes. Other examples involve nonallelic genes affecting the same process in different ways, such as the combined effects of structural hemoglobin genes and thalassemia genes in producing hematological disorders (Chapter 5).

Genetic Background. In inbred animals such as laboratory mice there are many examples of mutant genes expressed with different degrees of severity in animals of different genetic backgrounds. Analysis of gene action against different backgrounds is not practical in man, but it is logical to expect that the effect of genetic background would be important. For example, normal stature is believed to have multifactorial inheritance, that is, to be determined by multiple genes of minor effect, together with environmental factors such as nutrition. On a genetic background for tall stature, a gene for Marfan syndrome may have a striking effect, whereas against a genetic background for short stature its expression may not be apparent. Again, though mental retardation is a standard observation in fragile X syndrome, the degree of retardation varies, and one factor in the variation might be the genetic background of the patient with respect to intelligence.

GENERAL REFERENCES

Most of the General References to Chapter 1 also apply to this chapter. The following list is a sampling of more specialized references, dealing with specific conditions or malformations.

Cohen MM. The child with multiple birth defects. New York: Raven Press, 1982.
Dubowitz V. Muscle disorders in childhood. Philadelphia: W. B. Saunders, 1978.
Goodman RM, Gorlin RJ, The malformed infant and child. An illustrated guide. New York: Oxford University Press, 1983.
Harper PS. Myotonic dystrophy. Philadelphia: W. B. Saunders, 1979.
Hayden MR. Huntington's chorea. New York: Springer-Verlag, 1981.
Smith DW. Recognizable patterns of human malformation, 3rd ed. Philadelphia: W. B. Saunders, 1982.

PROBLEMS

1. Draw a three-generation pedigree of your own family, showing dates of births, dates and causes of deaths, miscarriages, stillbirths and any genetic disorders. Use standard symbols.

2. A woman, Ann, has oculocutaneous albinism (autosomal recessive). Ann's sister's daughter Brenda marries Ann's brother's son Bob and they have a child Charles.
   a) Draw the pedigree.
   b) What is the probability that Charles has albinism?
   c) If Charles has albinism, what is the risk that the next child of Brenda and Bob will also have albinism?

3. Don and his maternal grandfather Barry both have classic hemophilia. Don's partner Diane is his mother's sister's daughter. Don and Diane have one hemophilic son Edward, two hemophilic daughters Elise and Emily, and one unaffected daughter Enid.
   a) Draw the pedigree.
   b) Why are Elise and Emily hemophilic?
   c) What is the probability that Elise's son will be hemophilic?
   d) What is the risk that Enid's son will be hemophilic?

4. George, who has muscular dystrophy, is married to his first cousin, Gertie. Their mothers are sisters. The type of muscular dystrophy is unknown; onset was at age 16, and it is slowly progressive. There are no other known cases of muscular dystrophy in the family, but the children of George and Gertie include 2 affected sons, 1 unaffected son and 2 unaffected daughters.
   a) Draw the pedigree.
   b) On the basis of this pedigree, what pattern(s) of single-gene inheritance can be ruled out?
   c) You now learn that Gertie's two brothers have the same disease. Her three sisters are unaffected. What is now the most likely pattern of inheritance? What other pattern of inheritance cannot be ruled out?
   d) If George's affected son Harry asks for genetic counseling, what risk would you give that Harry would have an affected child?

5. Hedy, aged 25, comes to the Genetics Clinic with the following problem. Her father Gary had retinitis pigmentosa (RP), an eye disease. His mother Florence also has RP; she and Gary's father Fred are first cousins.

   RP is genetically heterogeneous. About 5 percent of cases are autosomal dominant, the great majority are autosomal recessive, and there are X-linked forms. As a further complication, the onset age varies, though visual impairment may be detected by school age in many genetic forms of RP and was present in early childhood in Gary.
   a) Draw the pedigree.
   b) What is the risk of RP in Hedy? In her future children?
   c) You investigate further and find that Gary's sister and her son Herb also have RP, with onset in childhood. How does this new information affect the risk estimate?
   d) Hedy is given an eye examination and is reported to have no sign of visual impairment. How does this finding affect your genetic counseling?

6. Linda, now 18 years old, consults you about her family history of Huntington disease (HD). A paternal aunt, born in 1930, died in 1973 of HD. A paternal uncle, born in 1936, has the same diagnosis. Two paternal aunts, aged 51 and 52, and Linda's father, aged 41, have recently been examined and show no sign of HD. Linda's paternal grandmother, born in 1913, died in 1960 of HD. How would you counsel Linda concerning the risk that she herself or her child might develop HD?

7. A color-blind man with limb-girdle muscular dystrophy and a woman with normal color vision and normal muscle function produce a daughter who is color-blind and has limb-girdle muscular dystrophy. State the genotypes of the parents and the child.

8. Figure 4–20 is a pedigree of congenital deafness, discussed in the text.
   a) State the probable genotypes of the children in Generation IV.
   b) If one of these boys mates with a daughter of III–1, what is the probability of deafness in their child?

9. A man has dentinogenesis imperfecta, his wife is normal. Both are tasters of PTC, but both of their mothers were nontasters.
   a) Give the genotypes of the couple.
   b) What is the chance that they will have a daughter who does not have dentinogenesis imperfecta and is a nontaster?

10. In a certain family, John has polydactyly, with an extra digit on each hand and each foot. His sister Jane and brother Jerry are unaffected. His father is normal; his mother Inez has polydactyly of both hands but normal feet. Inez has four sibs, three of whom have some expression of polydactyly. Their father also had polydactyly. Assume that polydactyly is inherited as an autosomal dominant trait with 80 percent penetrance, and with variable expressivity.
    a) What is the probability that John will have an affected child?
    b) What is the probability that John's brother Jerry will have an affected child?

5

HUMAN BIOCHEMICAL GENETICS

Many different proteins are synthesized in body cells. These proteins, which may be either enzymes or structural components, or even serve both functions, are responsible for all the developmental and metabolic processes of the organism. The fundamental relationship between genes and proteins is that the coding sequence of bases in the DNA of a given gene specifies the sequence of amino acids in the corresponding polypeptide chain. Alteration of the base sequence in the gene may result in the synthesis of a variant polypeptide with a correspondingly altered amino acid sequence. Proteins are composed of one or more polypeptide chains. Hence, a gene mutation may lead to the formation of a variant protein, which may have altered properties as a consequence of its changed structure.

The phenotypic changes produced by gene mutations are numerous, varied and frequently unexpected. If a mutation results in an amino acid substitution in a so-called structural protein, such as hemoglobin, the phenotypic effect, if any, will depend on how the alteration in amino acid sequence affects such properties of the hemoglobin molecule as its affinity for oxygen or its tendency to sickle. If the amino acid sequence of an enzyme polypeptide is altered, the enzyme synthesized by the mutant gene may have altered enzymatic activity. Most of the variant enzymes known have less activity than the normal forms; some are completely inactive. Rarely, a mutation may produce excessively high activity. Occasionally a mutation leads to synthesis of an unstable polypeptide chain that is rapidly destroyed in vivo, or to other types of change, some of which are described in the subsequent pages.

Not all variant proteins are clinically abnormal. On the contrary, many proteins exist in two or more relatively common, genetically distinct and structurally different "normal" forms. Such a situation is known as a **polymorphism**.

The subject matter of this chapter falls into three main areas: (1) hemoglobins, normal and abnormal, which have done more than any other type of protein to clarify the relationship between gene, protein and disease; (2) human biochemical disorders caused by metabolic errors that in turn are the result of gene mutations; and (3) examples of genetic variations in response to drugs (pharmacogenetics).

The Hemoglobins

The human hemoglobins occupy a unique position in medical genetics for many reasons. They have taught us more about the molecular basis of human and medical genetics than any other system. They are historically important for their part in the demonstration of the relationship between genetic information and protein structure. They also illustrate mechanisms of forming new genes other than by point mutation, cast light on the process of evolution at both the molecular and the population level, and provide a model of gene action during development. Furthermore, hemoglobin variants are clinically important as causes of a variety of genetic disorders of blood, some of which are very common. Molecular analysis of the hemoglobins has given us important new insights into the organization of the human genome and the kinds of genetic changes that can produce genetic disease. These insights are already being exploited clinically and in studies of human populations. Consequently, the genetic background of the human hemoglobins merits examination in some detail.

The first step in understanding the genetics of the hemoglobinopathies was taken in 1949 by Neel, who showed that patients with the blood disorder known as sickle cell disease were homozygous for a gene that produced a similar but much milder abnormality, sickle cell trait, in both parents, who were heterozygous. Shortly afterward Pauling and colleagues designated sickle cell disease as the prototype of the "molecular diseases," in which an abnormal hemoglobin molecule was the basic defect. Ingram then discovered that the abnormality in the hemoglobin in sickle cell disease constituted a replacement of only one of the 287 amino acids of the hemoglobin half-molecule; this was the first demonstration *in any organism* that a mutation in a structural gene could cause an amino acid substitution in the corresponding protein.

THE STRUCTURE AND FUNCTION OF HEMOGLOBIN

Hemoglobin is the respiratory carrier in vertebrate red blood cells and is also found in some invertebrates and in the root nodules of legumes. The molecule is tetrameric. Each subunit has two parts: a polypeptide chain, globin, and a prosthetic group, heme, which is an iron-containing pigment that combines with oxygen and gives the molecule its oxygen-transporting ability. The heme portion is alike in all forms of hemoglobin, genetic variation being restricted to the structure of the globin portion only. Even within the globin polypeptide there is some restriction on the kinds of amino acid substitutions that are acceptable in terms of natural selection, since impairment of the oxygen-carrying function of the molecule cannot be tolerated.

The hemoglobin molecule typically consists of two each of two different types of polypeptide chain. In normal adult hemoglobin (Hb A) these globin chains are designated α and β. The four chains are folded and fitted together to form a roughly globular molecule with a molecular weight of about 64,500 (Fig. 5–1). The "formula" for the composition is $\alpha_2^A\beta_2^A$, which translates to "two α chains typical of those in Hb A plus two β chains typical of those in Hb A." The

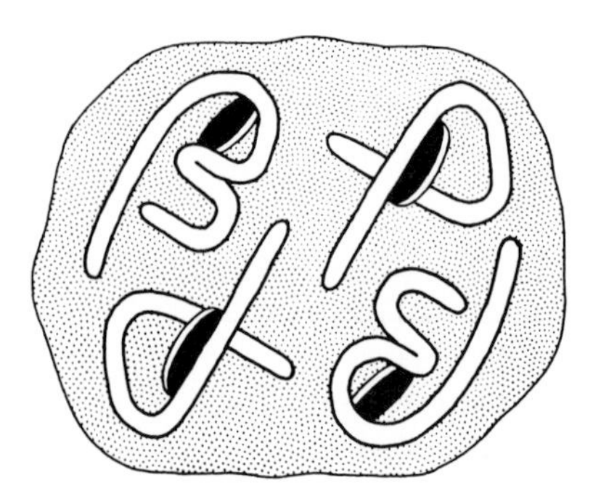

Figure 5–1. Representation of a molecule of normal adult hemoglobin. There are two α and two β chains, each associated with a heme moiety (black disc). After Ingram, Nature 1959; 183:1795–1798.

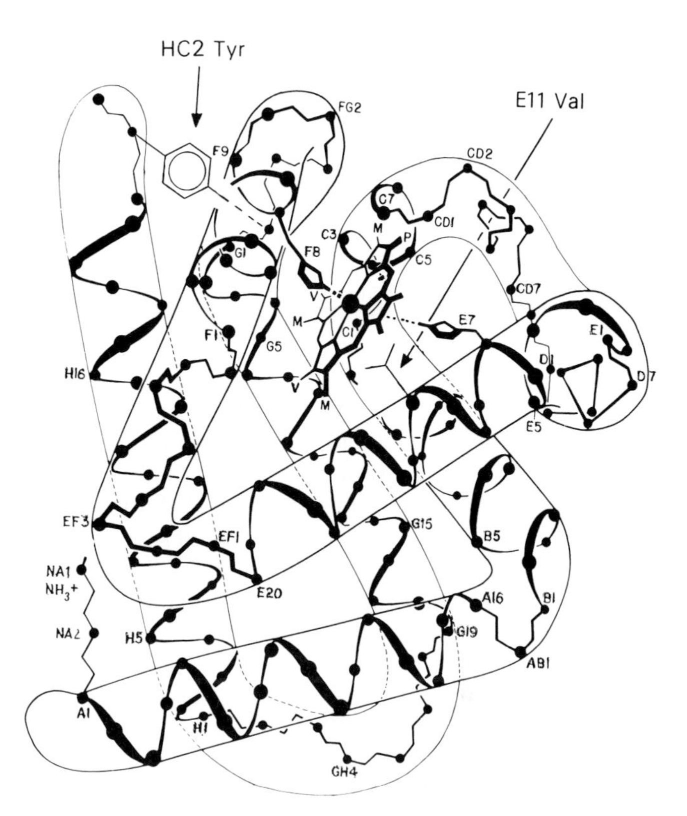

Figure 5–2. The hemoglobin molecule. The various hemoglobins and myoglobin differ only slightly. The molecule has eight helical regions, designated A to H; the amino acids are numbered from the N-terminal end of each segment. The letters, M, V and P refer to methyl, vinyl and propionate side chains of the heme. Note the bond between F8 (histidine) and the iron atom of the heme, which is the only covalent link between heme and globin. From Perutz MF. Structure and mechanism of haemoglobin. Br Med Bull 1976; 32:195–208. Reproduced by permission of The Medical Department, The British Council.

formula is often written $\alpha_2\beta_2$, omitting the superscript. The two kinds of chains are almost equal in length, the α chain having 141 amino acids and the β chain 146. Because the α and β chains are encoded by genes at separate loci (also designated α and β, on chromosomes 16 and 11 respectively) a mutation affects one chain or the other but not both. The chains resemble one another markedly both in amino acid sequence (primary structure) and in three-dimensional configuration (tertiary structure). They also resemble myoglobin, the respiratory pigment of muscle, though less closely; the myoglobin molecule has only a single polypeptide chain, but similarities in amino acid sequence and tertiary structure indicate that the hemoglobin and myoglobin molecules have evolved from a common precursor.

The details of the structure of hemoglobin are shown in Figure 5–2. Some points of the structure should be emphasized. In all hemoglobins, the heme and the amino acid to which the iron is linked (a histidine) are the same, the porphyrin is "wedged into its pocket" by a phenylalanine, and about 35 other sites along the chain are occupied by nonpolar residues. Otherwise, the sequence of amino acids along the chain is not rigidly specified but shows considerable variation between species, a fact that has allowed the hemoglobins to be used in studies of evolution.

THE HEMOGLOBIN LOCI

As noted above, the Hb A molecule is a tetramer of four globin chains coded by two gene loci designated α and β. The general model fits all the known types of normal human hemoglobin. Each has two α or α-like chains, and two non-α chains. The several types are summarized in Table 5–1. The minor hemoglobin component, Hb A2, which constitutes only 2 percent of the total, has the same

Table 5–1. TYPES OF NORMAL HUMAN HEMOGLOBIN

Adult	Hb A	$\alpha_2\beta_2$
	Hb A2	$\alpha_2\delta_2$
Fetal	Hb F	$\alpha_2\gamma_2$
Embryonic	Hb Portland	$\zeta_2\gamma_2$
	Hb Gower I	$\zeta_2\epsilon_2$
	Hb Gower II	$\alpha_2\epsilon_2$

two α chains as Hb A, but its non-α chains are δ chains coded by a δ locus. Other hemoglobins are formed in early development through the activity of several additional loci. In fetal hemoglobin (Hb F), the non-α chain is designated the γ chain; there are two different γ chains, differing only at a single site, $^{G}\gamma$ having glycine and $^{A}\gamma$ having alanine at position 136 of the chain.

Another type of non-α globin is the epsilon (ε) globin found in early embryonic life. Finally, an α-like globin, the zeta (ζ) type, combines with either γ or ε globins to form other types of embryonic hemoglobin.

The structure of the α and β globin gene complexes has already been described (Fig. 3–4). There are two α globin loci, designated α1 and α2, instead of a single one per chromosome. Evidence for the duplication came originally from individuals with two separate α chain mutations who nevertheless formed about 50 percent normal hemoglobin. The gene complex also contains a nonfunctional pseudo-α (ψα) gene and two α-like ζ genes, only one of which is functional. The β globin gene complex on chromosome 11 also shows extensive duplication. It includes the β locus, the γ locus, the $^{G}\gamma$ and $^{A}\gamma$ loci, the ε locus and an inactive pseudo-β locus $\psi\beta_1$.

In both the α and β gene complexes, the arrangement of the genes on the chromosome is in the sequence in which they are switched on during development. This seems unlikely to be coincidental but has not yet been explained.

EVOLUTIONARY ASPECTS OF THE GLOBIN GENES

In Chapter 3, a globin gene was used as a model of gene structure, transcription and processing. This section will briefly review the similarity in structure of the various globin genes.

Though the α and β genes now reside on different chromosomes, they show much structural hemology in overall length, amino acid sequence and exon-intron arrangements. In both genes the first intron begins between amino acids 30 and 31, and is about 130 nucleotides long. The second occurs at almost but not exactly the same site in both types of globin, though it is much longer in the β genes. The origin of the two genes (and other α-like and β-like genes) from a common ancestor seems obvious.

Within the β gene complex, a close homology exists between the β and δ globins, which differ in only 10 of their 146 amino acids. The two γ globins differ from one another in just one amino acid, and from the β chain in 38.

Hemoglobin Pseudogenes. Pseudogenes are DNA sequences that closely resemble known genes but are nonfunctional. A ψα globin and ψβ globin gene have been identified, and it has been suggested that the δ globin gene may be on its way to becoming a nonfunctional relic, as it is present but nonfunctional in

Old World monkeys and forms only a small proportion of human adult hemoglobin. If pseudogenes retain any function, it is unknown.

The Evolution of Hemoglobin

The extensive molecular homology among the various globin genes and their chromosomal arrangement demonstrates the role of gene duplication and divergence in evolution.

The evolutionary model proposed by Ingram (1963) and by Epstein and Motulsky (1965) on the basis of amino acid sequence is still in accordance with the newer information provided by molecular analysis of the globin genes (Fig. 5–3). According to this model the primitive α gene arose by duplication of a primitive gene for a myoglobin-like molecule. Mutation gradually brought about divergence of the two primitive genes. Further gene duplication led to the formation of "new" genes that in turn diverged from one another to result eventually in the amino acid sequences of the globin chains of present-day man. Some of the genes apparently evolved into nonfunctional pseudogenes. Meanwhile, development of the tetrameric configuration allowed much more efficient oxygenation.

There is a limit to the kind of amino acid substitution that is acceptable in terms of natural selection. Only those substitutions that do not impair function can be maintained; it is of interest that the sites of heme attachment are invariant in the different human globin chains and in those of other vertebrate species. Another constraint is that those amino acids which are classified as polar (hydrophilic) tend to occupy the surface of the hemoglobin molecule, where they are exposed to water, whereas those which are nonpolar (hydrophobic) are in the interior, where they play a role in maintaining the structure of the molecule. Consequently, substitutions are not likely to be preserved if they do not have the right type of polarity. The fact that many sites are alike in all four chains indicates that selection places rigid constraints on some types of changes in specific proteins. As a generalization, many proteins have "sensitive areas,"

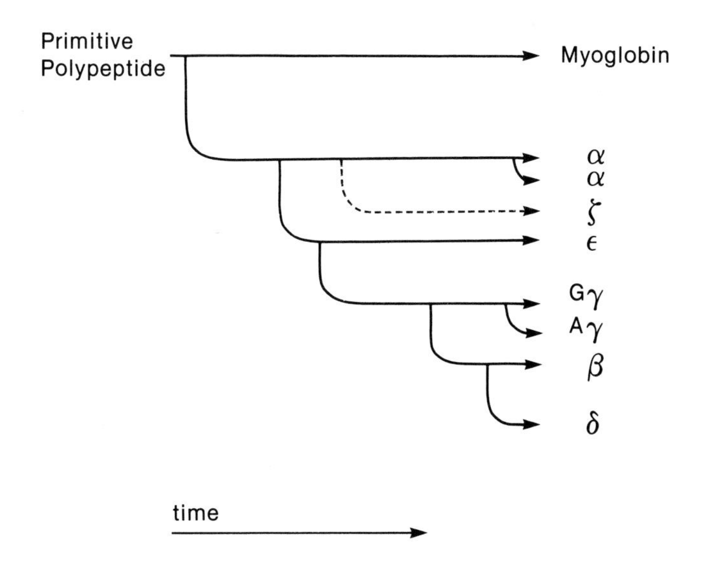

Figure 5–3. An interpretation of the evolution of hemoglobin by gene duplication and divergence. For discussion, see text. Adapted from Epstein CJ, Motulsky AG. Evolutionary origin of human proteins. Prog Med Genet 1965; 4:85–127, and from Giblett ER. Genetic markers in human blood. Oxford: Blackwell Scientific Publications, 1969, p. 402, by permission.

in which mutations cannot occur without affecting function, and "insensitive areas," where mutations are tolerated more freely.

HEMOGLOBIN VARIANTS

Most of the hemoglobin variants result from point mutations in the structural genes that code for the amino acid sequence of one of the types of chain of the molecule, but some are formed by other molecular mechanisms. More than 300 abnormal hemoglobins have been described. Their clinical manifestations, if any, result from altered oxygen affinity (either higher or lower than normal), instability of the molecule or, in a few cases, the presence of iron in the oxidized state, resulting in methemoglobinemia.

In **methemoglobinemia**, the iron of the heme group of the molecule is in the ferric state, unable to combine reversibly with oxygen. Cyanosis is the most obvious clinical sign of the condition. All the hemoglobin variants resulting in methemoglobinemia have an amino acid substitution in either the α or the β chain very close to the attached heme and are able to bond to it so that normal reduction by methemoglobin reductase cannot take place. Methemoglobinemia can also originate by quite a different mechanism, through a deficiency of the enzyme methemoglobin reductase itself; in this case the inheritance is autosomal recessive rather than autosomal dominant.

Sickle Cell Disease

The first abnormal hemoglobin to be detected and a very important one clinically is sickle cell hemoglobin (Hb S), the molecular defect of sickle cell disease.

Sickle cell disease is a severe hemolytic disease characterized by a tendency of the red cells to become grossly abnormal in shape (sickled cells) under conditions of low oxygen tension (Fig. 5–4). The clinical manifestations include

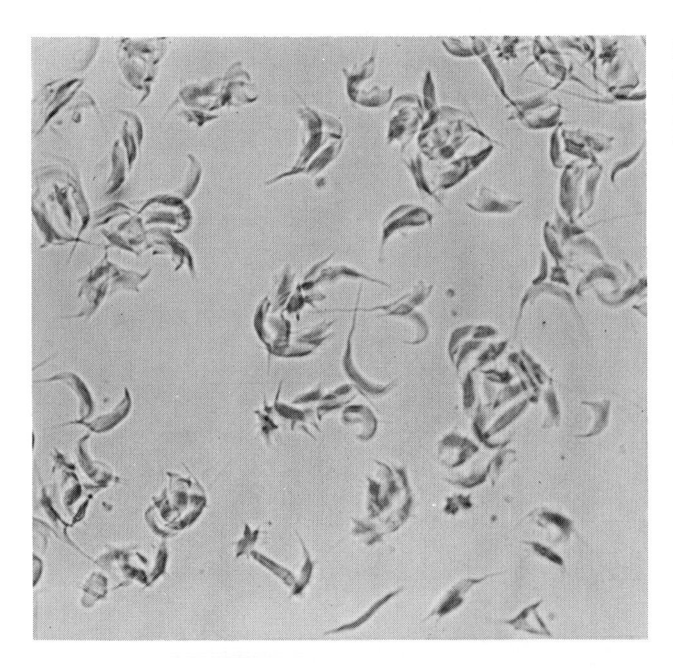

Figure 5–4. The red cell phenotype in sickle cell anemia. Homozygotes for sickle cell disease form only Hb S, no Hb A. The distorted shape of the red cells is due to aggregation of the abnormal hemoglobin molecules under conditions of reduced oxygen tension. Photomicrograph courtesy of J. H. Crookston.

anemia, jaundice and "sickle cell crises," marked by impaction of sickled cells, vascular obstruction and painful infarcts in various tissues such as the bones, spleen and lungs. The disease has a characteristic geographic distribution, occurring most frequently in equatorial Africa, less commonly in the Mediterranean area and India and in countries to which people from these regions have migrated. About 0.25 percent of American blacks are born with this disease, which is often fatal in early childhood though longer survival is becoming more common. Sickle cell disease results from homozygosity for a mutant gene that determines an abnormal hemoglobin, Hb S.

The parents of affected children, though usually clinically normal, have red cells that sickle when subjected to very low oxygen pressure in vitro (that is, they show a positive sickling test). Occasions when this might happen in vivo are very unusual. The heterozygous state, known as sickle cell trait, is present in approximately 8 percent of American blacks.

The physicochemical abnormality of sickle cell hemoglobin was identified in 1949 by Pauling and associates by electrophoresis. They found that Hb A and Hb S were readily distinguishable from one another by their mobility in an electrical field, and concluded that their globin molecules were different. They also found that the hemoglobin of persons with sickle cell trait behaved like a mixture of normal and sickle cell hemoglobin. The structure of sickle cell hemoglobin was, in their words, "a clear case of a change produced in a protein by an allelic change in a single gene involved in synthesis."

The precise nature of the change in the protein molecule predicted by these workers was later identified by Ingram by means of "fingerprinting." This is a technique developed by Sanger for determining the structure of proteins by breaking them down with trypsin, then separating the resulting small peptides by electrophoresis in one direction and by chromatography at right angles to it (Fig. 5–5). Ingram demonstrated that the difference between normal and sickle cell hemoglobin lay in the β chain and involved only one of the 146 amino acids in the chain, the amino acid sixth in position from the N-terminal end of the chain. At this position, the amino acid valine has replaced the glutamic acid of

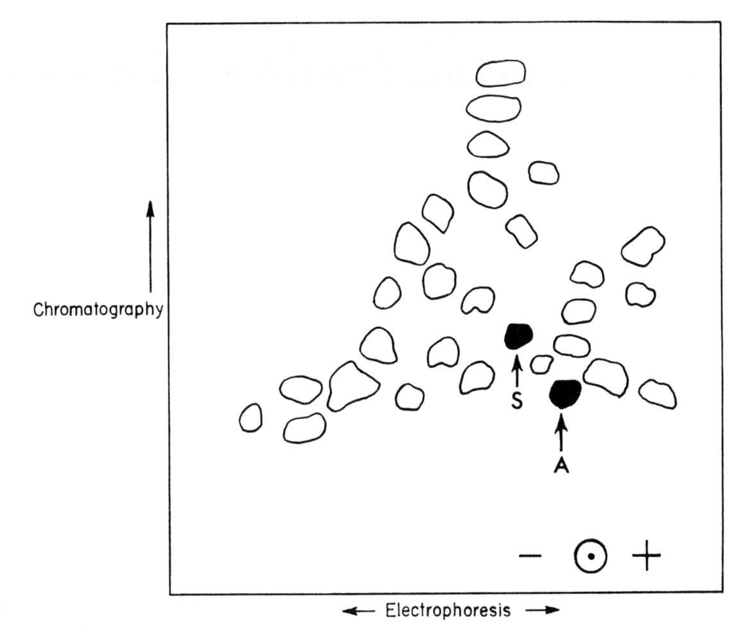

Figure 5–5. Representation of the "fingerprints" of Hb A and Hb S. Each spot represents a tryptic peptide. Hb A and Hb S differ in only a single peptide. See text.

normal hemoglobin. The sequences in the corresponding sections of the two hemoglobins are as follows:

Hb A: val-his-leu-thr-pro-*glu*-glu-lys

Hb S: val-his-leu-thr-pro-*val*-glu-lys

Note that only a single substitution in the second position of the triplet coding for glutamic acid alters it to code for valine:

Amino acid	mRNA code
Glutamic acid	GA (A or G)
Valine	GU-(U, C, A or G)

The substitution of a valine for a glutamic acid in each of the two β chains of the hemoglobin molecule is the only biochemical difference between Hb A and Hb S. This difference, which results in an altered electrical charge, explains the difference in electrophoretic mobility of the two hemoglobins. All the clinical manifestations of sickle cell hemoglobin are consequences of this relatively minor change. The physical basis of the sickling phenomenon is the tendency of the abnormal hemoglobin molecules to aggregate, forming rod-like masses that distort the red cells into sickle shapes under conditions of low oxygen tension.

Because the abnormality of Hb S is localized in the β chain, the formula for sickle cell hemoglobin may be written $\alpha_2^A\beta_2^S$ or simply $\alpha_2\beta_2^S$. This symbol indicates that the α chain is normal, but that the β chain is of the sickle cell hemoglobin type. A heterozygote has a mixture of the two types of hemoglobin, A and S. The relationships of clinical status, hemoglobin types and genes can be summarized as follows:

Clinical Status	Hemoglobin	Hemoglobin Composition	Genotype
Normal	Hb A	$\alpha_2^A\beta_2^A$	$\alpha\alpha/\alpha\alpha\ \beta\beta$
Sickle cell trait	Hb A and Hb S	$\alpha_2^A\beta_2^A$ and $\alpha_2^A\beta_2^S$	$\alpha\alpha/\alpha\alpha\ \beta\beta^S$
Sickle cell disease	Hb S	$\alpha_2^A\beta_2^S$	$\alpha\alpha/\alpha\alpha\ \beta^S\beta^S$

In addition to Hb S over 150 other β chain variants have been discovered, almost all of which are point mutations. Substitutions are known for many positions other than position 6. Though all the mutations that produce β chain variants are allelic in that they affect the amino acid sequence of the same chain, the mutation may be at the same site as the Hb S mutation or a different site, affecting the same or a different codon. The second variant to be found, Hb C, also has a substitution in position 6 of the β chain, a change from glutamic acid to lysine; a change in the first base of the triplet from G to A accounts for this variant.

The allelism of the β chain variants has been demonstrated both genetically and biochemically. For example, Figure 5–6 shows a family in which a person with two different mutations at the β locus, one for Hb S and one for Hb C, has parents with one abnormality or the other, but not both.

Lepore Hemoglobins: Fusion Genes

In the Lepore hemoglobins, the non-α chain has a segment homologous to the N-terminal end of a normal δ chain followed by a segment homologous to the C-terminal end of a normal β chain, which together form a new δβ fusion

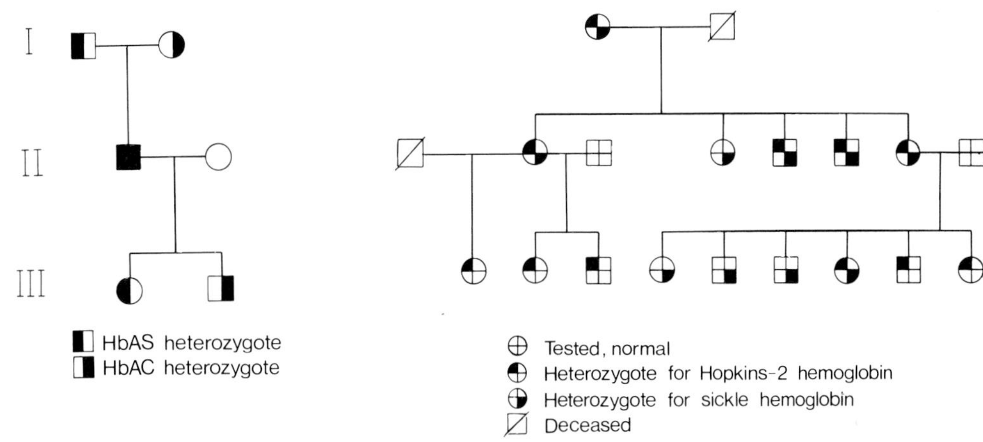

Figure 5–6. Allelism and non-allelism in the hemoglobinopathies. Left: a pedigree to demonstrate allelism of the β chain variants Hb S and Hb C. See text for details. Right: a pedigree to demonstrate independent assortment of Hb Hopkins-2, an α-chain variant, and Hb S, a β-chain variant. From Smith EW, Torbert JV. Bull Johns Hopkins Hosp 1958; 102:38–45.

chain. The explanation of the origin of these variants is shown in Figure 5–7. The δ and β genes have extensive homology, differing at only 10 of their 146 sites. Misalignment between the δ gene of one chromatid and the β gene of the homologous chromatid, occurring during meiosis, could happen as a relatively rare accident. Crossing over would then be possible between the two mismatched genes, resulting in the formation of two products: a new gene beginning as a δ and ending as a β and a reciprocal product with a normal δ and a normal β gene but also a fusion product coded for a β segment followed by a δ segment. The first is a Lepore gene, the second an "anti-Lepore." This mechanism is known as homologous but unequal crossing over, and is thought to be significant in the evolution of many genes that have duplicated and undergone some divergence.

The anti-Lepore gene is inevitably associated with normal β and δ genes, so it has no deleterious effect on Hb A synthesis. However, Lepore hemoglobin has clinical and hematological consequences representing a form of thalassemia (see below).

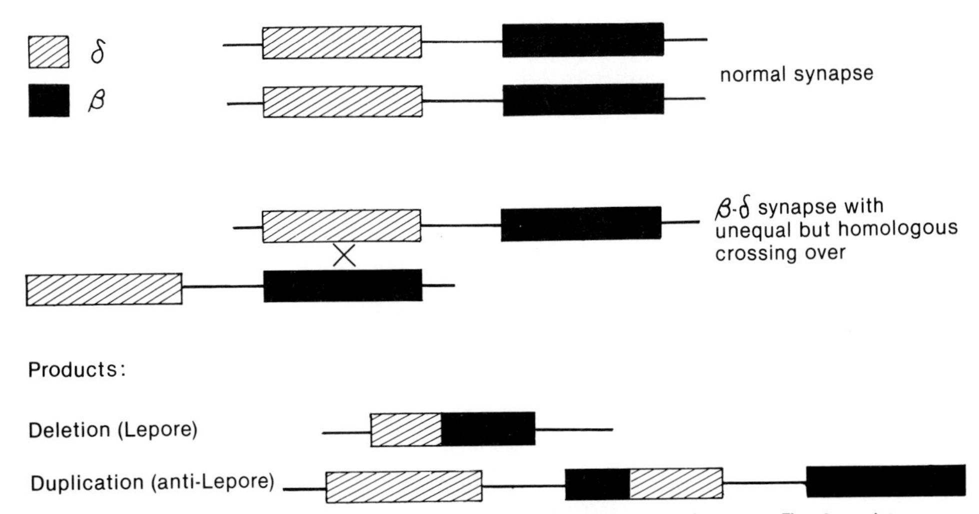

Figure 5–7. Model for the origin of a Lepore gene by unequal crossing over. The δ and β genes are very closely linked and their sequences differ at only 10 sites. If mispairing occurs, followed by intragenic crossing over, two hybrid genes result: one with a deletion of part of each locus (a Lepore gene), the other with a corresponding duplication (an anti-Lepore gene).

Alpha, Delta and Gamma Chain Variants

An alpha chain mutation alters all the hemoglobins that have α chains (A, A2, F and Gower 2). Alpha globin variants are less common than β globin variants, possibly because an α globin mutation leads to abnormal fetal hemoglobin. The first to be discovered is known as Hb Hopkins-2. In a family with both Hb Hopkins-2 and Hb S, it was found that the members with both abnormal hemoglobins also had large quantities of Hb A and had children with either one or both of the abnormal hemoglobins as well as Hb A (see Fig. 5–6). Thus it was clear that the two mutations were not allelic. As all three main hemoglobins were affected, the mutation was obviously in an α chain.

A mutation in the gene coding for the δ chain leads to production of an abnormal Hb A2, and a γ gene mutation affects fetal hemoglobin only.

The Thalassemias

The thalassemias are a heterogeneous group of disorders of hemoglobin synthesis in which the basic defect is not in molecular structure, but in a reduced rate of synthesis of either the α or β chains. This leads to imbalance of globin chain synthesis, precipitation of the globin chains that are produced in excess and consequent impairment of erythrocyte maturation and survival. Worldwide, thalassemias are the commonest of the human single-gene disorders. Thalassemia carriers are sufficiently common in Canada and the United States to pose the important problem of differential diagnosis from iron deficiency anemia, and to be a relatively common source of referral for homozygote detection in prenatal diagnosis. There is a characteristic distribution of the thalassemias in a band around the Old World—in the Mediterranean, the Middle East, parts of Africa, India and the Orient. The name is derived from the Greek word for sea, *thalassa*, and signifies that the disease was first discovered in persons of Mediterranean stock. The erythrocytes typically have a target cell appearance.

In recent years much progress has been made in clarifying the molecular and genetic basis of the thalassemias. Two main groups are defined: the α thalassemias, in which α chain synthesis is reduced or absent, and the β thalassemias, in which β chain synthesis is impaired. A genetic and clinical complication is that it is not unusual for genes for both types of thalassemia as well as for structural hemoglobin abnormalities to coexist in an individual and to interact. Since there are many such genes, they account for a wide variety of complex hematological problems. Not all the thalassemias are well characterized; this brief discussion risks oversimplification by mentioning only some of the most common and best known types.

The Alpha Thalassemias

Genetic disorders of α globin production affect the formation of both fetal and adult hemoglobins. They appear to be due chiefly to deletion of α genes, though other mechanisms are sometimes involved; for example, the variant hemoglobin Hb Constant Spring, in which a chain termination mutation allows synthesis of an elongated α globin chain, leads to α thalassemia.

As noted earlier, each chromosome 16 carries two identical genes arranged in tandem. Thus there are four α genes per individual, each apparently responsible for the synthesis of 25 percent of the total α globin. Deletion or other alteration of one, two, three or all four of these genes causes a correspondingly severe hematological abnormality, as tabulated below:

Genotype	α Chain Production	Clinical Expression
αα/αα	100%	Normal
αα/α-	75%	"Silent" carriers
α-/α- or αα/--	50%	α thalassemia trait (mild microcytosis and anemia)
α-/--	25%	Hb H disease: moderately severe hemolytic anemia
--/--	0	Hydrops fetalis with Hb Bart's

Homozygous α thalassemia, in which all four α chains are deleted, results in fetal hydrops, a lethal condition in which fluid accumulates throughout the body of the newborn. (Fetal hydrops also occurs in other conditions, for example, Rh hemolytic disease of the newborn.) The parents of children with this condition have the αα/-- genotype, and usually have only mild microcytosis and anemia. The homozygous deletion type of α thalassemia leading to hydrops fetalis is almost restricted to East Asians, though in many other populations α thalassemia trait and the silent carrier state are quite common. Gene mapping allows ready detection of the absence of α genes in --/-- fetuses.

The Beta Thalassemias

Genetic disorders of β globin production affect the synthesis of Hb A only. The β thalassemias are a heterogeneous group of disorders of β chain production which are usually due to defective β globin genes rather than to deletions. There is such great variety in β thalassemia genes that persons carrying two such genes, though commonly called homozygotes, are more likely to be compounds with two different β mutations, unless they have consanguineous parents.

As a general rule individuals with two β thalassemia genes ("homozygotes") have thalassemia major, a condition characterized by severe anemia and requiring lifelong medical management. Until recently, survival into adult life was unusual. The onset of the disorder is usually not apparent until β globin production replaces γ globin production a few months after birth.

At present some 30 different mutations causing β thalassemia have been defined. Some of these mutations are within the coding sequences of the β globin gene, but others are in intervening sequences or in regulatory sequences upstream of the gene. A second group of β thalassemias results from deletion of an extensive segment of DNA from the β gene complex (Fig. 5–8). The deletion thalassemias include that associated with Hb Lepore (the consequence of a δβ fusion gene, described earlier) and other forms due to more extensive deletions encompassing the γ globin loci as well as the δ and β loci. If the γ globin loci are intact but the β and δ loci are deleted, the consequence is hereditary persistence of high fetal hemoglobin (HPFH), a mild disorder in which fetal hemoglobin compensates at least partly for absent Hb A production.

Whether the genetic basis of a case of β thalassemia is a mutation or a deletion, it is typically clinically mild in heterozygotes. Homozygotes have clinical pictures of varying severity, depending on the nature of the specific defect present. The most common type involves mutation at the β locus, and since the δ gene is intact, this form is characterized clinically by a high level of Hb A2. There may be no synthesis of β globin (the β^0 form) or synthesis of a reduced amount of β globin (the β^+ form).

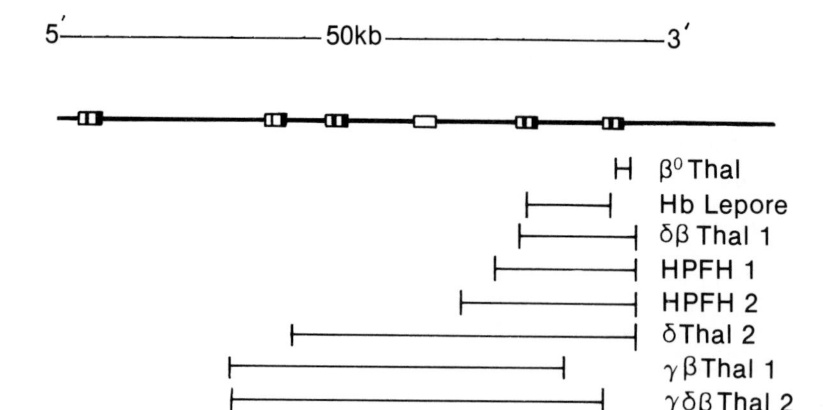

Figure 5–8. Examples of deletions resulting in β thalassemia. Based on Kan YW. The thalassemias. In: Stanbury JB, Wyngaarden, JB, Fredrickson DS, Goldstein JL, Brown MS. The metabolic basis of inherited disease, 5th ed. New York, McGraw-Hill, 1983.

PRENATAL DIAGNOSIS IN β THALASSEMIA

Because of the high frequency of β thalassemia in certain populations and the possibility of prenatal detection, it is considered good medical practice to identify carrier couples from high-risk ethnic groups during early pregnancy so that prenatal diagnosis can be offered. Homozygous β thalassemia can be detected by fetal blood sampling at the eighteenth week of gestation, followed by measurement of the ratio of β globin synthesis to γ globin synthesis by the fetal cells. Methods using linked restriction fragment length polymorphisms are also used in suitable families (see Chapter 11). In some areas, such as Cyprus or Sardinia, where β thalassemia reaches a heterozygote frequency of 10 percent or more, prenatal diagnosis has become widely used and has reduced the number of births of homozygous β thalassemics.

POLYMORPHISM ADJACENT TO THE β GLOBIN GENE

In a landmark paper, Kan and Dozy (1978) described polymorphism in a DNA sequence about 5000 nucleotides beyond the 3′ end of the β globin gene. In most people the restriction enzyme *HpaI* cuts the DNA strand at the site, generating a fragment 7.6 kb in length containing the β gene. In some Africans the site was altered and no longer cut by *HpaI*; instead, a fragment 13 kb in length was obtained, and the 13 kb fragment was so frequently associated with the sickle hemoglobin gene, rather than the normal allele, that it could be used with some success in prenatal diagnosis (Fig. 5–9). About 60 percent of black Americans who have the sickle mutation have the altered *HpaI* site. This was the first example of the potential of RFLP analysis for clinical application.

Subsequently it has been found that the restriction enzyme *MstII* recognizes the specific site of the sickle mutation, consequently providing a direct test for use in prenatal diagnosis.

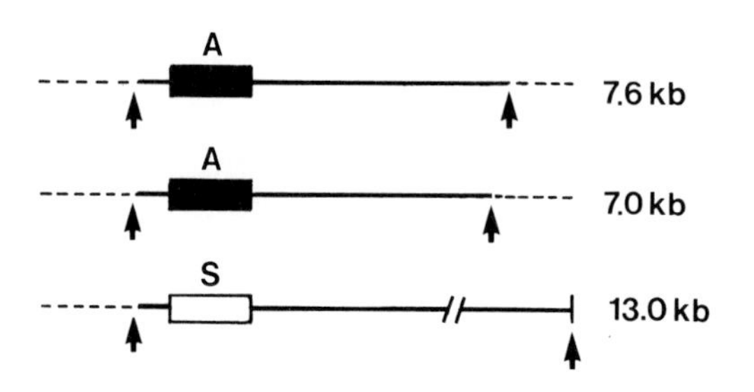

Figure 5–9. The *HpaI* restriction fragment length polymorphism adjacent to the sickle cell gene, demonstrated by Kan YW, Dozy Am. Proc Natl Acad Sci USA. 1978; 75:5631–5635. See text.

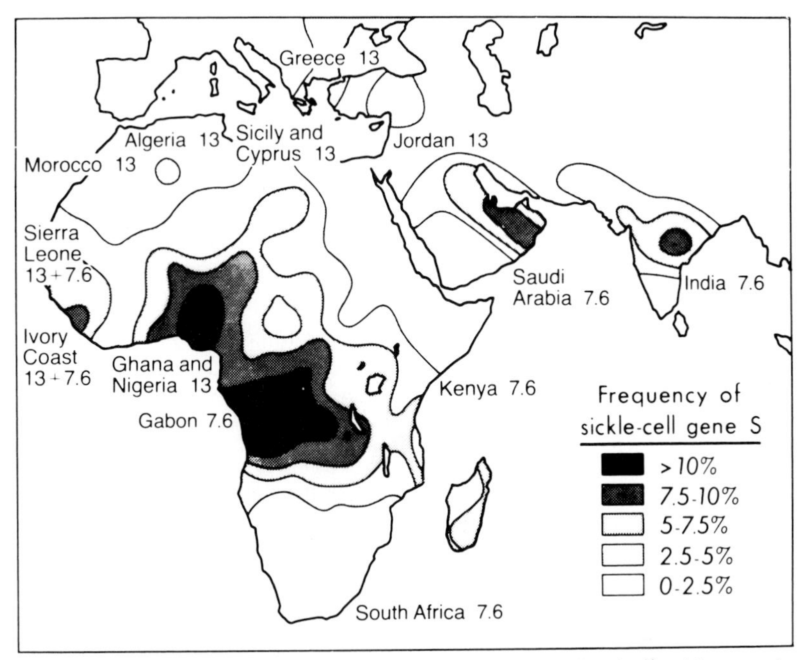

Figure 5–10. The distribution of the sickle-cell gene in relation to *HpaI* fragments 7.6 kb and 13 kb in length. The mutation associated with the 13-kb *HpaI* fragment originated in West Africa and spread from there. The mutation associated with the 7.6-kb fragment arose separately and probably had multiple origins. From Kan YW. Hemoglobin abnormalities: molecular and evolutionary studies. In The Harvey Lectures, Series 76. New York: Academic Press, 1982, pp 75–93, by permission.

Not all ethnic groups in which the sickle mutation is present have the gene in a 13 kb *HpaI* fragment as do West Africans. In India, Saudi Arabia, South Africa, Gabon and Kenya the mutation is typically associated with a 7.6 kb fragment (Fig. 5–10). These findings imply that the sickle mutation occurred in West Africa on a chromosome that already had a mutant *HpaI* site, and that it arose independently at least once elsewhere. The protection the sickle cell trait confers against malaria accounts for the high frequency the gene has reached in malarial areas of the world.

CLINICAL MANAGEMENT OF GENE EXPRESSION IN β THALASSEMIA

The first step in what has been described as "bringing molecular biology to the bedside" was an attempt at therapy of homozygous β thalassemia by selective activation of the γ globin genes (Ley et al., 1982). The patient was treated with 5-azacytidine, a cytidine analogue known to be capable of selectively increasing γ globin synthesis. Within a week the patient responded with an increase in γ globin synthesis that temporarily allowed synthesis of greatly increased quantities of *Hb* F. Although the general usefulness of the approach and its risk, in particular the possible carcinogenic effect of the agent 5-azacytidine, have yet to be determined, the results are encouraging for β thalassemia and sickle cell anemia. Before this report appeared, many physicians would have expected that gene therapy required a surgical approach with excision of the genetic lesion and transplantation of a normal DNA sequence. It is because the β allele has a fetal counterpart and persistence of fetal hemoglobin is known to be a harmless genetic anomaly that the experiment was possible. Manipulation of gene expression may not be feasible for many lesions, but it is a promising line for future efforts in gene therapy.

Human Biochemical Disorders

Human biochemical disorders, often called inborn errors of metabolism, are conditions in which a specific genetically determined enzyme defect produces a metabolic block that has pathological consequences. The concept of inborn errors of metabolism as causes of disease was first proposed by the eminent physician Garrod in 1902, shortly after the rediscovery of Mendel's laws. Garrod noted that in alcaptonuria and several other disorders the patients appeared to lack the ability to perform one specific metabolic step. (In the case of alcaptonuria, this step is the breakdown of the benzene ring of the amino acid tyrosine. Failure to complete the reaction leads to the excretion in urine of an abnormal constituent, homogentisic acid or alcapton.) Garrod also observed that these disorders showed a striking familial pattern of distribution, in that two or more sibs might be affected though the parents and other relatives were usually normal, and that the parents themselves were often consanguineous. Through discussions with Bateson, a leading biologist and pioneer geneticist, Garrod came to interpret this pattern as exactly what would be expected on the basis of Mendelian recessive inheritance, and thus described in alcaptonuria the first example of autosomal recessive inheritance in man.

Garrod's concept, that an inborn error of metabolism was a genetically determined enzyme defect leading to interruption of a metabolic pathway at a specific point, was far in advance of its time. The first actual demonstration of a specific enzyme defect in an inborn error of metabolism was not provided until 1952, when Gerty Cori showed that von Gierke's disease (glycogen storage disease Type I) is due to loss of activity of the enzyme glucose-6-phosphatase. The specific deficiency of the enzyme homogentisic oxidase, which is the cause of alcaptonuria, was not directly demonstrated until 1958, by La Du and colleagues.

The majority of inborn errors of metabolism are inherited as autosomal recessives. Several, including some of great clinical or genetic interest, are X-linked; for example, the glucose-6-phosphate dehydrogenase (G6PD) variants and hypoxanthine-guanine phosphoribosyl transferase (HPRT) deficiency (Lesch-Nyhan syndrome). Some are autosomal dominants; these are likely to involve regulators of metabolic pathways such as membrane receptors or rate-limiting enzymes which when deficient lead to impaired feedback inhibition.

A large number of biochemical defects that produce inborn errors of metabolism are now known. In the following sections some of the most common examples of the variety of clinical and genetic problems related to specific enzyme deficiencies are discussed.

MECHANISMS OF BIOCHEMICAL DISEASES

In the last analysis, the actual anatomical site of any single-gene disorder is an alteration in DNA sequence that accounts for synthesis of a defective protein. This defective protein in turn, if it is an enzyme, impairs some intracellular metabolic function; if it is a nonenzymatic (structural) protein, it may disrupt the normal structure and role of the cell or tissue of which it is a component. Since the structural and functional proteins encoded by normal genes are many and varied, the clinical disorders that represent the remote consequences of mutation in these genes are correspondingly numerous and diverse in their manifestations. They range in severity from those that are relatively harmless to those that cause death in early infancy. All are, of course, compatible with life at least to the time of birth and consequently must be milder clinically than the unknown mutations that are genetic lethals. Examples of the variety of pathogenetic mechanisms by which abnormal proteins cause disease are given both in this section and elsewhere in the book.

Most of the disorders classed as inborn errors of metabolism are due to defective enzymes, but a number of disorders due to defective nonenzymatic proteins have also been defined. Usually the defective enzymes have reduced enzymatic activity as compared with their normal counterparts. Occasionally the activity is completely lost; rarely it may be even higher than normal. Reduced activity can result from reduced affinity for substrate or cofactor, or from instability of the enzyme molecule.

Metabolism is performed as a stepwise series of reactions, each step catalyzed by a different enzyme. Mutation in one enzyme of the sequence results in accumulation of precursor and in a reduced quantity of product, either of which may have pathological consequences. For example, in phenylketonuria, phenylalanine is accumulated because of reduced activity of the enzyme phenylalanine hydroxylase. Since it cannot be metabolized normally, it is broken down to toxic products that bring about brain dysfunction. Similarly in the lysosomal storage disorders deficiency of a specific lysosomal enzyme leads to accumulation of a complex substance in lysosomes. An example of deficiency of the product of an enzymatic reaction is the impairment of cortisol synthesis in congenital adrenal hyperplasia.

Not all biochemical genetic disorders are due to disruption of metabolic pathways. A number of other mechanisms have been identified. For example, in I-cell disease the defect is in posttranslational modification of various hydrolase enzymes; the consequence is that these enzymes fail to enter lysosomes and therefore fail to become functional. In familial hypercholesterolemia, the defect has been traced to defective cell surface receptors for plasma LDL. Thus LDL-bound cholesterol fails to enter cells, or enters in reduced amounts. As a result the normal feedback regulation of cholesterol synthesis is impaired. In Lesch-Nyhan syndrome, an X-linked disorder of purine metabolism, the overproduction of end products also depends on failure of feedback inhibition.

Osteogenesis imperfecta and sickle cell anemia may both be regarded as disorders in which genetic defects lead to molecular defects in a nonenzymatic (structural) protein.

PHENYLKETONURIA AND OTHER HYPERPHENYLALANINEMIAS: DISORDERS OF AMINO ACID METABOLISM

Phenylketonuria

Classic phenylketonuria (PKU) is an autosomal recessive disorder of metabolism of the amino acid phenylalanine, which Jervis (1953) found to be caused by a defect in phenylalanine hydroxylase, the enzyme that normally converts phenylalanine to tyrosine as the first step in its degradation (Fig. 5–11). (Note that a statement such as this implies that the structure of the normal gene product is determined by the normal allele at the locus. The role of the normal gene is revealed by observing the consequences of its absence.) The discovery of PKU by Følling in 1934 marked the first demonstration of a genetic defect as a cause of mental retardation. Because patients with PKU cannot convert it to tyrosine, phenylalanine is degraded through an alternate pathway, producing

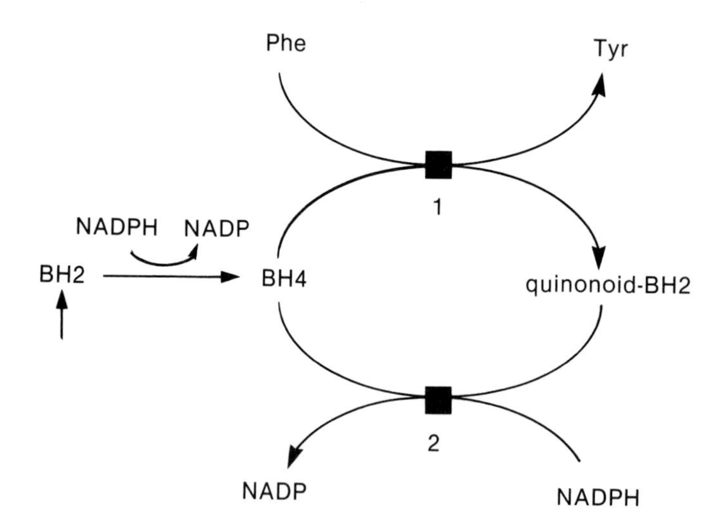

Figure 5–11. Scheme of defects in phenylalanine metabolism. 1. A deficiency of phenylalanine hydroxylase blocks the phe → tyr pathway. 2. A defect in dihydropteridine reductase leads to deficiency of the phenylalanine hydroxylase coenzyme, tetrahydrobiopterin (BH4).

phenylpyruvic acid and other abnormal metabolites, which are excreted in the urine.

Children with PKU are normal at birth and become retarded only when they ingest phenylalanine; thus PKU is the prototype of those inborn errors of metabolism for which mass screening of the newborn population is appropriate. The usual screening test, the Guthrie bacterial inhibition assay, makes use of the high serum phenylalanine (about 30 times normal) and the principle that the growth of bacteria on agar containing β-2-thienylalanine, a competitive inhibitor of phenylalanine, is a function of the phenylalanine added to the medium. A droplet of blood obtained by a heel prick and dried on filter paper is obtained a few days after birth, usually before the infant leaves the hospital, and sent to a central laboratory for assay. There is also a fluorimetric method for direct determination of phenylalanine in blood after elution from filter paper; this method is also applicable to mass screening.

The consequences of the metabolic block in PKU may be partly circumvented by reducing the amount of precursor, that is, by giving the patient a diet low in phenylalanine. Phenylketonuric children are normal at birth because the maternal enzyme protects them during prenatal life. The results of treatment are best when the diagnosis is made soon after birth and treatment is begun promptly. If the child is fed phenylalanine for some time, irreversible mental retardation occurs.

Other Disorders of Phenylalanine Metabolism

Decreased rather than absent phenylalanine hydroxylase produces a relatively benign form of persistent hyperphenylalaninemia in which, typically, no treatment is required. In a transient mild form, maturation of the enzyme is delayed but there are no clinical consequences. Although these children have a positive screening test, a phenylalanine-free diet is not required and in fact is harmful.

Though originally it was thought that all children with hyperphenylalaninemia lacked phenylalanine hydroxylase, it is now clear that absence or deficiency of two other enzymes, dihydropteridine reductase or dihydrobiopterin synthetase, can also lead to hyperphenylalaninemia.

Dihydropteridine reductase deficiency is a rare and serious defect of an enzyme involved in the synthesis of the co-factor, tetrahydrobiopterin (BH4) (see Fig. 5–11). (The name biopterin dates back to the discovery of substances called "pterins" because they were found in butterfly wing pigments.) The affected children do not respond to a low phenylalanine diet, remaining hyperphenylalaninemic. Neurological problems develop during the first year of life. The biochemical defect leads to impaired synthesis of the neurotransmitters dopa, noradrenalin, adrenalin and serotonin. A different entity, dihydrobiopterin synthetase defect, has much the same clinical course and serious outcome. Like classical PKU these disorders are autosomal recessives.

Maternal Phenylketonuria

PKU can now be treated successfully enough to allow homozygotes an independent life and almost normal prospects for marriage and parenthood. Because phenylketonurics who have been effectively managed from birth can function adequately in many ways, it is disconcerting that almost all the children of female phenylketonurics are mentally retarded. Most of these children are heterozygotes (a few are homozygotes, having heterozygous fathers). Their retardation is apparently due not to their own genetic constitution but to their intrauterine development in mothers who have unduly high plasma phenylalanine levels. About 25 percent of the children of phenylketonuric mothers (vs. about 3 percent of controls) also have congenital malformations of various types. Currently considerable progress is being made in improving the palatability of low phenylalanine diets; it is considered reasonable to continue to maintain females on such diets until the end of their childbearing years. Screening of all primigravid mothers for hyperphenylalaninemia has been proposed as a public health measure.

Prenatal Diagnosis of PKU

A 1983 development in molecular genetics by Savio Woo and his colleagues has provided a means of prenatal diagnosis. These workers cloned the PKU gene by isolation of messenger RNA coding for phenylalanine hydroxylase from rat liver, using it to synthesize complementary DNA, and using the rat cDNA as a probe to allow cloning of the corresponding human cDNA. Three different restriction fragment polymorphisms generating several haplotypes within or very close to the gene were then identified. It is estimated that at present about 75 percent of the population can be shown to be heterozygous for haplotypes linked closely to PKU. In these families, after the birth of a child with PKU (but not before), DNA analysis of the affected child, the carrier parents and the fetus could allow the genotype of the fetus to be determined.

GALACTOSEMIA: A DISORDER OF CARBOHYDRATE METABOLISM

Galactosemia results from inability to metabolize galactose, a monosaccharide that is a component of lactose (milk sugar). The inheritance is autosomal recessive. Affected infants completely lack the enzyme galactose-1-phosphate uridyl transferase (gal-1-PUT), which normally catalyzes the conversion of galactose-1-phosphate to uridine diphosphogalactose.

Infants with galactosemia are usually normal at birth but begin to develop gastrointestinal problems, cirrhosis of the liver and cataracts as soon as they are given milk. Galactosemia can, however, be the cause of abnormal liver function and failure to thrive in newborns. Pregnant women who are known heterozygotes at risk of having homozygous children can be placed on galactose-free diets to prevent problems arising prenatally. If untreated, galactosemia is usually fatal, though in older children an alternative pathway for galactose metabolism eventually develops. Complete removal of milk from the diet can protect against the harmful consequences of the enzyme deficiency.

In addition to the normal allele Gt^{+} and the galactosemia allele *gt*, there is another well-known allele Gt^{D} that when homozygous produces the "Duarte variant," with only half the normal enzyme activity but without associated clinical problems. The approximate frequencies, enzyme activities and phenotypes of the six possible genotypes are as follows:

Genotype	Frequency (%)	Enzyme Activity (%)	Phenotype
$Gt\ Gt$	91.2	100	Normal
$Gt\ Gt^{D}$	7.6	75	Normal
$Gt^{D}\ Gt^{D}$	0.16	50	Normal
$Gt\ gt$	0.96	50	Normal
$Gt^{D}\ gt$	0.04	25	Borderline
$gt\ gt$	0.0025	0	Galactosemia

Multiple allelism at a locus resulting in a relatively high frequency of compound genotypes is thought to be an important source of clinical heterogeneity for many disorders. Unless the affected person is the product of a consanguineous mating, he or she may have a compound genotype rather than a homozygous one. However, if the parents are consanguineous, the probability is that the patient has received the identical mutation in double dose, by two separate lines of descent from one carrier ancestor.

Heterozygotes and compounds for the various alleles at the galactosemia locus are identified at present chiefly by assaying for the enzyme in red cells or cultured skin fibroblasts. Because the enzyme is expressed in cultured cells, prenatal diagnosis is feasible. Mass screening for galactosemia in newborns is also feasible and is routinely used in a number of areas though less widely and somewhat less successfully than PKU screening.

GALACTOKINASE DEFICIENCY

Galactokinase catalyzes the first step in the pathway of galactose metabolism, whereas galactose-1-phosphate uridyl transferase catalyzes the second step. Galactokinase deficiency is a rare cause of galactosemia. Among the patients reported, cataracts developing in later childhood are common, but liver disease and mental retardation are variable.

LYSOSOMAL STORAGE DISEASES

Tay-Sachs Disease

There are many different disorders in which the characteristic defect is the accumulation of complex substances that in normal cells are degraded to their constituents by specific hydrolytic enzymes segregated within lysosomes. The sphingolipidoses are a group of these lysosomal storage disorders in which the stored substances are sphingolipids. Gangliosides are a form of sphingolipid in which the basic structure is a ceramide linked to a polysaccharide chain, as shown in Figure 5–12. Tay-Sachs disease (G_{M2} gangliosidosis or infantile amaurotic idiocy) is a lysosomal storage disease in which the stored substance is a ganglioside.

Tay-Sachs disease is one of several different forms of amaurotic familial

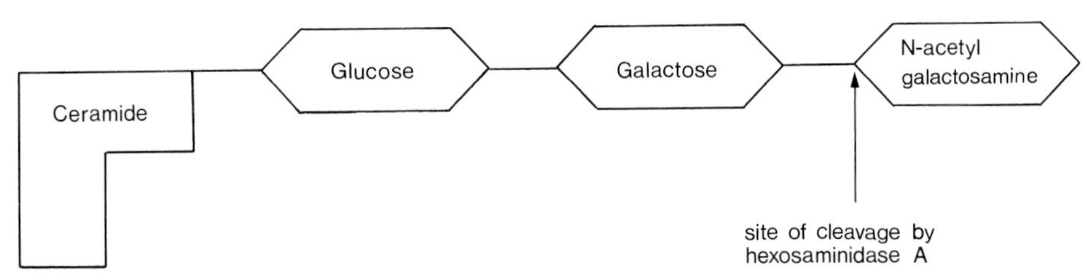

Figure 5–12. Representation of the structure of G_{M2} ganglioside. The ceramide subunit consists of sphingosine linked to a long-chain fatty acid.

idiocy, not all of which are gangliosidoses. The term amaurotic is derived from *amauros* (dark or obscure) and referred originally to the obscure cause of the progressive blindness that is one of the features of the disorder.

The enzyme lesion in Tay-Sachs disease is a marked deficiency of hexosaminidase A (hex A) in a wide variety of tissues. The function of hex A is to cleave the terminal N-acetyl-β-galactosamine residue from the polysaccharide chain of the ganglioside molecule; in its absence the ganglioside accumulates, especially in brain tissue. The disorder is particularly common in Ashkenazi Jews. The frequency is about 1 affected child in 3600 births among the Ashkenazi, and about 1 in 360,000 in other populations (including non-Ashkenazi Jews). The frequency of heterozygotes is about 1 in 30 among the Ashkenazi, 1 in 300 in other populations.

Heterozygotes can be detected by screening blood samples for hexosaminidase activity in serum; the disease can be diagnosed prenatally by biochemical analysis of cultured amniocytes. Because of these characteristics (high frequency in a particular population group, feasibility of mass screening for heterozygotes and possibility of prenatal detection), Tay-Sachs disease has become the prototype of those inborn errors in which matings of two heterozygotes can be identified and each pregnancy of such matings can be monitored in order to identify potentially affected fetuses, which in many jurisdictions can then be aborted if the parents so choose. In other words, Tay-Sachs disease has become in theory an almost entirely preventable disease. Many screening programs have already been carried out, both in adults and in high school students, but the long-range effects of these programs on the frequency of the disease and the reproductive decisions of identified carriers have not yet been assessed.

The Mucopolysaccharidoses

Acid mucopolysaccharides (glycosaminoglycans) are normal constituents of many body tissues, especially connective tissues. The mucopolysaccharidoses (MPS) are a heterogeneous group of lysosomal storage diseases in which the stored substances are partially degraded acid mucopolysaccharides, especially breakdown products of dermatan sulfate and heparan sulfate. They are of particular genetic interest because as a group they include examples both of mutations at different loci and of more than one mutation at a single locus. Several types are summarized in Table 5–2.

The common feature of the structure of mucopolysaccharides is that they are large polymers with a protein core to which are attached extensive polysaccharide branches. In patients with a particular MPS, the protein portion of the molecule has been completely hydrolyzed, but the polysaccharide portion only partly so. Characteristic breakdown products are stored within the lysosomes and appear in large quantities in the urine.

The first two mucopolysaccharidoses to be recognized were the X-linked

Table 5–2. EXAMPLES OF MUCOPOLYSACCHARIDOSES

Syndrome	Enzyme Deficiency	Genetics	Mucopolysaccharides Stored/Excreted
Hurler (MPS IH)	α-Iduronidase	Homozygous for autosomal recessive gene	Dermatan and heparan sulfates
Hunter (MPS II–XR)	Iduronidate sulfatase	Hemizygous for X-linked gene	Dermatan and heparan sulfates
Scheie (MPS IS)	α-Iduronidase	Homozygous for autosomal recessive gene	Dermatan and heparan sulfates
Hurler/Scheie compound (MPS IH/S)	α-Iduronidase	Genetic compound (2 different autosomal recessive alleles)	Dermatan and heparan sulfates
Sanfilippo A (MPS IIIA)	Heparan N-sulfatase	Homozygous for autosomal recessive gene	Heparan sulfate
Sanfilippo B (MPS IIIBV)	N-acetyl-α-glucosamin-idase	Homozygous for autosomal recessive gene	Heparan sulfate

Data from McKusick VA, Neufeld EF. The mucopolysaccharide storage diseases. In Stanbury et al., eds. The Metabolic Basis of Inherited Disease, 5th ed. New York: McGraw-Hill, 1983, pp. 751–777.

Hunter syndrome (in three young Winnipeg brothers) in 1917 and the more severe autosomal recessive Hurler syndrome in 1919. Originally each of these conditions were called "gargoylism" because of the coarseness of the facial features (Fig. 5–13). Affected children are mentally retarded and skeletally

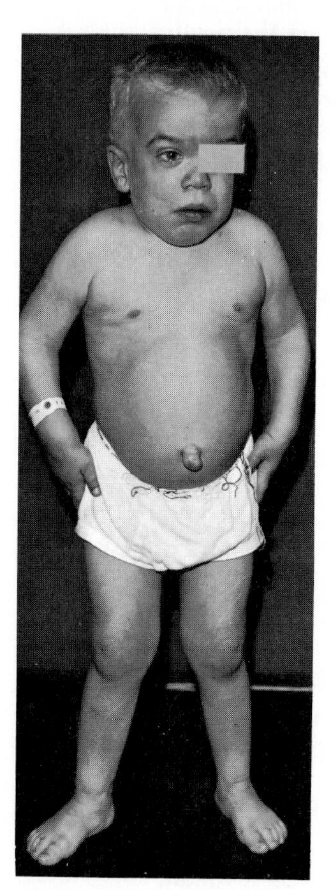

Figure 5–13. A child with Hurler syndrome (mucopolysaccharidosis I). At five years of age, the child is only as tall as a typical three-year-old. From Smith DW. Recognizable patterns of human malformation, 3rd ed. Philadelphia: W. B. Saunders Company, 1982.

abnormal with short stature. The corneal clouding that develops in the Hurler but not in the Hunter syndrome is a useful distinguishing sign. Both disorders are progressive, and Hurler syndrome is usually fatal in childhood. Mild and severe forms of Hunter syndrome, possibly allelic, are now distinguished.

The difference in the pattern of inheritance of the two conditions indicated a biochemical difference; this was demonstrated clearly by Fratantoni, Hall and Neufeld (1970), who showed that in cell culture, although fibroblasts from patients of either type accumulated mucopolysaccharides in the culture medium, the accumulation could be corrected by co-cultivation (that is, by growing cells of both types together in the same culture vessel). This experiment is a type of complementation test, a test to see whether more than one abnormal protein is involved. The correction factors proved to be lysosomal enzymes, α-iduronidase in the case of Hurler syndrome and sulfoiduronate sulfatase in the case of Hunter syndrome. It has been possible to effect at least temporary clinical improvement by transfusion of normal plasma or leucocytes as a source of the missing enzyme.

Scheie syndrome is clinically much milder than Hurler syndrome, so the finding (by failure to cross-correct in co-cultivation) that the same enzyme deficiency was involved in both was unexpected. The Scheie and Hurler genes appear to be alleles at the same locus but with different effects on α-iduronidase activity. The Hurler-Scheie compound has a condition intermediate in severity between the homozygous Hurler and the homozygous Scheie phenotypes.

In contrast, the Sanfilippo syndrome, which was earlier regarded as a single entity, has been split into several nonallelic disorders on the basis of their different enzyme deficiencies and ability to cross-correct. Clinically, they remain indistinguishable. The intelligence is severely impaired, but the physical abnormalities are very mild.

I-Cell Disease: A Disorder of Lysosomal Function

I-cell disease is an autosomal recessive condition phenotypically resembling Hurler syndrome, though with earlier onset, in which cultured skin fibroblasts contain numerous inclusions throughout the cytoplasm. The term I-cell refers to these inclusions, which are composed of membranous material.

In I-cell disease, not one but many enzymes are affected—the acid hydrolases that normally function within lysosomes. These enzymes are grossly deficient within cells but are present in excess in the culture medium.

It appears that in I-cell disease many acid hydrolases fail to undergo normal posttranslational modification. A typical hydrolase is a glycoprotein with mannose residues that become phosphorylated. The mannose-6-phosphate residues are essential for lysosomal localization of the hydrolases. A defect in the phosphorylation pathway results in failure of binding of the hydrolases to receptors that deliver them to the lysosomes. The location of the defect is shown in Figure 5–14. As the figure shows, the unphosphorylated precursors of the mature hydrolases are sialylated, then secreted into the extracellular medium instead of being inserted into lysosomes.

Not all lysosomal enzymes are affected. For unknown reasons, acid phosphatase and β-glucosidase are normal in amount and distribution. Some types of cells, especially neurons and hepatocytes, are unaffected, again for unknown reasons.

The enzyme defect has a range of pleiotropic effects, involving facial features, skeletal changes, severe growth retardation and mental retardation. The expected survival is about five years.

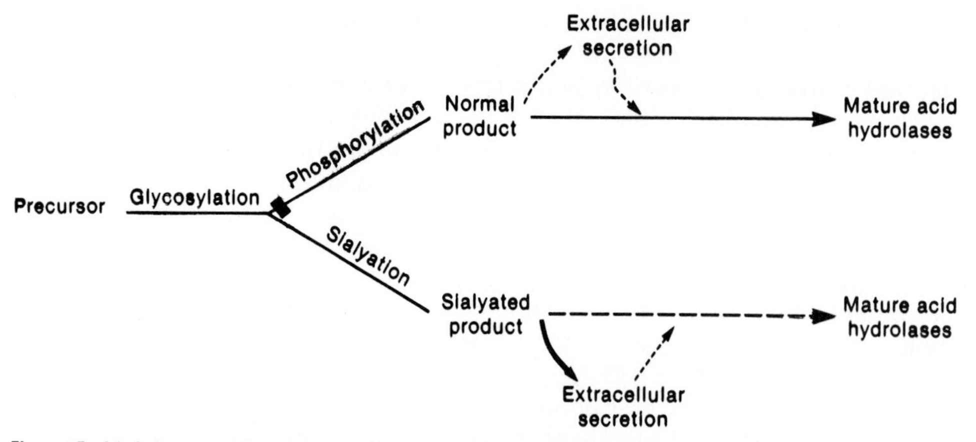

Figure 5–14. Scheme of the abnormality of phosphorylation that leads to multiple lysosomal enzyme abnormalities in I-cell disease. See text.

THE GENETIC HYPERLIPOPROTEINEMIAS: DISORDERS OF LIPID METABOLISM

Among the various genetic disorders of lipid metabolism, the genetic hyperlipoproteinemias are of special interest and clinical significance because of their role in myocardial infarction, a major cause of death and disability. Hyperlipoproteinemias are characterized by elevated levels of plasma lipids (cholesterol and triglycerides) and specific plasma lipoproteins. A number of distinct single-gene forms have been defined, differing in their biochemical and clinical phenotypes, though in some cases the phenotypes are not yet completely defined. These single-gene disorders are believed to result from mutations in single autosomal genes at different loci. At each locus there may well be more than one mutant allele.

To complicate the genetic picture further, it is known that not all hyperlipoproteinemias are caused by single-gene mechanisms; on the contrary, epidemiological studies have shown that only a minority of persons with significantly elevated plasma lipid and lipoprotein levels have a specific genetic disorder. In the general population, the background levels are continuously distributed and appear to be determined by a complex interaction of environmental and minor genetic factors. The variable expression of the major genes responsible for the genetic hyperlipoproteinemias against this multifactorial background makes it difficult to define the phenotypes of the various genetic forms until the basic defect in each is known.

Only one of these disorders will be mentioned here. Familial hypercholesterolemia (one of several disorders grouped as familial type 2 hyperlipoproteinemia) is an autosomal dominant disorder characterized by elevation of plasma cholesterol carried in the low-density lipoprotein (LDL) fraction. Heterozygotes have premature coronary heart disease, xanthomas (cholesterol-containing tumors) and early development of arcus corneae (deposits of cholesterol around the edges of the cornea). Homozygotes have exceptionally high plasma cholesterol levels and may have clinically significant coronary heart disease in childhood. The homozygous form is rare, but the heterozygous form, with a population frequency of at least 1 in 500, may be among the most common human single-gene disorders.

The basic defect has been found through analysis of its expression in cultured fibroblasts; the discovery has cast much light on the normal process of cholesterol regulation in the cell. Brown and Goldstein (1974) showed that fibroblasts from homozygotes may exhibit either absence or profound deficiency of the cell surface receptors that normally bind LDL; that is, they are either completely receptor-negative or receptor-defective. A third, rare mutation leads to receptors that can bind normally but are unable to internalize the lipoprotein. Heterozygotes have half the normal number of receptors. In normal LDL metabolism, the lipoprotein is taken up from the plasma by LDL receptors, undergoes endocytosis and is hydrolyzed in the lysosomes, with release of free cholesterol from the cholesteryl component of LDL. Cholesterol not required for cellular metabolism may be re-esterified for storage. The free intracellular cholesterol suppresses the activity of the microsomal enzyme 3-hydroxy 3-methylglutaryl coenzyme A reductase (HMG CoA reductase), thus causing reduction of cholesterol synthesis and also activating an enzyme that re-esterifies cholesterol for storage (Fig. 5–15). In familial hypercholesterolemia, since no LDL (or only a small amount) enters the cell, feedback regulation cannot take place; cholesterol synthesis is not suppressed, nor are cholesteryl esters formed. Thus there is overproduction of cholesterol, much more severe in homozygotes than in heterozygotes.

Prenatal diagnosis of homozygous familial hypercholesterolemia is possible by assay of LDL receptor activity in cultured amniocytes.

LESCH-NYHAN SYNDROME: A DISORDER OF PURINE METABOLISM

Lesch and Nyhan (1964) were the first to describe this rare X-linked disorder, often referred to as HPRT deficiency. Its features include uric aciduria, cerebral palsy, mental retardation and a compulsive behavior of self-mutilation by gnawing the lips and fingers. The affected boys completely lack the enzyme

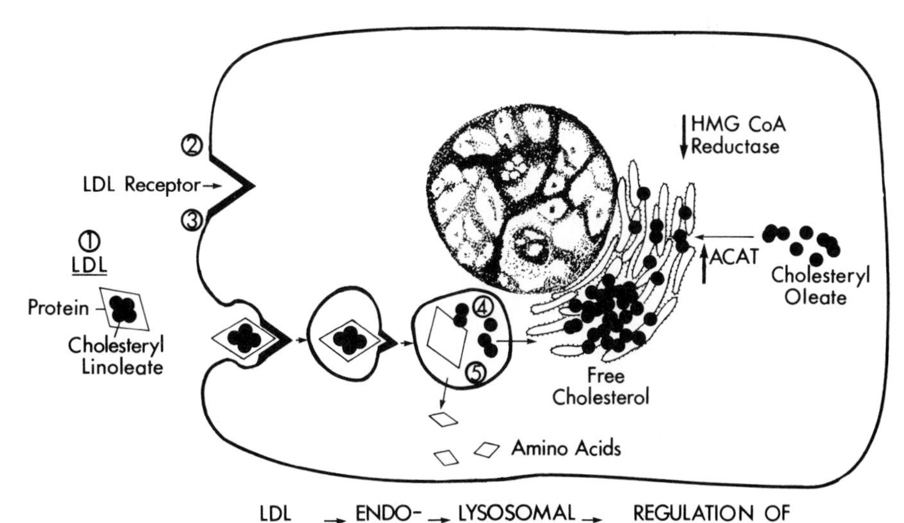

Figure 5–15. The pathways of LDL metabolism in cultured human fibroblasts. The numbers indicate specific known mutations in the pathway: 2, a form of familial hypercholesterolemia lacking LDL receptors; 3, a form having defective receptors; 1, 4 and 5, other single-gene defects. ACAT, fatty acyl CoA: cholesterol acyltransferase; other abbreviations in text. From Fredrickson DS, Goldstein JL, Brown MS. The familial hyperlipoproteinemias. In Stanbury JB, Wyngaarden JB, Fredrickson DS, eds. The metabolic basis of inherited disease, 4th ed. New York: McGraw-Hill Book Company, 1978.

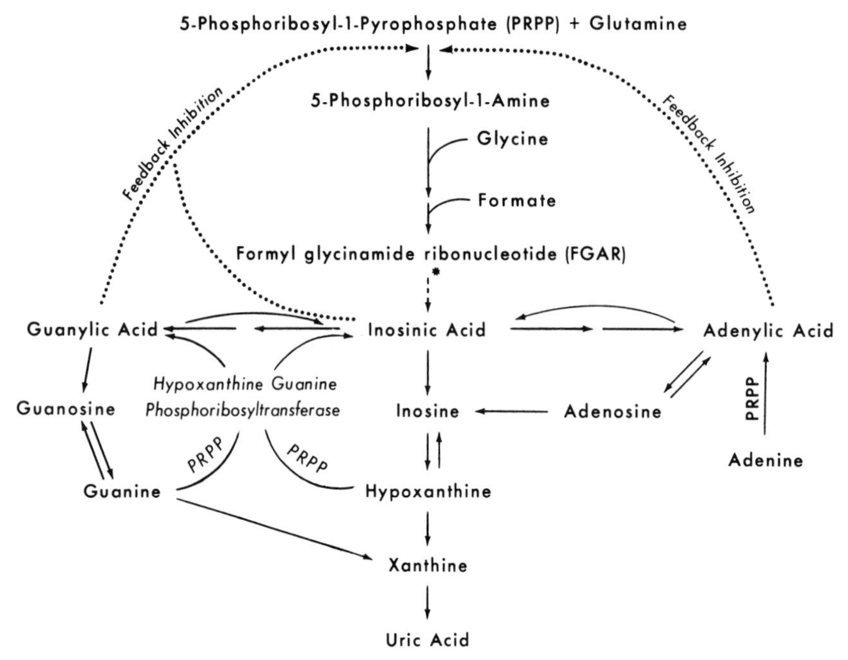

Figure 5–16. Feedback control system for regulation of purine synthesis. From Seegmiller JE, Rosenbloom FM, Kelley WN. Enzyme defect associated with a sex-linked human neurological disorder and excessive purine synthesis. Science 1967; 115:1682–1684. Copyright © 1967 by the American Association for the Advancement of Science.

hypoxanthine-guanine phosphoribosyltransferase (HPRT), which plays a role in the regulation of purine synthesis (Fig. 5–16), converting guanine and hypoxanthine to their respective nucleotides. Deficiency of HPRT leads to overproduction of purines and consequently excessive excretion of uric acid. Not all forms of HPRT deficiency completely lack the enzyme; a much less severe disorder results if even a very small amount of enzymatic activity remains. HPRT deficiency is of particular significance because of the usefulness of the mutation in somatic cell genetics (discussed in Chapter 10).

Many patients who survive for several years with HPRT deficiency develop gouty arthritis. Gout is characterized by recurrent and disabling inflammation, usually involving the first metatarsophalangeal joint. The underlying biochemical finding is hyperuricemia; not all hyperuricemics develop clinical gout, but perhaps 25 percent of them have urate crystals deposited in and around the joints, in the cartilage of the outer ear (tophi) and sometimes in the kidney as kidney stones. The hyperuricemia of gout has a heterogeneous group of causes, but many cases appear to be due to genetic defects involving purine and pyrimidine metabolism.

Heterozygotes for HPRT can be detected, and the defect can also be identified in early prenatal life by assaying for the enzyme in cultured amniotic fluid cells. Cloning of the gene itself has been accomplished.

Pharmacogenetics

Pharmacogenetics is the special area of biochemical genetics that deals with variation in drug response and the contribution of genetics to such variation. Broadly speaking, pharmacogenetics can be said to encompass any genetically determined variation in response to drugs; for instance, the effect of barbiturates in precipitating attacks of porphyria in individuals with the gene for acute intermittent porphyria, or the effect of alcohol use by pregnant women on the incidence of birth defects in the progeny. In a narrower sense, pharmacogenetics can be restricted to those genetic variations that are revealed *only* by response to drugs. Some classic examples of pharmacogenetic variations are described briefly in this section.

The origin of polymorphisms for drug response and the mechanisms by which they are maintained pose a problem. Obviously they have not developed in response to drugs, since they antedate the drugs concerned. The handling of and response to drugs require many specific biochemical reactions, and the enzymatic sequences involved may be used in the metabolism of ordinary food substances as well as drugs. For example, a potato extract, solanine, is an inhibitor of serum cholinesterase; it may be that an atypical serum cholinesterase phenotype conferred some selective advantage in the remote past. At least one abnormal drug response, primaquine sensitivity (glucose-6-phosphate dehydrogenase deficiency), is known to be associated with protection against malaria.

Pharmacologists recognize that there is normal variation in response to drugs by defining the "potency" of a drug as that dose that produces a given effect in 50 percent of the population. For genetic traits, continuous variation is usually best explained on the basis of multifactorial inheritance or by a combination of genetic and environmental factors. But response to drugs can also show discontinuous variation, with sharp distinctions between different degrees of response. The finding of a bimodal or trimodal population distribution of activity of a drug-metabolizing enzyme may indicate that the enzyme is coded by genes at a polymorphic locus.

GENETIC PROBLEMS IN ANESTHESIA

Serum Cholinesterase and Succinylcholine Sensitivity

Serum cholinesterase is an enzyme of human plasma that has the property of hydrolyzing choline esters, such as acetylcholine, to form free choline and the corresponding organic acid. Its older name of pseudocholinesterase came about because its hydrolytic action on acetylcholine is slow, compared with the speed with which "true" cholinesterase, a red cell enzyme, destroys acetylcholine at the neuromuscular junction.

The function of serum cholinesterase is obscure. It may be a "protective enzyme" because it can hydrolyze some choline esters that in high concentrations inhibit acetylcholinesterase. A low level of serum cholinesterase, or even its complete absence, is fully compatible with normal development and health; hence it cannot play a major physiological role.

Succinylcholine (suxamethonium) is a drug widely used as a muscle relaxant in anesthesia and in connection with electroconvulsive therapy. Chemically it is made up of two molecules of acetylcholine, and it is rapidly hydrolyzed by serum cholinesterase. The rapidity of its hydrolysis effectively reduces the amount of succinylcholine that reaches the motor endplates; this hydrolysis is allowed for in the dosage given to the average patient. Occasionally, however, a patient will respond to the administration of succinylcholine by developing prolonged apnea lasting from one to several hours.

An abnormal response to succinylcholine is not always genetic. Some 50 percent of patients who respond abnormally have an abnormal genotype, but in the remaining 50 percent the underlying problem is a nongenetic, pathological change, or even a technical problem in administration of the anesthetic.

GENETICS OF SERUM CHOLINESTERASE VARIANTS

The activity of cholinesterase in the plasma is determined by two codominant alleles, known as the "usual" *(U)* and "atypical" *(A)* alleles. Cholinesterase alteration occurs in persons who are homozygous for the atypical allele; the enzyme produced by homozygotes is qualitatively altered and has lower activity than the usual type.

Serum cholinesterase phenotypes cannot be determined with certainty on the basis of cholinesterase levels in serum, because the values thus obtained show considerable overlap. Normal and abnormal phenotypes can be distinguished by using an inhibitor of cholinesterase, dibucaine (Nupercaine), which is a well-known local anesthetic. The "dibucaine number" (DN) of a serum sample expresses its percent inhibition by dibucaine. The following relationship exists:

DN	Phenotype	Genotype	Approximate Frequency (Canadian Data)
About 80	Usual	*UU*	0.9625
About 60	Intermediate	*UA*	0.0370
About 20	Atypical	*AA*	0.0005

As shown above, about 1 person in 2000 is homozygous for the atypical allele and sensitive to succinylcholine.

Total absence of cholinesterase activity is a rare defect thought to result from homozygosity for a "silent" allele. Still another allele determines the structure of a type of cholinesterase that is unusually resistant to inhibition by sodium fluoride. The various compound genotypes of the atypical, silent or fluoride-resistant alleles all have some reduction in activity of the enzyme.

Malignant Hyperthermia

Malignant hyperthermia is an autosomal dominant condition in which there may be a dramatic adverse response to the administration of various anesthetic drugs and muscle relaxants, chiefly halothane and succinylcholine, with development of a very high temperature and muscular rigidity. The condition is potentially fatal. The precise biochemical defect is not yet precisely defined, but a disturbance of calcium metabolism is involved, and dantrolene is effective in preventing or reducing the severity of the response. The importance of identifying and warning the relatives of affected members and the need for special precautions when at-risk persons require anesthesia are obvious.

THE ACETYLATION POLYMORPHISM

Isoniazid is a drug used in the treatment of tuberculosis. The rate of inactivation of isoniazid after a test dose shows polymorphism, either in the quantity of free isoniazid excreted in the urine during the following 24 hours or in the plasma concentration after 6 hours. "Slow inactivators" are homozygous for a recessive gene, and "rapid inactivators" are normal homozygotes or heterozygotes. It is now clear that the slow and rapid isoniazid inactivation phenotypes and their consequences are due to polymorphism of the enzyme hepatic N-acetyl-transferase, which also acetylates other drugs, for example sulfamethazine and hydralazine.

The isoniazid acetylation phenotype appears to have no influence upon the response of patients with tuberculosis to isoniazid treatment when treatment is on a daily or twice-weekly basis, but if it is on a weekly basis the failure rate is higher for rapid acetylators. Thus a dosing schedule that is adequate for both rapid and slow acetylators is required.

The alleles for rapid and slow inactivation show marked differences in geographic distribution. The "slow" allele is very rare in Eskimos, for example, but up to 30 percent of blacks and of some European populations are "slow" homozygotes. In Canada, a high proportion of the tuberculosis cases are in the native population; members of this population are rapid inactivators who require isoniazid treatment at intervals of less than a week.

The isoniazid acetylator phenotype has a bearing on the development of isoniazid-induced peripheral neuropathy, a side effect of isoniazid treatment for which slow acetylators are at increased risk. Slow acetylators develop toxic reactions at the dosage necessary to maintain adequate blood levels in rapid acetylators. They are also at risk of drug-induced lupus in response to a number of other drugs that are acetylated by the same enzyme as isoniazid. On the other hand, rapid acetylators are at greater risk of hepatotoxicity. In general terms, the acetylator polymorphism is an example of a significant pharmacogenetic problem in which rational therapy must take into account wide, genetically determined individual differences in response.

The role of drug metabolism in causing mutagenesis, carcinogenesis, teratogenesis, cytotoxic damage or autoimmune diseases has recently attracted interest. Family histories of patients with toxic reactions to drugs should include questions about such toxicities, both to attempt to uncover genetic abnormalities of drug handling and to allow for appropriate genetic counseling about potential risks of certain drugs.

GLUCOSE-6-PHOSPHATE DEHYDROGENASE VARIANTS

Primaquine is an antimalarial drug that is active against both the *vivax* and the *falciparum* forms of malaria. (*Plasmodium vivax* is the usual cause of malaria in malarial regions of the world; *P. falciparum* is more restricted, occurring chiefly in jungle areas.) From the time primaquine was first introduced, it was known to be capable of inducing hemolytic anemia in some patients, especially in black males. Further investigation of this phenomenon showed that the red cells of primaquine-sensitive subjects are deficient in glucose-6-phosphate dehydrogenase (G6PD), a ubiquitous X-linked enzyme involved in glucose metabolism.

Favism, a severe hemolytic anemia in response to ingesting the broad bean *Vicia faba*, has been known since ancient times in parts of Italy, and its hereditary nature has been recognized. The basis of favism, like that of primaquine sensitivity, is G6PD deficiency, but a different variant is involved.

The normal and common abnormal variants of G6PD are listed in Table 5–3. The normal is known as type B, which is the conventional label given to normal electrophoretic variants of a number of genetic markers. About 20 percent of American black males have the faster, nearly normal A variant. The two deficient

Table 5–3. COMMON G6PD PHENOTYPES

G6PD Type	Gene Symbol	Electrophoretic Mobility	Enzyme Activity (% normal, approx.)	Approximate Population Distribution
B	Gd^B	Normal	100	Normal
B−	Gd^{B-}	Normal	4	Common in Mediterranean areas
A	Gd^A	Fast	90	20% of American black males
A−	Gd^{A-}	Fast	15	10% of American black males

types migrate electrophoretically at the same rate as A and B but have much lower activity, and so are called A− and B− respectively. Because the G6PD locus is on the X chromosome, males have only a single *Gd* gene and a single band for the corresponding protein, but heterozygous females (e.g., Gd^AGd^B heterozygotes) show two bands. The electrophoretic patterns are shown in Figure 5–17.

The G6PD locus is near the tip of Xq, within measurable distance of the loci for hemophilia A, hemophilia B, deutan and protan color blindness, adrenoleukodystrophy and the fragile X mental retardation syndrome. Because it is highly polymorphic, with well over 200 known variants, it is an especially useful X-linked marker.

Many of the variants of G6PD are associated with some degree of deficiency of activity of the enzyme. About 100 million people throughout the world have G6PD deficiency. For a deleterious gene to reach such a high population frequency, a selective advantage must be postulated; it is believed that many

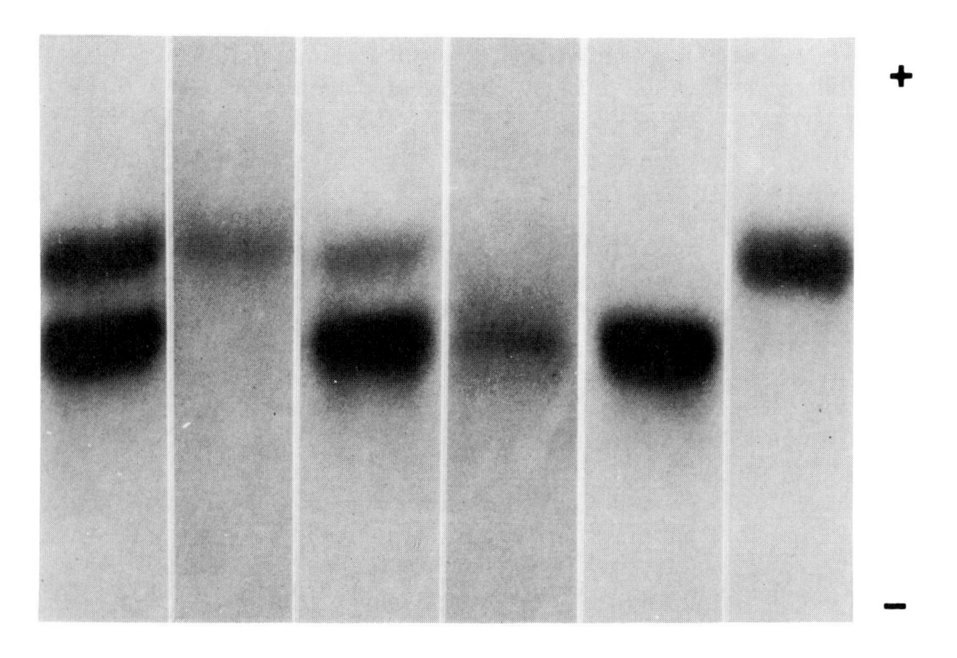

Figure 5–17. Starch gel electrophoretic patterns of six G6PD phenotypes in red blood cell hemolysates. From Giblett E. Genetic markers in human blood. Oxford: Blackwell Scientific Publications, 1969, by permission.

G6PD variants, like sickle cell hemoglobin and thalassemia, confer some protection against malaria.

Many of the G6PD variants are associated with a high risk of hemolytic crises on exposure to certain drugs that stress the shunt pathway. These drugs include sulfanilamide, trinitrotoluene, quinidine, primaquine, naphthalene (moth balls) and many others. When an infection or some other precipitating factor is present, an even longer list of drugs can cause hemolysis—for example, acetylsalicyclic acid. It is noteworthy that the A− type of deficiency is much less susceptible to hemolysis than the B− type; for example, favism is common in the Mediterranean region but almost unknown in blacks. Deficient infants of the Mediterranean type can even develop hemolytic crises if dressed in clothes that have been stored in moth balls and not thoroughly laundered. Congenital nonspherocytic hemolytic anemia associated with very low enzyme activity is a consequence of some of the rare variants, especially in whites.

Though G6PD deficiency is far more common in males, it is not impossible for females to inherit two abnormal alleles and consequently to be affected. About 2 percent of American black females are genetically *A−A−* and clinically susceptible to drug-induced hemolysis.

GENERAL REFERENCES

Bondy PK, Rosenberg LE, eds. Metabolic Control and Disease, 8th ed. Philadelphia: W.B. Saunders, 1980.

Giblett E. Genetic markers in human blood. Oxford: Blackwell, 1969.

Harris H. The principles of human biochemical genetics, 3rd ed. Amsterdam: Elsevier/North–Holland, 1980.

Spielberg SP. Pharmacogenetics. In MacLeod SM, Radde IC, eds. Textbook of pediatric clinical pharmacology, Chapter 24. Littleton, Massachusetts: PSG Inc., 1984.

Stanbury JB, Wyngaarden JB, Fredricksen DS, Goldstein JL, Brown MS, eds. The metabolic basis of inherited disease, 5th ed. New York: McGraw-Hill, 1983.

Weatherall DJ. The thalassemias. Edinburgh: Churchill Livingstone, 1983.

PROBLEMS

1. A man is heterozygous for Hb M Saskatoon, a hemoglobinopathy in which *his* is replaced by *tyr* at position 63 of the β chain. His partner is heterozygous for Hb M Boston, in which *his* is replaced by *tyr* at position 58 of the α chain. Heterozygosity for either of these alleles produces methemoglobinemia. Outline the possible genotypes and phenotypes of the offspring.

2. A child has a paternal uncle and a maternal aunt with sickle cell disease. What is the probability of sickle cell disease in the child?

3. A woman has sickle cell trait and her mate is heterozygous for Hb C. What is the probability that their child has no abnormal hemoglobin?

4. A woman had a brother who died in childhood of a disorder described as "gargoylism." Her husband's brother has Scheie syndrome (MPS IS).
   a) If the first child of this couple is a male with Hunter syndrome, what are the most likely genotypes of the child, his parents and his maternal uncle?
   b) If the child has a disorder resembling Hurler syndrome, but less severe, with demonstrable α-iduronidase deficiency, how would this change your judgment of the most likely genotypes?

5. a) Suppose that all fetuses with Tay-Sachs disease are aborted. What would be the expected effect on the population frequency of the Tay-Sachs gene?
   b) Suppose that carrier couples opt for termination of pregnancy if the fetus has Tay-Sachs disease but compensate by having another child. What then would happen to the frequency of the Tay-Sachs gene?

6. Match

A.	B.
—Rare X-linked defect of purine synthesis	1) Four
—Number of α globin genes per haploid genome	2) Compound
—Chromosome that carries the β globin locus	3) Pseudogene
—Number of α globin genes missing in hydrops fetalis with Hb Bart's	4) I-cell disease
	5) Two
—Used in prenatal diagnosis of sickle cell disease	6) Lesch-Nyhan syndrome
—Two different mutant alleles at a locus	7) Malignant hyperthermia
—Nonfunctional member of a gene family	8) Eleven
—Single gene mutation responsible for many abnormal enzymes	9) Restriction enzyme *Mst*II
	10) Three
—Cell surface receptor deficiency	11) Familial hypercholesterolemia
—Number of α globin genes missing in Hb H disease	

6

CHROMOSOMAL ABERRATIONS

A new era in medical genetics opened in 1959 with the demonstration by Lejeune and his colleagues that "enfants mongoliens" (children with mongolism or, as it is now commonly called, Down syndrome) have 47 chromosomes instead of the usual 46 in their body cells. Since then, chromosomal abnormalities have become well-defined causes of maldevelopment. Chromosomal aberrations are much more frequent and varied than was originally anticipated. They are a significant cause of birth defects and fetal loss, present in an estimated 0.7 percent of live births, half of all spontaneous abortions and about 8 percent of all conceptuses. Sixty or more different disorders are known to be caused by chromosomal abnormalities. There are many excellent reviews of medical cytogenetics, some of which are listed in the General References of this chapter.

Patients with chromosomal aberrations usually have characteristic phenotypes and often have a closer resemblance to other patients with the same karyotype than to their own brothers and sisters. The phenotypic anomalies result from genetic imbalance, which perturbs the normal course of development; nevertheless, little is known about how any specific imbalance produces its phenotypic consequences. Developmental retardation and multiple dysmorphic features are common to all the autosomal aberrations, regardless of which chromosome is involved and whether the chromosomal material is deleted or present in excess. Balanced structural rearrangements, in which the genetic material is all present but abnormally arranged, are usually associated with normal phenotypes; however, there is an excess of de novo balanced rearrangements in mentally retarded populations.

We each carry an assortment of variant genes, responsible either for common polymorphisms or for rare mutations. Consequently, the general phenotypic similarity of subjects with a specific chromosome abnormality is modified by individual differences determined by the individual genotype. If the abnormality is the loss of part or all of a chromosome, the subject is then hemizygous for all the gene loci that have been deleted; any recessive mutants on the chromosome or chromosome segment homologous to the deletion may be expressed. This may explain why, in general, monosomy is more damaging than trisomy.

Classification of Chromosomal Aberrations

Abnormalities of the chromosomes may be either numerical or structural, and may involve one or more autosomes, sex chromosomes or both simultaneously. A given abnormality may be present in all body cells, or there may be two or more cell lines, one or more of which are abnormal; the latter situation is termed **mosaicism**. In this section the more common types of chromosomal aberration are defined, and the terms used to describe them are introduced.

ABERRATIONS OF CHROMOSOME NUMBER

Numerical changes arise chiefly through the process of **nondisjunction** (failure of paired chromosomes or sister chromatids to disjoin at anaphase, either in the first or second meiotic division or in mitosis). **Anaphase lag**, which occurs in anaphase of mitosis when one or both daughter chromosomes lag behind and fail to reach either pole, is a type of nondisjunction that can result in one or both members of the pair failing to be included in either daughter cell.

Every species has a characteristic chromosome number (in man, $2n = 46$ and $n = 23$). Any number that is an exact multiple of the haploid number is **euploid.** Euploid numbers need not be normal; 3n (triploid) and 4n (tetraploid) chromosome numbers are known in man, though only a few triploids have been born alive and tetraploids have been seen only in early abortuses. Chromosome numbers such as 3n and 4n, which are exact multiples of n but greater than 2n, are said to be **polyploid**. Polyploidy can arise by a variety of mechanisms. Triploidy probably results from failure of one of the maturation divisions, in either ovum or sperm. Tetraploids are always XXXX or XXYY, suggesting that tetraploidy results from failure of completion of an early cleavage division of the zygote.

Any number that is not an exact multiple of n is **aneuploid**. Some types of aneuploids are trisomic, with $2n+1$ chromosomes and three members of one particular chromosome, as in Down syndrome; monosomic, with $2n-1$ chromosomes and only one member of some chromosome; and doubly trisomic ($2n+1+1$), with two different additional chromosomes.

Any chromosome number that deviates from the characteristic n and 2n is **heteroploid**, whether it is euploid or aneuploid.

Aneuploidy

As noted above, the chief cause of aneuploidy is nondisjunction in a meiotic division, leading to unequal distribution of one pair of homologous chromosomes to the daughter cells so that one daughter cell has both and the other has neither chromosome of a pair. Other mechanisms not described here can also cause aneuploidy (Bond and Chandley, 1983).

If nondisjunction of chromosome 21 occurs at meiosis, the gametes formed have either an extra chromosome 21 ($n+1=24$ chromosomes in all) or one too few ($n-1=22$). Fertilization of the 24-chromosome gamete by a normal gamete produces a 47-chromosome zygote, trisomic for chromosome 21.

Nondisjunction can take place at either the first or the second meiotic division. The consequences are rather different (Fig. 6–1). If nondisjunction occurs at meiosis I, the gamete with $n+1$ chromosomes will contain both the paternal and the maternal representatives of that chromosome; if it involves the two chromatids of a single chromosome at meiosis II, the gamete with $n+1$ chromosomes will contain a double complement of either the paternal or the maternal chromosome. (This simplified explanation omits the effect of meiotic crossing over on the gene content of the chromosomes.) Nondisjunction can occasionally occur at successive meiotic divisions, or in both male and female gametes, so that zygotes with bizarre chromosome numbers may be formed; these "multisomics" have been described only rarely except with respect to the X chromosome. Mitotic nondisjunction also produces multisomy in some malig-

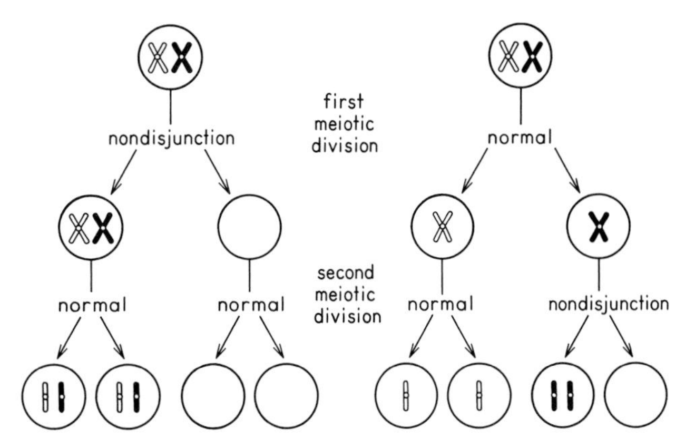

Figure 6–1. Nondisjunction occurring at the first and second meiotic divisions. Nondisjunction at meiosis I results in gametes with both members of a chromosome pair or with neither member. Nondisjunction at meiosis II results in gametes containing (or lacking) two identical chromosomes both derived from the same member of the homologous pair.

nant cell lines and some cell cultures. Double aneuploidy (trisomy for two different chromosomes at once) is not uncommon.

Nondisjunction can also occur at mitosis after formation of the zygote, in which case the nondisjoining objects are the chromatids of a single chromosome, as in meiosis II. If this happens at an early cleavage division, a trisomic and a monosomic cell line are established; the trisomic one might persist, but the monosomic one usually does not. Again the X chromosomes are an exception, as lines with a single X are viable.

Aberrations of Chromosome Structure

Much of our knowledge of human structural aberrations is based on work with other organisms, especially the fruit fly and certain plants. Structural rearrangements result from chromosome breakage, followed by reconstitution in an abnormal combination. Chromosome breaks occur normally at a low frequency, but may also be induced by a wide variety of breaking agents (clastogens) such as ionizing radiation, some viral infections and many chemicals.

The changes in chromosome structure resulting from breakage may be either stable (that is, capable of passing through cell division unaltered) or unstable. The stable types of aberration are deletions, duplications, inversions, translocations, insertions and isochromosomes. The unstable types, which fail to undergo regular cell division, are dicentrics, acentrics and rings.

The only chromosomal anomalies that are likely to be transmitted from parent to child are structural rearrangements of the inversion or translocation types (see later in this chapter).

DELETION

Deletion is loss of a portion of a chromosome, perhaps terminally following a single chromosome break, though usually interstitially between two breaks, with a minute terminal segment remaining. The deleted portion, if it lacks a centromere, is an acentric fragment. Because it has no centromere, the fragment fails to move on the spindle and is usually lost at a subsequent cell division. The structurally abnormal chromosome lacks whatever genetic information was present in the lost fragment. The common example of deletion in humans is the cri du chat syndrome, in which part of the short arm of chromosome 5 is deleted (see later in this chapter).

Figure 6–2 shows a chromosome in which the order of the genes is given as ABCDEFGH. A terminal deletion might produce a chromosome lacking H. A

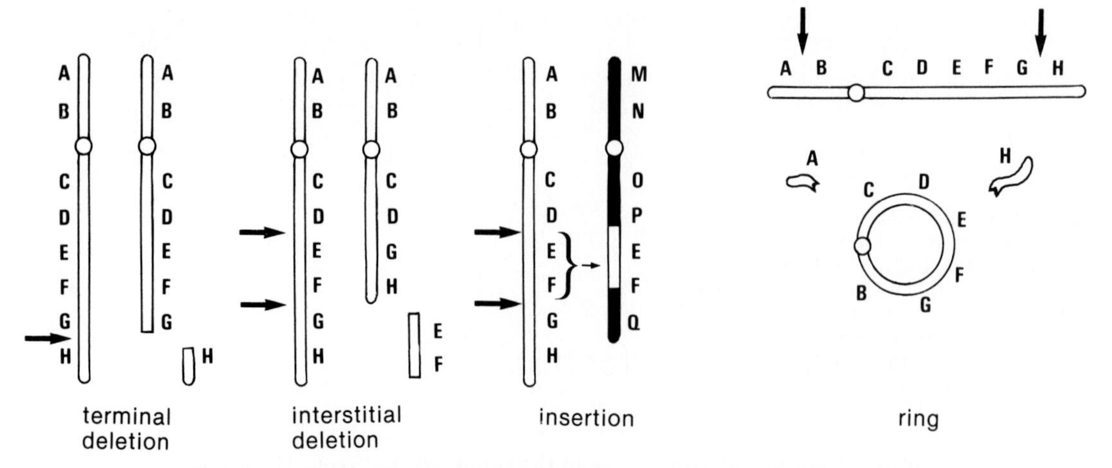

Figure 6–2. Structural rearrangements of chromosomes. For details, see text.

deletion produced by two breaks, between D and E and between F and G with loss of the fragment EF between them, might produce a chromosome ABCDGH. If the deleted portion does not involve the centromere, the chromosome replicates and divides in the normal way at subsequent divisions, but the acentric fragment is usually lost. A fragment EF deleted from one chromosome might be inserted into a nonhomologous chromosome.

A ring chromosome is a special type of deletion chromosome in which both ends have been lost and the two broken ends have reunited to form a ring. If it has a centromere a ring chromosome can pass through cell division, but it may undergo alteration in structure (Fig. 6–3). Ring chromosomes have been identified for many different human chromosomes.

DUPLICATION

Duplication is the presence of an extra segment of a chromosome. Duplications are more common and much less harmful than deletions. In fact, small duplications identifiable by molecular techniques appear to provide an evolutionary mechanism for the acquisition of new genes, which may then evolve into genes with quite different functions from the genes from which they originated. Duplications of whole genes such as the globin genes, or, much less frequently, of parts of genes as in Lepore hemoglobin or haptoglobin are well known and are discussed elsewhere. Recall also the repetitive structure of numerous DNA sequences.

Gross duplication of microscopically detectable parts of chromosomes may occur as a consequence of various structural rearrangements. For example, if an individual is a balanced translocation carrier, he or she may form unbalanced

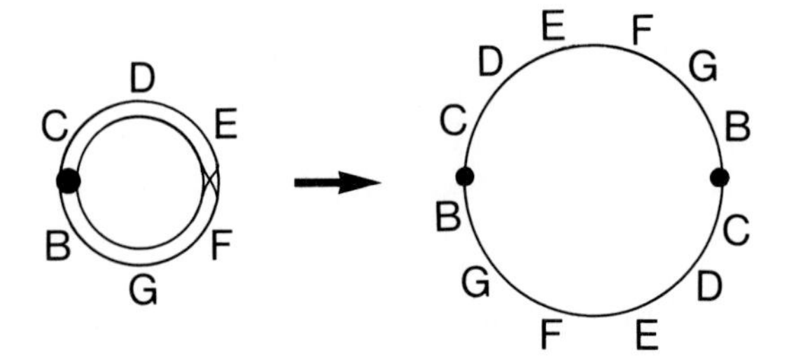

Figure 6–3. Origin of a double-sized dicentric ring after sister chromatid exchange in mitosis at X between E and F. Ring opens up to form dicentric ring, which may be broken at anaphase.

gametes which have, in effect, a duplication of one segment and deletion of another. Partial duplications also result from crossing over in inversion heterozygotes or from isochromosome formation, as shown below.

INVERSION

Inversion involves fragmentation of a chromosome by two breaks, followed by reconstitution with inversion of the section of the chromosome between the breaks (ABCdefGH might become ABCfedGH). If the inversion is in a single chromosome arm it is **paracentric** (beside the centromere), but if it involves the centromere region it is **pericentric** (around the centromere). Because paracentric inversion does not lead to a change in arm ratio, paracentric inversions were rarely identified until chromosome banding came into use. Pericentric inversions may change the proportion of the chromosome arms.

An inversion alone does not appear to lead to an abnormal phenotype in man, though theoretically it could do so if either break is within a gene or its regulatory sequences. The medical significance of inversions is for the subsequent generation, and arises from the consequences of crossing over between a normal chromosome and one with a pericentric inversion.

The consequences of the two kinds of inversion are shown in Figure 6–4. For the homologous chromosomes to pair at meiosis I, one of them must form a loop in the region of the inversion.

Paracentric Inversion. Here the centromere lies outside the loop. When a crossover occurs within the loop, a dicentric chromatid and an acentric fragment are formed, as well as a normal and an inverted chromatid. Both the dicentric and the acentric are unstable, so only gametes with the normal chromosome or with the inverted segment are formed.

Pericentric Inversion. Here the centromere lies within the loop. If a crossover now takes place, each of the two chromatids involved in the crossover has both a duplication and a deletion. If gametes are formed with these abnormal chromosomes, the resulting progeny will have a deficiency of one part of the chromosome and a duplication of another part.

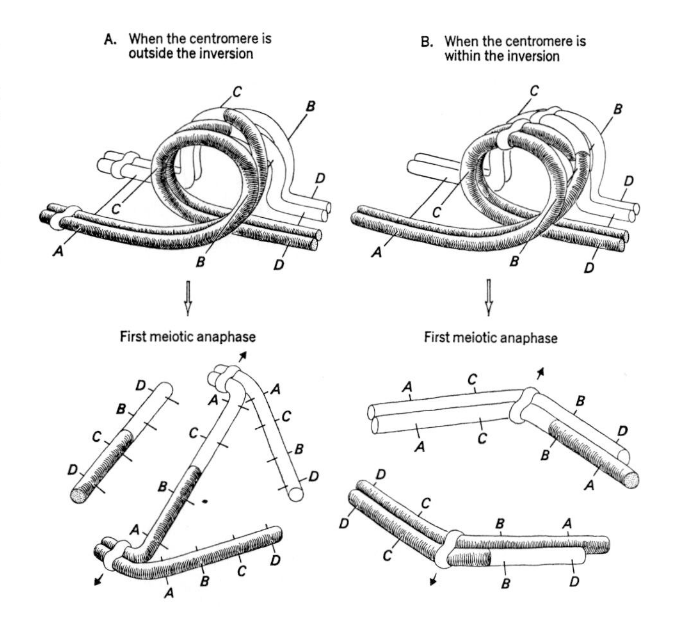

Figure 6–4. Crossing over within the loop formed at meiosis I in inversion heterozygotes. A. Paracentric inversion. The only gametes formed are normal or balanced. B. Pericentric inversion. The gametes formed may be normal, balanced or unbalanced with both duplications and deficiencies. See Figure 6–5 for a consequence of a pericentric inversion. From Srb AM, Owen RD, Edgar RS. General genetics, 2nd ed. San Francisco: W. H. Freeman and Company, 1965.

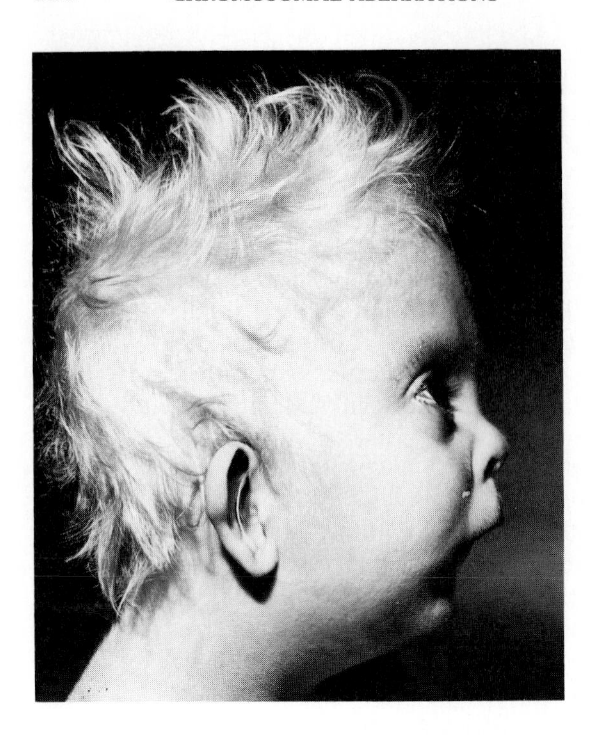

Figure 6–5. A child with an abnormal karyotype, chromosome 3 duplication q21 → qter, deletion p25 → pter; one parent carried a pericentric inversion of chromosome 3. From Allderdice et al., 1975, by permission. See text for discussion.

A very large pedigree has been reported from Newfoundland in which there are numerous carriers of a pericentric inversion of the long arm of chromosome 3 described as inv(3) (p25q21), which has segregated in the family for at least 6 generations (Allderdice et al., 1975). A number of children in families in which one parent carries the inversion have a syndrome of congenital defects including hirsutism, micrognathia, heart and renal anomalies and mental retardation (Fig. 6–5). The karyotype of these children includes a recombinant chromosome with a duplication of part of the long arm and a deficiency of part of the short arm of chromosome 3. The identical duplication-deficiency chromosome in phenotypically similar children has been reported from several other centers; in some of these cases relationship to the Newfoundland kindred could be traced.

TRANSLOCATION

Translocations are of two main types, reciprocal and Robertsonian. **Reciprocal translocation** is the exchange of blocks of chromatin between two nonhomologous chromosomes. The process requires breakages of both chromosomes, with repair in an abnormal arrangement. A balanced translocation does not necessarily lead to an abnormal phenotype; however, translocations, like inversions, can lead to the formation of unbalanced gametes and therefore carry a high risk of abnormal offspring. **Robertsonian translocation** involves two acrocentric chromosomes, which fuse at the centromere region and lose their heterochromatic short arms. A carrier of a balanced Robertsonian translocation has 45 chromosomes including the translocation chromosome. The significance of Robertsonian translocation in Down syndrome is described below.

Insertion

This is a rare non-reciprocal type of translocation that involves three breaks, with a segment removed from one chromosome, then inserted into a broken region of a non-homologous chromosome.

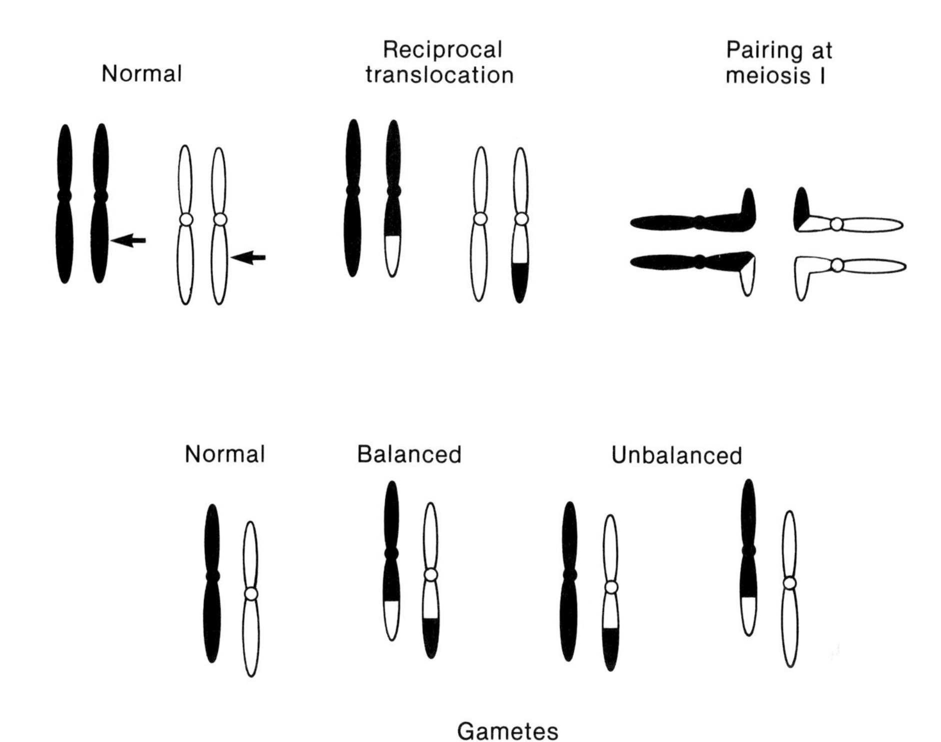

Figure 6–6. Reciprocal translocation and its consequences. Above, origin by breakage of two nonhomologous chromosomes and reconstitution with the broken-off ends interchanged. At meiosis, a cross-shaped figure is formed. Below, chief types of gametes formed by this translocation heterozygote. Alternate segregation produces gametes with normal and balanced chromosome complements. Adjacent I segregation produces two types of unbalanced chromosome complements. The less common adjacent II segregation (in which the chromosomes with identical centromeres segregate together) is not shown. Normal or balanced gametes result in phenotypically normal progeny, but unbalanced gametes result in zygotes that are partially trisomic and partially monosomic and consequently develop abnormally.

Gametogenesis in Carriers of Reciprocal Translocations

Because translocations interfere with normal chromosome pairing and segregation at meiosis I, they can lead to unbalanced gametes and unbalanced offspring. The consequences of gametogenesis in an individual carrying a reciprocal translocation are shown in Figure 6–6. The two normal and the two translocated chromosomes synapse as a cross-shaped figure, which may open up into a ring or chain unless the arms of the chromosomes are held together by chiasmata.

The most frequent types of gametes formed include a normal combination, an abnormal but balanced combination and two abnormal unbalanced combinations. The first two types can lead to normal progeny, but the last two can produce unbalanced progeny with a duplication and a deletion of parts of chromosomes. There is also an increased risk of nondisjunction in translocation heterozygotes.

Gametogenesis in Carriers of Robertsonian Translocations

Robertsonian translocations arise either by mutation or by segregation in the offspring of a balanced carrier. The clinical significance of this phenomenon is that the carriers of a translocation involving chromosome 21 and another acrocentric chromosome have a high risk of producing offspring with Down

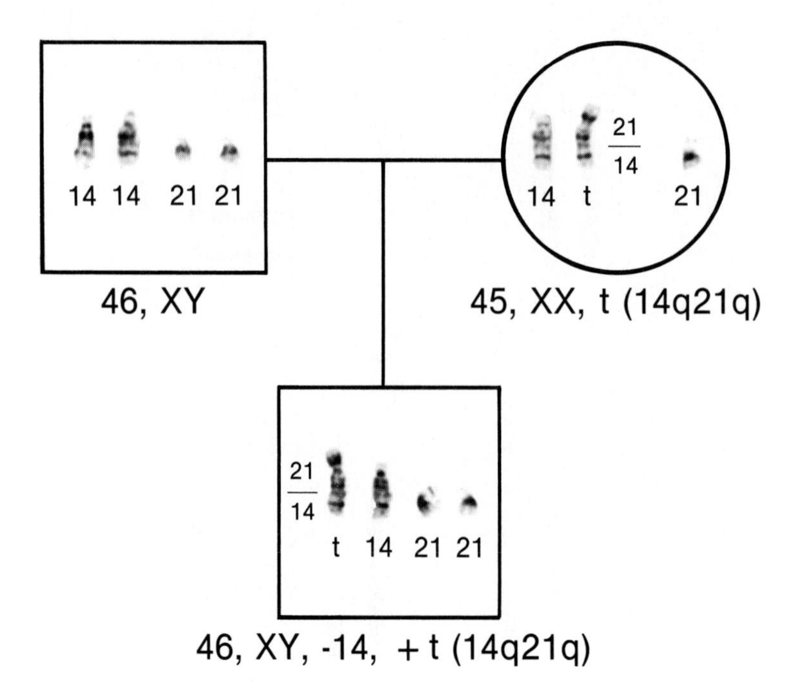

Figure 6–7. Translocation Down syndrome transmitted from a carrier mother to her child. The father's chromosomes are normal. Only chromosomes 14, 21 and t(14q21q) are shown. G-banding. Courtesy of R. G. Worton.

syndrome (Fig. 6–7). In these carriers, one of the chromosomes involved in the translocation is always a 21; the other may be any of the acrocentrics, though chromosome 14 is the most common. The theoretical risk is that one-third of the offspring of such translocation carriers have Down syndrome, but the observed risk is lower (see later). The homologous 21q21q translocation is a special case, because all the offspring inevitably have either Down syndrome or monosomy 21, which is usually lethal in early development.

Figure 6–8 shows gametogenesis in a balanced carrier of a Robertsonian translocation involving 14q and 21q. The carrier has 45 chromosomes including t(14q21q) and is phenotypically normal. Theoretically, he or she forms six types of gametes in equal proportions:

A1 Balanced, 22 chromosomes including t(14q21q)
A2 Normal, 23 chromosomes
B1 Abnormal, 23 chromosomes including t(14q21q) and 21
B2 Abnormal, 22 chromosomes with no 21
C1 Abnormal, 23 chromosomes including t(14q21q) and 14
C2 Abnormal, 22 chromosomes with no 14

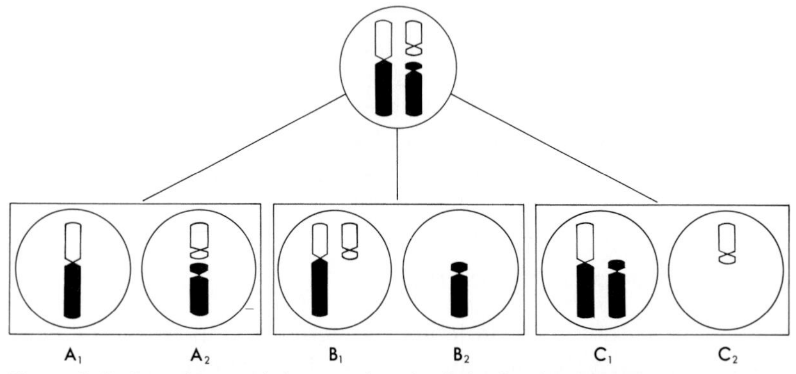

Figure 6–8. Gametogenesis in a carrier of a Robertsonian translocation involving the long arms of chromosome 14 (black) and chromosome 21 (outline). For details, see text.

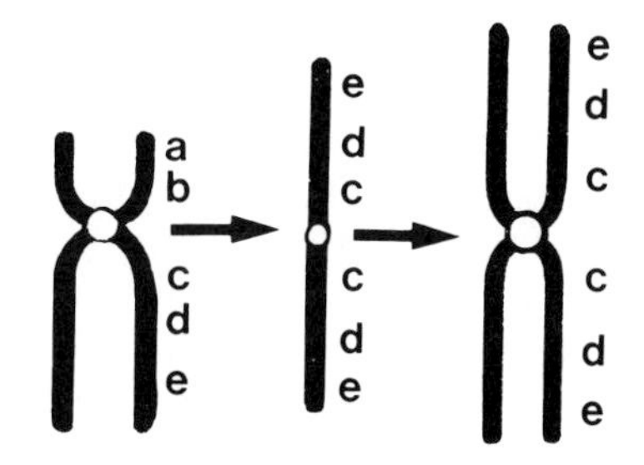

Figure 6–9. Possible mechanism of formation of an isochromosome of the long arm of the X chromosome. The centromere divides transversely rather than longitudinally; the long arm may then replicate to form a metacentric chromsome, which contains a duplication of the normal long arm (Xq) and a deficiency of the normal short arm (Xp). A female with a normal X and i(Xq) thus is trisomic for Xq and monosomic for Xp.

Two of these types (B2 and C2) lack rather large amounts of chromosomal material. Assuming union with a normal gamete, they would produce zygotes monosomic for chromosome 21 and chromosome 14 respectively. These are inviable (except for rare instances of monosomy 21 in liveborn infants). C1 would lead to a zygote with trisomy 14, which also seems to be inviable but occurs in many spontaneous abortions.

The three remaining types of gametes can produce viable offspring. Type A2 is entirely normal. Type A1 leads to a phenotypically normal balanced translocation carrier, like the parent. Type B1 produces translocation Down syndrome, that is, a child with 46 chromosomes, one of which is a 14q21q translocation. The karyotype is fully described as 46, XX or XY, −14, +t(14q21q), indicating that there are 46 chromosomes in all, a chromosome 14 is missing, and a Robertsonian translocation of the long arms of a 14 and a 21 is present.

ISOCHROMOSOMES

During cell division the centromere of a chromosome sometimes mistakenly divides so that it separates the two arms rather than the two chromatids (Fig. 6–9). The chromosomes so formed are isochromosomes. The most common kind of isochromosome involves the long arm of the X and is designated i(Xq). (An isochromosome for the short arm of the X does not occur.) A woman with a normal X and an i(Xq) is monosomic for the genes on the short arm of the X and trisomic for the genes on the long arm. About 15 to 20 percent of women with Turner syndrome have this karyotype (see Chapter 7).

Mosaicism

If nondisjunction occurs at an early cleavage division of the zygote rather than during gametogenesis, an individual with two or more cell lines with different chromosome numbers is produced. Such individuals are termed mosaics.

A chromosomal mosaic has at least two cell lines, with different karyotypes, derived from a single zygote. The alterations in the karyotype may be either numerical or structural. Many different mosaics have been described, most of which have cell lines with different sex chromosome constitutions. About 1 percent of patients with Down syndrome have a mixture of 46-chromosome and 47-chromosome tissues.

The proportion of normal and abnormal cells may vary from tissue to tissue within the same patient, as well as from patient to patient. It may also change during development; patients are known in whom mosaicism was apparent in lymphocyte cultures at birth but had disappeared a few months later, apparently through selection against the abnormal cell line. On the average mosaics are less severely abnormal than their nonmosaic counterparts.

There are many practical difficulties in the investigation of mosaicism. Even if mosaicism cannot be demonstrated, it cannot definitely be ruled out. Because normal and abnormal cells may survive and multiply at different rates in culture, the relative proportions in a chromosome culture may not reflect the proportions in the patient or the proportions present during the critical early stage of development. Mosaicism may arise independently in vitro; this is a common cause of difficulty in interpretation of karyotypes from amniotic fluid cell cultures.

Causes of Chromosomal Aberrations

Although by now the mechanisms that produce aberrations of chromosome number and structure are understood in a general way, little is known about the predisposing genetic and environmental factors. Most studies aimed at elucidation of underlying causes have involved Down syndrome, since it is the most common chromosomal disorder. Because the recurrence risk after the birth of a Down child to a young mother is about 1 percent, much higher than the baseline risk of only about 1 in 2000 for mothers under the age of 30, it seems clear that certain people or certain families are predisposed to have Down children; however, the reason has been extraordinarily difficult to find.

Late maternal age is a major factor (perhaps the only factor of real significance) in the etiology of Down syndrome and, to a lesser extent, of the other trisomies, but the reason for the correlation of late maternal age with the nondisjunctional event that underlies Down syndrome is unknown. Paternal age probably has no effect upon nondisjunction, though some studies show a very slight effect. Any explanation of the maternal age effect must take into account the observation that the majority of Down syndrome cases involve nondisjunction at meiosis I in the mother.

Normally homologous chromosomes are held together in meiosis I by chiasmata. One role of chiasmata is to orient the paired homologues to opposite poles of the cell. If fewer chiasmata are formed, or if they terminalize prematurely, nondisjunction could result. In mouse oocytes there are fewer chiasmata and a higher incidence of univalents (unpaired chromosomes) with increasing maternal age.

Genes predisposing to nondisjunction may exist in man, since such genes are known in other organisms, but evidence for their presence in man is scant. The risk of an aneuploid child does not seem to increase in inbred populations or when a parent might be homozygous for a "nondisjunction gene" through consanguinity of his or her parents. Close relatives of mothers of Down children do not seem to be at increased risk of having affected children, unless of course they carry a familial translocation.

Autoimmune disease seems to have some role in the pathogenesis of nondisjunction, in view of an observed correlation between high thyroid autoantibody levels in mothers and Down syndrome in their children.

Radiation has been postulated as a cause of nondisjunction in man, and is known to increase the rate of nondisjunction in the fruit fly. Uchida (1977) has reported experimental data and reviewed a number of epidemiological studies of the association between maternal radiation, late maternal age and nondisjunction. In her experiments, when oocytes in metaphase of meiosis II from irradiated mice were compared with those of unirradiated age-matched controls, the frequency of trisomy was four times as high in the irradiated series as in the controls. Finally, of eleven epidemiological studies of the radiation histories of the mothers of Down patients as compared with mothers of controls, nine showed an increase in radiation exposure prior to conception in the mothers of the Down patients, though the difference was significant in only four. Though it may be premature to conclude that radiation, as a cause of nondisjunction, increases the frequency of trisomy 21, it seems logical to avoid unnecessary radiation exposure.

There seems to be no evidence that **viruses** or **teratogenic agents**, including chemotherapeutic agents, cause nondisjunction in the offspring of exposed parents.

Chromosome abnormality, such as mosaicism or balanced translocation in a parent, can lead to abnormal chromosomes in the offspring, but this would account for only a small proportion of chromosomally abnormal children.

It is disappointing that so little is known about how to identify mothers at high risk of having Down children. The usefulness of prenatal diagnosis would be greatly enhanced if risk factors other than late maternal age could be demonstrated. A recent observation that is potentially useful is that the α-fetoprotein level in the serum of a pregnant woman is unusually low if her fetus has Down syndrome.

Clinical Aspects of Autosomal Disorders

There are several well-defined chromosome disorders in which an abnormality of either an autosome or a sex chromosome is present. There are also many very rare chromosomal syndromes, usually involving loss or gain of only a segment of a chromosome arm instead of a whole chromosome. In this section the best known autosomal disorders are described and some of the less common syndromes involving parts of chromosomes or chromosome breakage are briefly noted. The sex chromosome disorders are discussed in Chapter 7.

TRISOMY 21 (DOWN SYNDROME)

Down syndrome is by far the most common and best known of the chromosome disorders. It was first described by Langdon Down in 1866, but its cause remained a deep mystery for nearly a century. Two noteworthy features of its population distribution drew attention: the late maternal age and the peculiar pattern within families—concordance in all monozygotic twins but almost complete discordance in dizygotic twins and other relatives. It was suggested by Waardenburg in 1932 that a chromosomal anomaly could explain these observations; in 1959 this was verified when it was found that children with Down syndrome have 47 chromosomes, the extra member being a small acrocentric that has since been designated chromosome 21. (It is actually the smallest chromosome and should originally have been designated chromosome 22.)

The older name of mongolism, now falling into disuse, refers to the somewhat Oriental cast of countenance produced by the characteristic epicanthal folds, which give the eyes a slanting appearance.

Down syndrome can usually be diagnosed at birth or shortly thereafter by its phenotypic features (Fig. 6–10). Hypotonia is often the first abnormality noticed. Mental retardation is present, the intelligence quotient usually being in the 25–50 range when a child is old enough to be tested. There is a 15-fold increase in the risk of leukemia. The head is brachycephalic, with a flat occiput. The eyes have epicanthal folds, and the iris shows speckles (Brushfield spots)

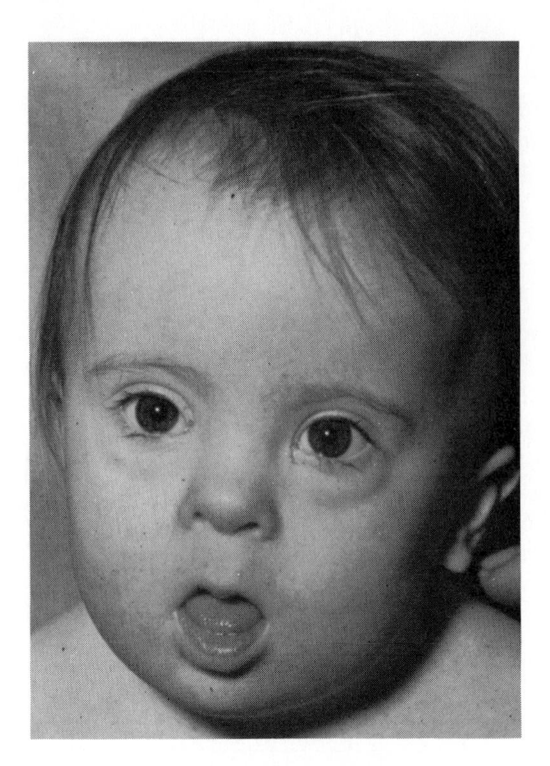

Figure 6–10. A child with Down syndrome. From Smith DW. Recognizable patterns of human malformation, 3rd ed. Philadelphia: W. B. Saunders Company, 1982.

around the margin. The nose has a low bridge. The tongue usually protrudes and is furrowed, lacking a central fissure. The hands are short and broad, usually with a simian crease and clinodactyly (incurving) of the fifth finger. There may be a single crease on the fifth finger. The dermal patterns are characteristic. On the feet, there is often a wide gap between the first and second toes and a furrow extending proximally along the plantar surface. In about half the patients the hallucal dermal pattern is an arch tibial, which is rare in normal persons. (See also Chapter 17.) About one-third of the patients have congenital malformations of the heart. Radiologically, the acetabular and iliac angles are decreased. The stature is below average. Often the diagnosis presents no particular difficulty, but karyotyping is nevertheless indicated for confirmation and for determining whether the child has the typical 47,XX or XY,+21 karyotype (95 percent of cases), a translocation (4 percent) or mosaicism (1 percent).

The Chromosomes in Down Syndrome

Though Down syndrome always involves trisomy for chromosome 21, in about 4 percent of the cases the extra chromosomal material is present not as a separate chromosome but as a translocation of 21q to another acrocentric chromosome, as described earlier in this chapter.

A child with a translocation, t(14q21q) for example, has 46 chromosomes in all, but the karyotype is effectively trisomic for chromosome 21. The phenotypic consequences are indistinguishable from those of standard trisomy 21.

About 1 percent of Down syndrome patients are mosaics, usually 46/47 mosaics (that is, with a mixture of 46- and 47-chromosome cells). It is likely that most mosaic Down patients derive from trisomy 21 zygotes. Such patients have relatively mild stigmata and are less retarded than the typical trisomies. Low-grade mosaicism in germinal tissue of a parent is a postulated cause of Down syndrome.

Risk of Down Syndrome

A frequent problem in genetic counseling is the assessment of the risk that a woman will have a Down child. This is especially important now that guidelines are used to select for prenatal diagnosis only those women in whom the risk of a Down child outweighs the risk of amniocentesis itself. The risk varies with the woman's age, her karyotype and that of her husband, and her reproductive history with respect to Down syndrome and perhaps also to other trisomies.

Estimates of the population incidence of Down syndrome in newborns currently average about 1 in 800. The incidence seems to have remained about the same in recent years despite some decline in mean maternal age. Affected persons survive to an average age of about 40 years, unless severe physical abnormalities lead to early death. Premature senility associated with the cortical atrophy, ventricular dilatation and neurofibrillar tangles characteristic of Alzheimer disease is common in older patients.

The average maternal age at the birth of a Down child is about 34 years, compared with the current mean maternal age in North America of about 26 years. Though the risk is lower at early maternal ages, there are so many more births at the younger maternal ages that the absolute number of Down babies born to young mothers is quite high. Mothers of translocation Down patients have the same age distribution as control mothers. Of all the Down children born to women under 30, about 9 percent have a translocation, whereas for mothers over 30 the proportion is only 1.5 percent.

The increasing risk of a Down child, or a Down fetus, with higher maternal age is shown in Table 6–1 and has already been referred to repeatedly. As shown, the incidence of the syndrome in fetuses karyotyped at the sixteenth week of gestation or near that time, especially when the mother is 40 or older, is higher than one would expect from the estimates for newborns. The difference in incidence does not seem to be fully accounted for by fetal loss later in pregnancy, so it appears that the only reasonable explanation is that in the past the diagnosis was often missed in newborns, and perhaps especially often in stillbirths. However, the incidence in newborn surveys, in which only infants dying within the first 24 hours after birth were missed, is still much lower than the figures obtained by prenatal diagnosis; the discrepancy is still not fully explained.

Table 6–1. RISK OF DOWN SYNDROME IN FETUSES AT AMNIOCENTESIS AND IN LIVE BIRTHS

	Frequency of Down Syndrome	
Maternal Age*	*Fetuses*	*Live Births*
–19	—	1/1550
20–24	—	1/1550
25–29	—	1/1050
30–34		1/700
35	1/350	1/350
36	1/260	1/300
37	1/200	1/225
38	1/160	1/175
39	1/125	1/150
40	1/70	1/100
41	1/35	1/85
42	1/30	1/65
43	1/20	1/50
44	1/13	1/40
45+	1/25	1/25

*Approximate (rounded) estimates chiefly from data of Hook EB. Rates of chromosome abnormalities at different maternal ages. Obstet Gynecol 1981; 58:282–285.

Table 6–2. PROGENY OF CARRIERS OF ROBERTSONIAN TRANSLOCATIONS

	Normal	Carrier	Unbalanced
Theoretical	0.33	0.33	0.33
Observed			
Mother t(14q21q) carrier	0.35	0.50	0.15
Father t(14q21q) carrier	0.39	0.61	—
Mother t(13q14q) carrier	0.44	0.56	—
Father t(13q14q) carrier	0.37	0.63	—

Data from: Boué A, Gallano P. A collaborative study of the segregation of inherited chromosome structural rearrangements in 1356 prenatal diagnoses. Prenat Diagn 1984; 4:45–67.

In the United States and Canada, 50 percent or more of pregnant women over the age of 35 (and many younger women) undergo amniocentesis for fetal chromosome analysis.

Risk to Offspring of Translocation Carriers. Most of the translocations seen in Down syndrome are of the Robertsonian type, either t(14q21q) or t(21q22q). As discussed earlier, the theoretical risk for the progeny of carriers of this type of translocation is 1/3 normal, 1/3 carrier and 1/3 unbalanced. The observed proportion of unbalanced offspring is much lower, especially for the offspring of male carriers, as shown in Table 6–2.

Previous Trisomic Child. On average, the risk of Down syndrome (or some other trisomy) after the birth of one such child appears to be about 1 percent regardless of maternal age. This represents a large increase in risk for mothers under the age of 30, but not for older mothers.

Other Abnormalities of Chromosome 21

Monosomy 21 ("antimongolism") is rare, having been reported only a few times. The infants are severely abnormal, retarded in growth and mental development. Partial monosomy and partial trisomy have both been described; the trisomies in particular have indicated that the part of the chromosome chiefly responsible for the phenotypic features of Down syndrome, when present in triplicate, is the light-staining band at the distal end of the long arm.

Source of the Abnormal Gamete in Down Syndrome

To understand the causes of Down syndrome, it would be helpful to know what proportion of Down children receive their extra chromosome from the father and what proportion from the mother, plus the relative frequency of nondisjunction at the first and second meiotic divisions. Analysis of heteromorphisms (variants) of chromosome 21 in Down children and their parents is beginning to provide some answers. About 20 percent of Down children result from paternal rather than maternal nondisjunction; in both sexes the accident is more likely to happen at meiosis I than at meiosis II (Hassold and Matsuyama 1979, Mikkelsen et al., 1980).

TRISOMY 18 (47,XX or XY, +18)

The syndrome caused by trisomy 18 was first described by Edwards and colleagues in 1960. In the older literature it is often called E trisomy, because

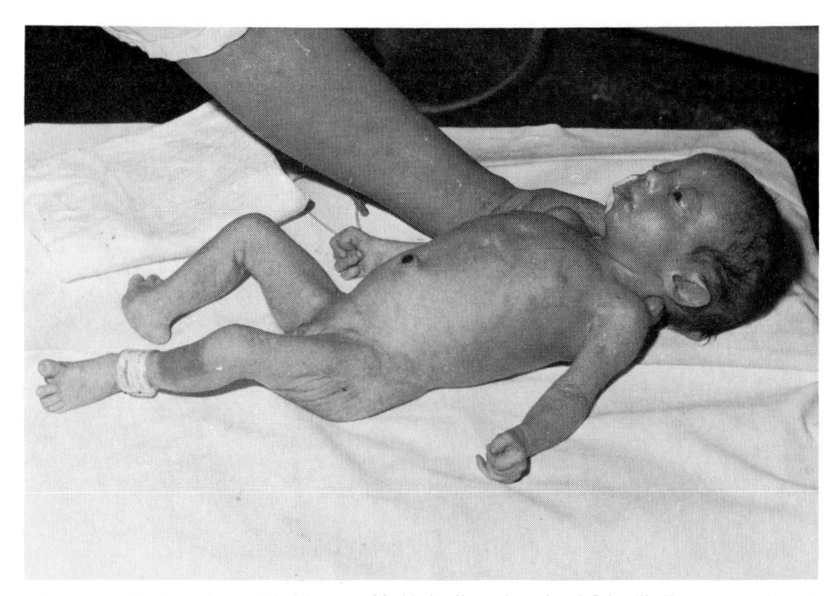

Figure 6–11. An infant with trisomy 18. Note the clenched fist with the second and fifth digits overlapping the third and fourth, rocker-bottom feet with prominent calcanei, dorsiflexion of the big toe and large low-set simple helix. Courtesy of D. H. Carr.

until chromosome banding came into use the specific chromosome concerned could not be positively identified. Its incidence is about 1 in 8000 newborns. Probably 95 percent of trisomy 18 fetuses abort spontaneously. Postnatal survival is also poor. Among those who are liveborn, the mean survival is only 2 months, though a few survive for 15 years or more. About 80 percent of the patients are female, perhaps because of preferential survival. As in most other trisomies, maternal age is advanced. Usually the cause is nondisjunction. About 10 percent of cases are mosaics; these display milder manifestations, survive longer and are born to mothers of normal age distribution.

The features of trisomy 18 are shown in Figure 6–11. Mental retardation and failure to thrive are always present. Hypertonia is a typical finding. The head has a prominent occiput, and the jaw recedes. The ears are low-set and malformed. The sternum is short. The fists clench in a characteristic way, the second and fifth digits overlapping the third and fourth. The feet are rocker-bottom, with prominent calcanei. The dermal patterns are very distinctive, with simian creases on the palms and simple arch patterns on most or all digits. The nails are usually hypoplastic. Severe congenital malformations of the heart are present in almost all cases.

Other Abnormalities of Chromosome 18. A variety of other anomalies of chromosome 18 have been identified as partial deletion of the short arm and both partial deletions and partial trisomies of the long arm. There are also ring 18's lacking part of each arm. It is difficult to generalize briefly about the variety of phenotypes associated with these different karyotypes. All share mental and growth retardation.

TRISOMY 13 (47,XX or XY, +13)

The striking phenotype of trisomy 13 is shown in Figure 6–12. Though the pattern of malformations characteristic of this disorder was probably recognized

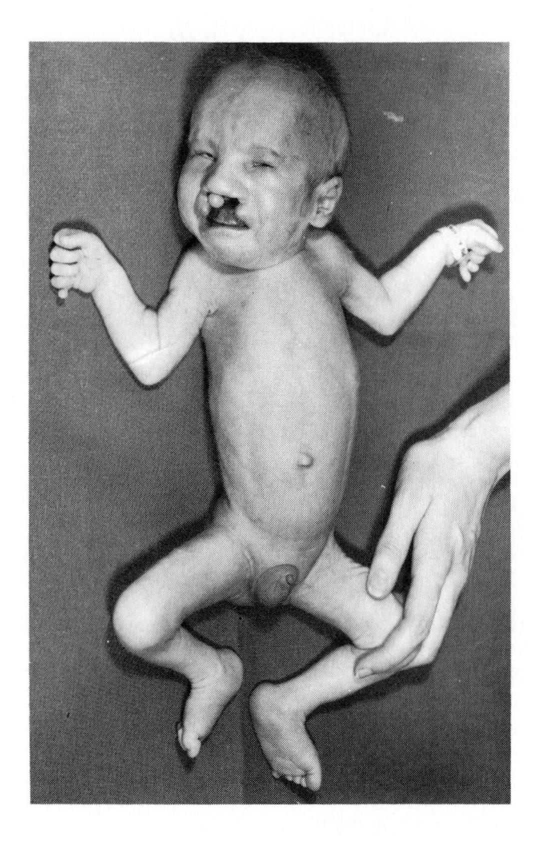

Figure 6–12. An infant with trisomy 13. Note particularly the bilateral cleft lip and polydactyly. Courtesy of P. E. Conen.

at least 300 years ago, the chromosome anomaly was first identified in 1960 by Patau and associates. Because the three pairs of D group chromosomes were originally indistinguishable, trisomy 13 is often referred to in the older literature as D or D1 trisomy.

Trisomy 13 is a severe disorder, lethal in about half the liveborn infants within the first month. It is very rare or unknown in first-trimester abortions and is not often seen in prenatal diagnosis even though the mean maternal age is advanced.

About 20 percent of the cases are caused by translocation, far above the frequency of translocation in Down syndrome. Even when one parent is a translocation carrier, the empirical risk of the same defect in a subsequent child seems to be below 2 percent.

The phenotype of trisomy 13 includes severe central nervous system malformations such as arhinencephaly and holoprosencephaly. Growth retardation and severe mental retardation are present. The forehead is sloping, there is ocular hypertelorism, and there may be microphthalmia, iris coloboma or even absence of the eyes. The ears are malformed. Cleft lip and cleft palate are often present. The hands and feet may show postaxial polydactyly and the hands clench with the second and fifth fingers overlapping the third and fourth, as in trisomy 18. The feet are rocker-bottom. The dermal patterns are unusual, with simian creases on the palms, distal axial triradii and distinctive hallucal patterns. Internally there are usually congenital heart defects of specific types and urogenital defects including cryptorchidism in males, bicornuate uterus and hypoplastic ovaries in females and polycystic kidneys. Of this constellation of defects, the most distinctive are the general facial appearance with cleft lip and palate and ocular abnormalities, plus polydactyly, the clenched fists and rocker-bottom feet.

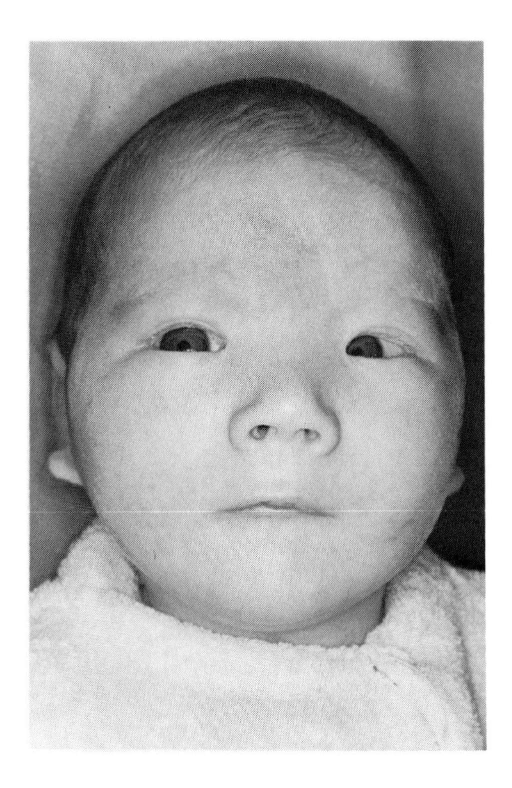

Figure 6–13. An infant with cri du chat syndrome, resulting from deletion of part of the short arm of chromosome 5 (5p –). Note characteristic facies with hypertelorism, epicanthus and retrognathia.

CRI DU CHAT SYNDROME (46,XX or XY, 5p –)

Deletion of part of the short arm of chromosome 5 results in a syndrome, originally described by Lejeune and colleagues, which has been named *cri du chat* because of the resemblance of the cry of an affected newborn to the mewing of a cat. The facial appearance (Fig. 6–13) is distinctive, with microcephaly, hypertelorism, antimongoloid slant of the palpebral fissures, epicanthus, low-set ears (sometimes with preauricular tags) and micrognathia. The dermal patterns of the palms, fingers and soles are also characteristic, with simian creases, a high total ridge count and a high frequency of thenar patterns.

Most cases are sporadic, but 10 to 15 percent are the offspring of translocation carriers.

Families with translocations involving 5p show that deletion of a chromosome segment can be much more harmful than duplication of the same segment. In one illustrative family, the 5p– patient had a retarded but physically completely normal sister whose karyotype showed an extra segment attached to chromosome 9p (9p+). The father had a normal karyotype, but the mother was found to be a balanced carrier of a translocation of a segment of 5p to 9p. In two later pregnancies, the mother underwent amniocentesis for prenatal diagnosis. The first fetus was found to have a 46,XY karyotype; the second was a balanced carrier like the mother (Fig. 6–14).

OTHER AUTOSOMAL ABNORMALITY SYNDROMES

In addition to the classic syndromes described above, numerous other autosomal abnormalities associated with malformations have been described; some of these are common enough to have reached the status of recognized syndromes. Examples include trisomy 8 (usually seen as mosaic trisomy 8), various abnormalities of chromosome 9 (full trisomy, mosaic trisomy 9, or

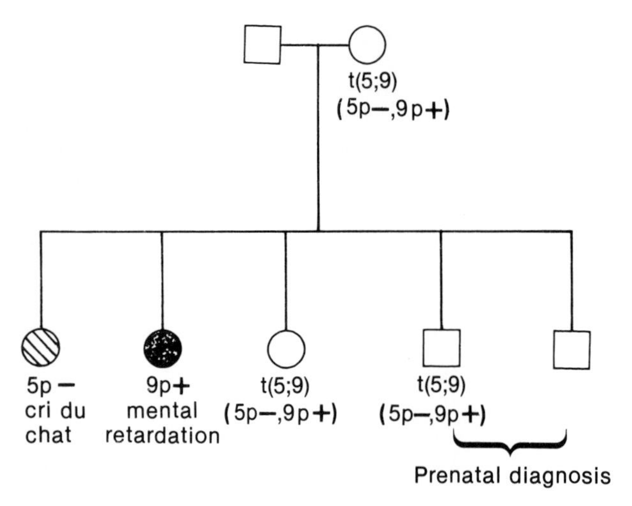

Figure 6–14. Pedigree of child with cri du chat syndrome whose mother was a carrier of a balanced translocation. The mother's karyotype showed that the short arm of chromosome 5 had lost material (5p −) which was translocated to the short arm of chromosome 9 (9p +). See text for further discussion. (Note: For simplicity, the complete formal descriptions of the karyotypes are not given.)

trisomy or monosomy of the short arm) and deletion of the short arm of chromosome 4. Though the pattern of anomalies is characteristic of the specific chromosome abnormality present, these defects are characterized in general by abnormal facies, multiple dysmorphic features and developmental and mental retardation. Chromosome analysis is an essential part of the diagnosis of any such patient, though in many cases no cytogenetic abnormality will be detectable; the cause of the syndrome will have to be sought elsewhere.

CHROMOSOMAL ABERRATIONS IN ABORTIONS AND NEWBORN SURVEYS

Though population cytogenetics encompasses a wide range of studies of cytogenetic abnormalities and variants—their phenotypic consequences, causation and segregation in families as well as their frequency—this discussion is restricted mainly to the incidence of the various abnormalities as found at different ages, especially in spontaneous abortions, 16-week fetuses and newborns.

Spontaneous Abortions

Chromosome abnormalities are an important cause of spontaneous abortion. About 15 percent of all recognized pregnancies end in spontaneous abortion, 80 percent of these during the first trimester. The overall frequency of chromosome abnormalities in spontaneous abortions is very close to 50 percent, and 1 in 13 conceptuses has a chromosome abnormality of some kind.

The kinds of abnormalities seen in abortuses (Table 6–3) differ in a number of ways from those seen in liveborn infants (Table 6–4). The single most common abnormality is 45,X (18 percent of chromosomally abnormal spontaneous abortions as compared with about 1 in 20,000 live births). It appears that more than 99 percent of all 45,X conceptuses abort spontaneously. Sex chromosome abnormalities other than 45,X are unusual in abortions. So far trisomies have been seen for all the autosomes except chromosome 1, but the distribution of types is quite different in abortions and live births. Trisomy 16, the most common trisomy in abortions, is not seen at all in live births. Trisomy 21 and trisomy 22 are about equally common in abortions, but in live births trisomy 21 is common and trisomy 22 very rare. If 9 percent of trisomic abortions (4.5 percent of all chromosomally abnormal abortions, 2.25 percent of all spontaneous abortions)

Table 6–3. RELATIVE FREQUENCIES OF TYPES OF ABERRATION IN CHROMOSOMALLY ABNORMAL ABORTUSES

Type		Frequency (%)
Trisomy		52
Trisomy 14	3.7	
15	4.2	
16	16.4	
18	3.0	
21	4.7	
22	5.7	
Other	14.3	
45,X		18
Triploid		17
Tetraploid		6
Unbalanced translocations		3
Other		4
Total		100

Data summarized from Carr and Gedeon, 1977.

and only 1 live birth in 800 are 21 trisomic, then about 75 percent of all Down syndrome conceptuses must abort spontaneously. Of the three D trisomies, trisomy 13 is seen in liveborn children, but both trisomy 14 and trisomy 15 rank high in frequency in spontaneous abortions, and are seen only rarely in liveborns, as mosaics. Trisomy 18 occurs in 5 or 6 percent of trisomic abortions and 1 in 8000 newborns, so it appears that over 95 percent of all trisomy 18's abort. Triploids and tetraploids are common in abortions, but triploids are rarely born

Table 6–4. INCIDENCE OF CHROMOSOMAL ABERRATIONS SEEN IN NEWBORN SURVEYS

	Number Seen	Approx. Incidence
Sex Chromosome Abnormalities in 37,779 Males		
XYY	35	1/1000 male births
XXY	35	1/1000 male births
Other	28	1/1300 male births
Sex Chromosome Abnormalities in 19,173 Females		
45,X	2	1/10,000 female births
XXX	20	1/1,000 female births
Other	7	1/3,000 female births
Autosomal Aberrations in 56,952 Babies		
+D (trisomy 13)	3	1/20,000 births
+E (trisomy 18)	7	1/8000 births
+G (nearly all trisomy 21)	71	1/800 births
Other trisomies	1	1/50,000 births
Rearrangements		
Balanced	110	1/500 births
Unbalanced	34	1/2000 births
Total Chromosomal Aberrations	353	1/160 births

Summarized from Hook EB, Hamerton JL. The frequency of chromosome abnormalities detected in consecutive newborn studies. Differences between studies. Results by sex and by severity of phenotypic involvement. In: Hook EB, Porter IH, eds. Population cytogenetics: studies in humans. New York: Academic Press, 1977.

alive and only a few who are mosaic with a diploid line survive any length of time; tetraploids have even worse survival, being seen most frequently in the very earliest abortions.

Fetuses at 16 Weeks of Development

Since most spontaneous abortions in which the fetus is chromosomally abnormal happen before the sixteenth week of pregnancy, the frequency of the chromosome abnormalities seen in prenatal diagnosis should be much the same as the frequency at birth. On the whole this is true, but there seems to be an exception for Down syndrome at maternal ages 40 to 44. For this maternal age range, as noted earlier, the incidence of Down syndrome is substantially higher in the 16-week fetus than at birth.

Newborn Surveys

The incidence of chromosome abnormalities in newborns has been estimated in several large studies, which included over 50,000 children, of whom two thirds were male. (In some series, only males were examined.) The findings are summarized in Table 6–4.

In summary, consider the fate of 10,000 conceptuses. About 800 are chromosomally abnormal. These include about 140 with 45,X; 112 with trisomy 16; 20 with trisomy 18; and 45 with trisomy 21.

Of these, 750 abort spontaneously as do 750 chromosomally normal fetuses. The abortions include about 139 with 45,X, all 112 with trisomy 16, 19 with trisomy 18, 35 with trisomy 21 and a variety of rarer types, chiefly trisomies.

The remaining 50 abnormals represent about 0.6 percent of the 8500 liveborns. The liveborn group includes about one each with 45,X and trisomy 18, 10 with trisomy 21, 15 with sex chromosome trisomy (XYY, XXY or XXX), and 20 with rearrangements, of which 16 are balanced and 4 unbalanced.

Other surveys in special populations have revealed a higher than normal incidence of chromosome abnormality in stillbirths (about 5 percent, almost 10 times the incidence in live births), in institutionalized retarded children, and in women with a history of repeated abortion or their spouses.

Conditions with Defects Demonstrable by High-Resolution Banding

The technique of high-resolution banding has made it possible to identify and precisely locate a number of previously undetectable chromosomal defects. Retinoblastoma and Prader-Willi syndrome are examples of conditions in which deletion of a very short segment of a chromosome is associated with an abnormal phenotype and, in the case of retinoblastoma, with oncogenesis.

RETINOBLASTOMA

Retinoblastoma (Rb) is an embryonic tumor of the retina that in some families is inherited but in other families is isolated, that is, occurs in one member only. In the hereditary form several tumors develop, usually in both eyes; there is a high risk of a second primary malignancy such as osteosarcoma. In the isolated form, only one tumor develops. The Rb locus has been assigned to chromosome 13 at band q14 (symbolized 13q14); there are some patients with retinoblastoma who have a deletion of this band.

In the hereditary form, the heterozygous presence of the Rb gene is not itself sufficient to produce retinoblastoma. In addition, a second event (for example, a somatic mutation in the normal allelic gene in some retinoblast) is required for the retinoblastoma to develop. However, because of the large number of retinoblasts, it is almost certain that this second event will occur in one or more of them.

Possible mechanisms for the first and second events have been proposed (Cavenee et al., 1983; Murphree and Benedict, 1984). The first would be a mutation, either germinal or somatic, or a deletion of band 13q14. The second would be mutation, deletion or some other chromosomal mechanism that would have the effect of producing homozygosity or hemizygosity of the retinoblastoma allele. Although retinoblastoma follows the rules of autosomal dominant transmission in many families, it is not expressed until a second somatic event has rendered the locus either homozygous or hemizygous.

PRADER-WILLI SYNDROME

Prader-Willi syndrome (PWS) is a relatively common syndrome of hypotonia, progressive obesity with compulsive eating tendencies, hypogonadism and small hands and feet, usually associated with mental retardation and not typically familial in its occurrence (Fig. 6–15).

By high-resolution banding, at least half the children with PWS are known to have a specific small deletion of chromosome 15, specifically of band 15q12 (Ledbetter et al., 1982). Various other alterations of chromosome 15 have also

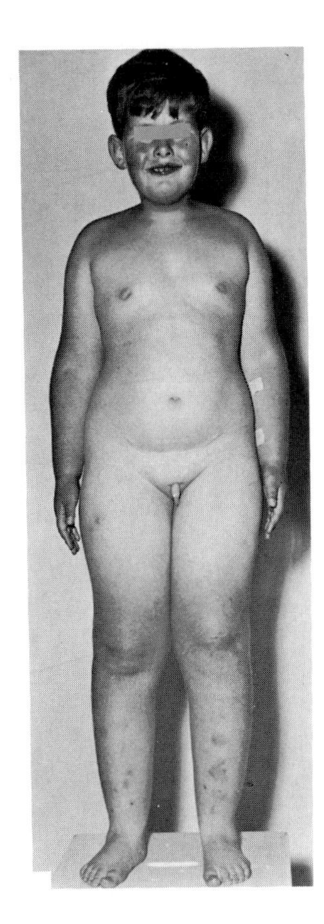

Figure 6–15. Prader-Willi syndrome. Note obesity, small hands and feet, small penis with cryptorchidism. The child is short for his age and is mentally retarded. From Smith DW. Recognizable patterns of human malformation, 3rd ed. Philadelphia: W. B. Saunders Company, 1982.

been seen. In the other half of the cases, however, no chromosome abnormality has been detected microscopically. It may be that PWS is etiologically heterogeneous, or that a submicroscopic deletion is present in children with apparently normal karyotypes. At present, finding a 15q deletion in a suspected case of PWS would confirm the diagnosis, but a normal karyotype would not rule it out.

Chromosomal Breakage Syndromes

A special group of disorders of cytogenetic interest comprises several rare autosomal recessive conditions in each of which there is excessive chromosome instability, defective DNA repair, impairment of growth and increased susceptibility to neoplasia.

Fanconi anemia is characterized primarily by pancytopenia and excess chromosomal breakage observed in cultured cells. Experimentally, fibroblast cultures from patients show increased breakage with mitomycin C. Most patients have prenatal and postnatal growth retardation. Many have hypoplastic or missing thumbs, occasionally with more extensive arm defects including radial aplasia. There is brownish pigmentation of the skin, and bone marrow failure occurs in later childhood. The characteristic hematological findings include a high level of fetal hemoglobin. There is an increased incidence of malignancy, especially leukemia and lymphomas.

In Bloom syndrome (Fig. 6–16), there is a greatly increased incidence of sister chromatid exchange (very useful in diagnosis) and induction of breakage by ultraviolet light. Growth is severely retarded both prenatally and postnatally. Most but not all patients are of Ashkenazi Jewish descent.

Ataxia-telangiectasia, as the name implies, is associated with progressive ataxia and the presence of telangiectasia, easily seen in the bulbar conjunctiva and the auricles. Clinically this often presents as a defect of cellular immunity. The cells of patients have impaired ability to excise DNA that has been damaged by ionizing radiation.

Xeroderma pigmentosum, which is chiefly a skin disease often resulting in skin cancer, exists in two major forms and eight subgroups ("complementation groups"). Cultured cells from patients typically show chromosome changes only after exposure to ultraviolet light and chemical carcinogens, which produce an increased frequency of sister chromatid exchange and other chromosome abnor-

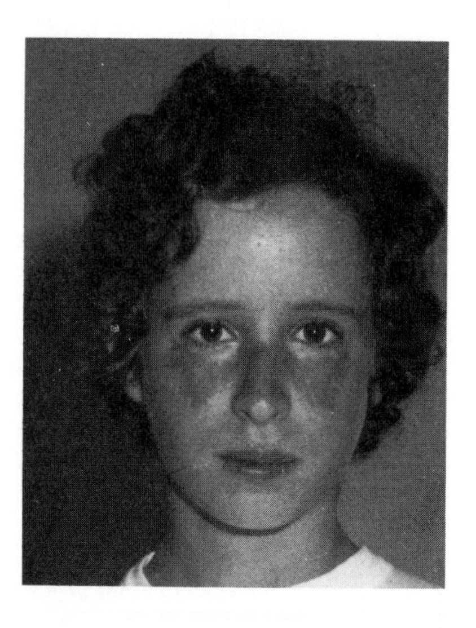

Figure 6–16. A child with Bloom syndrome. Note butterfly rash on midface. Courtesy of J. L. German.

malities. It is known to be associated with specific deficiency of one of several different enzymes required for excision of pyrimidine dimers from damaged DNA and repair replication.

Chromosomal Changes in Neoplasia

High-resolution banding has led to detailed knowledge concerning chromosomal changes in cancer (Rowley, 1983; Yunis, 1983).

The malignant cells of most neoplasias have chromosome defects; many of the defects are consistent. The alterations may involve deletion of a band, reciprocal translocation (in which one chromosome and its breakpoint are fixed but the other chromosome involved may vary) or, less often, trisomy for a specific chromosome.

The following are some examples of consistent chromosomal changes seen in human neoplasia:

1. The original and still the best-known example is a reciprocal translocation between chromosome 22q and 9q, symbolized as t(9q;22q) seen in about 95 percent of adults with **chronic myelogenous leukemia** and a much smaller proportion of patients with other types of leukemia (Fig. 6–17). The deleted chromosome 22 is called the **Philadelphia chromosome** because of its original discovery there.

2. In **Burkitt lymphoma**, a malignancy usually found in central Africa and typically characterized by an osteolytic lesion of the jaw, a high proportion of patients have a consistent reciprocal translocation between chromosomes 8 and 14, t(8q;14q). This translocation is also seen in Burkitt lymphoma of non-African origin and in a type of B-cell acute lymphocytic leukemia (ALL), indicating that Burkitt lymphoma and this type of ALL are probably manifestations of the same disease.

3. In **retinoblastoma**, as described earlier, there is often a deletion of a specific band of chromosome 13 (13q14).

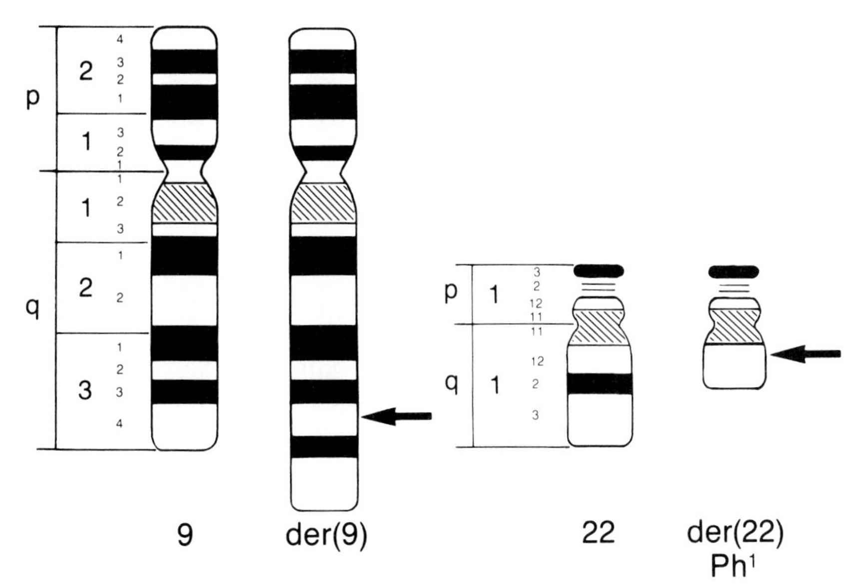

Figure 6–17. The defect in the Philadelphia chromosome, a translocation of part of the long arm of chromosome 22 to chromosome 9, compared with the normal chromosomes 9 and 22.

4. In **small cell carcinoma of the lung** there is deletion or translocation of a segment of chromosome 3 (p14–23).

5. In the **aniridia–Wilms tumor association** a band of chromosome 11 (11p15) may be deleted. Wilms tumor is a malignant tumor of the kidney, typically occurring in early childhood or even prenatally. Although both aniridia (absence of the iris) and Wilms tumor can occur independently, at least 50 cases of aniridia and Wilms tumor in association have been reported. Many of the patients also have other malformations, mental deficiency and growth retardation. The 11p– deletion is found in many but apparently not all patients with the aniridia–Wilms tumor association. It is noteworthy that a particular oncogene known as c-Ha-*ras* is located at the site of the deletion.

SECONDARY CHROMOSOMAL CHANGES

Distinctive nonrandom chromosomal defects may be acquired during development of neoplasia. For example, in chronic myelogenous leukemia a second Philadelphia chromosome and an additional chromosome 8 or isochromosome of the long arm of 17 may be found when the patient enters the terminal phase of the disease. The Y chromosome may be lost in males. These abnormalities are thought to confer proliferative advantage on the malignant clone.

Other changes often seen in solid tumors are the acquisition of homogeneously staining regions (HSR's) and double minutes (acentric fragments). These are thought to represent sites of gene amplification. The extra gene dosage may be important in allowing the tumor to increase its aggressiveness.

RELATION OF ONCOGENES TO CHROMOSOMAL DEFECTS

Oncogenes are a recently discovered group of some 20 structurally and functionally heterogeneous genes that are important in the transformation of cells to the malignant state. They exist in the cell in the form of **proto-oncogenes**, which become activated either by association with retroviruses or by mutational events.

The names given to oncogenes are acronyms based on their origin; for example, c-*myc* (see below) was originally found in B-cell avian myelocytoma. Oncogenes have been preserved throughout vertebrate evolution, and it is presumed that each of them exists in at least one copy in the genome of man (and the other vertebrates). Their role in normal cell function is only beginning to be elucidated, but there are already strong hints that they are involved in cell division and growth.

The best-known relation between an oncogene and a chromosome defect is that of c-*myc* to the Burkitt lymphoma 8;14 translocation (Fig. 6–18). In man, c-*myc* has been precisely localized to the break in band 8q24 involved in the translocation. The translocation transfers the c-*myc* gene to a region of chromosome 14q32 within the immunoglobulin heavy chain gene. (See Chapter 8 for a description of the immunoglobulin genes.) As a result of the translocation, transcription of c-*myc* increases up to 20 times normal in some patients, while in others an altered gene product is formed. In other forms of Burkitt lymphoma, the same breakpoint occurs in chromosome 8, but the reciprocal translocation is with chromosome 2 near the locus of the κ light chain immunoglobulin gene, or with chromosome 22, near the λ light chain locus. The mechanism by which the translocation of c-*myc* close to an immunoglobulin gene triggers malignant growth is still an enigma.

Concordance of the location of other human cellular oncogenes and the

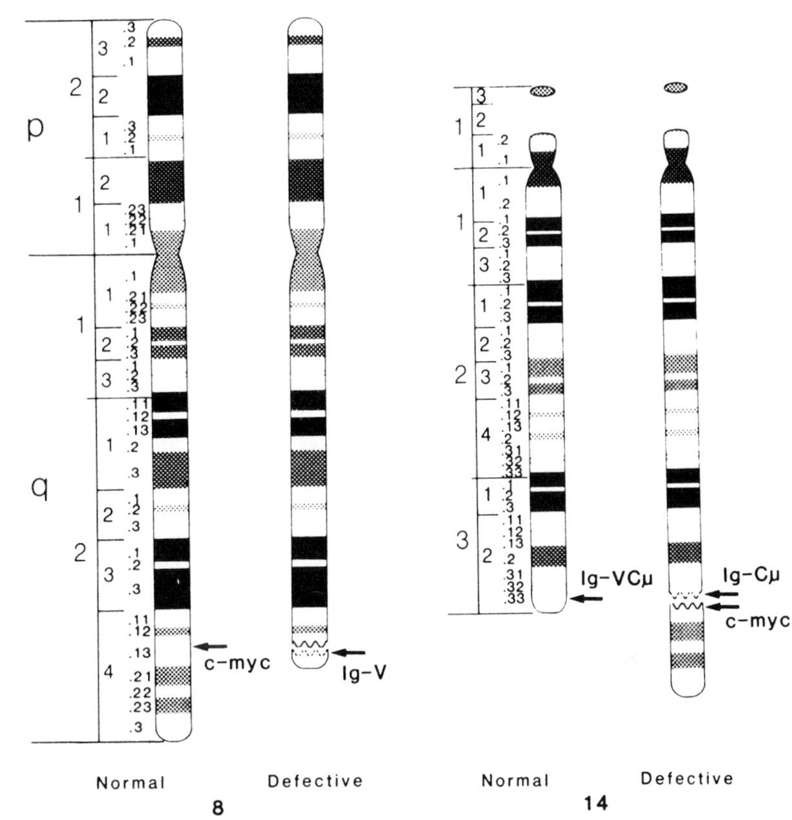

Figure 6–18. The chromosome defect in Burkitt lymphoma. From Yunis JJ. The chromosomal basis of human neoplasia. Science 1983; 221:227–236, by permission.

breakpoints involved in chromosome changes in other forms of cancer are becoming clear; for example, c-Ha-*ras* is located at 11p13, the site of the deletion in the aniridia–Wilms tumor association.

Malignant transformation is a multistep process; activation of a single cellular oncogene is not enough to convert a normal cell into a tumor cell. But it also appears that oncogenes as a group may only serve a few separate and perhaps sequential functions in carcinogenesis (Land et al., 1983).

GENERAL REFERENCES

Bishop JM. Cancer genes come of age. Cell 1983; 32:1018–1020.
Bond DJ, Chandley AC. Aneuploidy. Oxford: Oxford University Press, 1983.
Borgaonkar DS. Chromosomal variation in man: a catalog of chromosomal variants and anomalies, 4th ed. New York: Liss, 1984.
DeGrouchy J, Turleau C. Clinical atlas of human chromosomes, 2nd ed. New York: Wiley, 1984.
Hamerton JL. Human cytogenetics. Vol. I, General cytogenetics; Vol. II, Clinical cytogenetics. New York: Academic Press, 1971.
Hassold TJ, Jacobs PA. Trisomy in man. Ann Rev Genet 1984; 18:69–97.
Hook EB, Porter IH, eds. Population cytogenetics: studies in humans. New York: Academic Press, 1977.
Rowley JD. Human oncogene locations and chromosome aberrations. Nature 1983; 301:290–191.
Schinzel A. Catalogue of unbalanced chromosome aberrations in man. Berlin, New York: deGruyter, 1984.
Smith GF, Berg JM. Down's anomaly, 2nd ed. Edinburgh: Churchill Livingstone, 1976.
Yunis JJ, ed. New chromosomal syndromes. New York: Academic Press, 1977.
Yunis JJ. The chromosomal basis of human neoplasia. Science 1983; 221:227–236.

PROBLEMS

1. Describe the chromosome constitution for each of the following notations. State which represent a balanced and which an unbalanced karyotype.
   a) 46,XY/47,XY,+21
   b) 46,XY,−14, +t(14;21)(p11;q11)
   c) 46,XX, inv ins(5;2)(p14;q32q22)
   d) 46,XY,−7, + der(7), t(7;11)(q36;q21)mat
   e) 46,XX, t(2;5;7)(p21;q23;q22)

2. List the possible ways in which triploids could arise. Which is the most frequent origin?

3. A pair of twins consists of a 46,XY and a 47,XY,+21 member, They are identical for all the genetic markers tested. Explain how this could arise.

4. List three interstitial deletions that are frequently associated with certain malformation syndromes or disorders.

5. About 3 percent of trisomic abortuses have trisomy 13. Using information in this chapter, calculate the proportion of all trisomy 13 conceptuses that abort spontaneously.

6. What proportion of all human conceptuses are chromosomally abnormal?

7. You discover a 21q22q translocation in a baby with Down syndrome and in his father. The father's sister has a child with Down syndrome in an institution in a different city. Discuss the implications and how you would deal with the genetic, social and ethical problems of such a situation.

7

THE SEX CHROMOSOMES AND THEIR DISORDERS

Man and the other mammals have a sex determination mechanism in which the crucial factor is the presence or absence of the Y chromosome. Because the sex chromosome constitution underlies normal sexual development and many of its disorders, it is appropriate to begin this chapter with a brief review of the embryology of the human reproductive system.

Embryology of the Reproductive System

The primordial germ cells are wandering cells found in the fourth week of embryonic development in the endoderm of the yolk sac near its union with the allantois. By the sixth week of intrauterine life, they have migrated to the gonadal ridges, where they are surrounded by the sex cords, epithelial outgrowths of the genital ridges, to form a primitive, bipotential gonad. The outer cortex of the gonad normally becomes the ovary in females, the inner medulla becomes the testis in males. If there is a Y chromosome, even if more than one X is present, the gonad normally differentiates in the male direction, the medullary tissue forming typical testes with seminiferous tubules and Leydig cells which, under the stimulation of human chorionic gonadotropin from the placenta, become capable of androgen secretion. If no Y is present, the gonad differentiates in the female direction: the cortex develops, the medulla regresses, and oogonia form and multiply in the cortex. Before birth primary oocytes have differentiated and entered the first meiotic division.

The differentiation of the primitive gonad into a testis may be determined by the H-Y antigen, a cell surface component specified by a Y-linked gene and present in males but usually (not invariably) absent in females. The antigen was originally discovered in transplantation experiments with inbred mice, in which females of some strains reject grafts from males of the same strain. A variety of serological techniques for detection of the H-Y antigen were later developed. At present there is controversy over whether or not the serologically defined male antigen is the same as the histocompatibility antigen that accounts for rejection of male skin grafts by female mice. There is also disagreement as to whether the differentiation of the primitive gonad into a testis depends on interactions between autosomal genes and genes on the X and Y chromosomes, the H-Y histocompatibility antigen, and/or the serologically defined male antigen (reviewed by Zenzes and Reed, 1984).

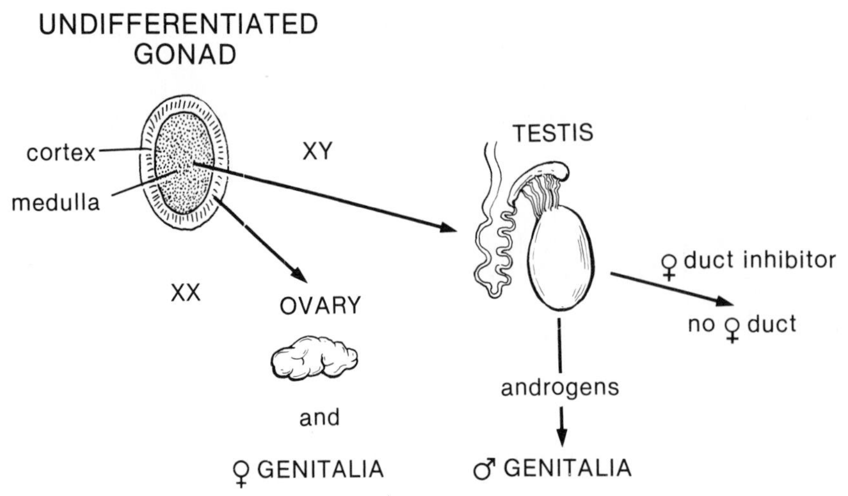

Figure 7–1. Development of the gonads and genital ducts, described in the text.

While the primordial germ cells are migrating to the gonadal ridges, thickenings in the ridges indicate the positions of the mesonephric (wolffian) and uterine (müllerian) duct systems. In the presence of androgen-secreting Leydig cells the mesonephric duct normally differentiates to form the typical male duct system. At the same time the Leydig cells produce a second factor, the müllerian inhibiting factor, which causes regression of the uterine duct system. If an ovary is present, or if the gonad remains undifferentiated, usually a typical female duct system develops and the mesonephric system regresses (Fig. 7–1).

In the early embryo the external genitalia consist of a genital tubercle, paired labioscrotal swellings and paired urethral folds. The genital tubercle can form either penis or clitoris. In the male, under the influence of androgens, the labioscrotal folds fuse to form the scrotum and the urethral folds fuse to form the urethra including its penile portion, while the genital tubercle takes on a male (penile) configuration. In the absence of a differentiated gonad, or in the presence of an ovary, the labioscrotal folds remain separate to form the labia majora, the urethral folds remain unfused to form the labia minora, and the genital tubercle assumes the female (clitoral) configuration.

All these developmental steps are under genetic control, and can be altered by gene mutation, chromosomal abnormalities or environmental effects.

The Y Chromosome

The short arm of the human Y chromosome is thought to be the segment chiefly responsible for male sex differentiation. Deletions of Yp lead to a female phenotype, streak gonads and absence of H-Y antigen. The long arm of the Y, in contrast, can show considerable variability in length without any phenotypic consequences, apparently because its distal portion is composed of highly repetitive DNA sequences. This part of Yq fluoresces so brilliantly with quinacrine staining that it can be seen even in interphase cells. About 10 percent of males have a fluorescent segment of Yq that is longer or shorter than normal; the difference is inherited (Fig. 7–2).

The maleness factor or factors on the Y chromosome appear to be localized to a segment of Yp very close to the centromere. A few studies have also indicated the presence of one or more male factors on Yq, also near the centromere (Davis, 1981). Several patients who lacked the paracentromeric part

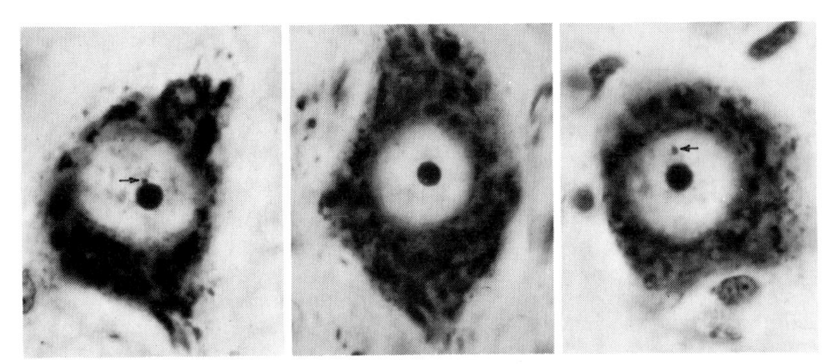

Figure 7–2. Sex chromatin (arrows) in nerve cells of a female cat. The center cell, from a male, lacks sex chromatin. From the original description of sex chromatin in Barr ML, Bertram EG. A morphological distinction between neurones of the male and female, and the behavior of the nucleolar satellite during accelerated nucleoprotein synthesis. Nature 1949; 163:676–677.

of Yp were phenotypic females with varying degrees of Turner syndrome, whereas other patients with a dicentric Y that contained a duplicated paracentromeric segment of Yp were males with testes.

The presence of an extra Y chromosome is quite a common observation, occurring in about 1 of every 1000 male births (see XYY syndrome, later in this chapter). Numerous different structural rearrangements of the Y have been reported, including deletions, dicentrics, duplications, inversions, isochromosomes, rings, satellited Y chromosomes or translocations involving either the X or an autosome. Mosaicism may be present, the second line having a normal XY karyotype or sometimes a single X. Since there are so many types of structural alteration, the phenotypic consequences are also variable. There may be normal differentiation and fertility or varying degrees of abnormal testicular development and ambiguity of the external genitalia.

The X Chromosome

SEX CHROMATIN

It has been known since 1921 that male and female cells differ in their sex chromosome complement, but it was not until 1949 that a sex difference in interphase cells was detected. In that year Barr and Bertram noted that a mass of chromatin in the nuclei of some nerve cells in cats was frequently present in females but not in males (Fig. 7–2). This mass is now known as the sex chromatin or Barr body. Cells (or people) are said to be chromatin-positive if sex chromatin is present and chromatin-negative if it is absent.

Barr and his students later found sex chromatin in the cells of most of the tissues of females of many species of mammals including humans. The Barr body can be seen in many cell types, but is most conveniently examined in the epithelium of the buccal mucosa. A buccal smear is made simply by scraping a few cells from the inside of the cheek, smearing them on a slide and staining. The slide is then examined microscopically to determine the percentage of cells that show Barr bodies (Fig. 7–3). At one time buccal smears were widely used in diagnosis of disorders of sexual development; today they have been replaced by complete karyotyping.

The association of the sex chromatin with the number of X chromosomes and its lack of association with the Y is shown in the following table. The

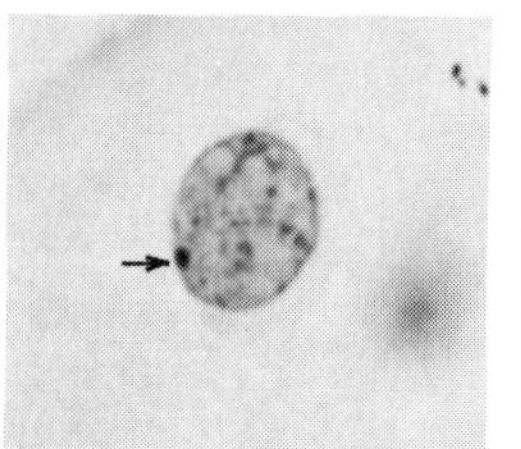
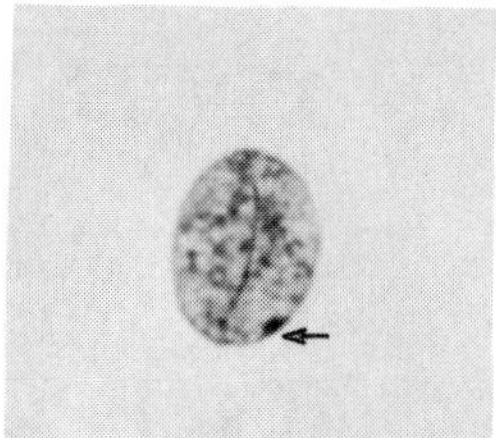
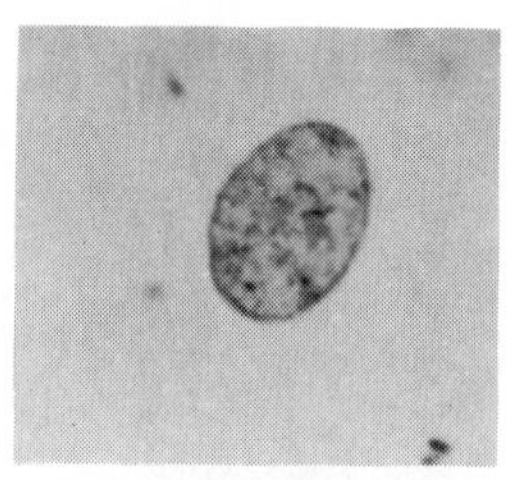

Figure 7–3. Sex chromatin in epithelial cells of human buccal mucosa. Arrows indicate the sex chromatin at the nuclear membrane of female cells. A male cell (right) lacks sex chromatin. From the original description of the buccal smear technique in Moore KL, Barr ML. Smears from the oral mucosa in the determination of chromosomal sex. Lancet 1955; 2:57–58.

phenotypes typical of some anomalous sex chromosome complements are described later in this chapter. Note that sex chromatin is present if there are two or more X chromosomes, and that the number of Barr bodies is always one less than the number of X's.

Chromosome Complement	Number of Barr Bodies
45,X 46,XY 47,XYY	0
46,XX 47,XXY 48,XXYY	1
47,XXX 48,XXXY 49,XXXYY	2
48,XXXX 49,XXXXY	3
49,XXXXX	4

The Barr body represents one of the two X chromosomes of female cells, which remains condensed and genetically inactive throughout interphase and is late-replicating (that is, completes its replication later than its homologue). This is true only of somatic cells. In the germ line of the female both X's remain active; in the germ line of the male the single X is inactivated.

X INACTIVATION

For many years the action of X-linked genes was a puzzle to geneticists. How could it be that females with two representatives of every gene on the X chromosome formed no more of the product of these genes than did hemizygous males with only a single X? And why were females homozygous for an X-linked mutant gene no more severely affected than were hemizygous males? Some mechanism of "dosage compensation" was indicated.

A hypothesis to explain dosage compensation in terms of a single active X chromosome was arrived at independently by several groups of workers; it is often called the Lyon hypothesis after Mary Lyon, who was the first to state it explicitly and in detail. Lyon (1962) based her hypothesis of X inactivation in part on genetic observations of X-linked coat color genes in the mouse and in part upon cytological data. She noted that in a female mouse heterozygous for X-linked coat color genes, the coat was mottled (that is, made up of patches of the two colors, random in arrangement and rarely crossing the midline), as shown in Figure 7–4. On the other hand, males never had this patchy phenotype but instead had coats of a uniform color. Mice with one X and no Y chromosome also expressed X-linked coat color genes as a uniform, not mottled, coat color.

The chief cytological observation on which Lyon based her hypothesis has already been mentioned: the number of Barr bodies in interphase cells is always one less than the number of X chromosomes seen at metaphase. Additional

Figure 7–4. Female mouse heterozygous for the X-linked coat color gene *Tortoiseshell.* The mosaic phenotype of such females provided one of the first lines of evidence for X inactivation (see text). From Thompson MW. Genetic consequences of heteropyknosis of an X chromosome. Can J Genet Cytol 1965; 7:202–213.

cytological evidence is found in the observation that at prophase one X chromosome is late-replicating and heteropyknotic (out of phase with the other chromosomes with respect to its condensation and staining properties). Thus it is clear that the sex chromatin is an X chromosome; only one X is active in cellular metabolism, and the second (or in abnormal cells with additional X chromosomes, any extra X) appears as a sex chromatin body. In the male, the single X is active; consequently there is no sex chromatin.

The Lyon hypothesis (now an accepted theory) states that:

1. In the somatic cells of female mammals, only one X chromosome is active. The second X is condensed and inactive, and appears in interphase cells as the sex chromatin.

2. Inactivation occurs early in embryonic life.

3. The inactive X can be either the paternal or maternal X (X^p or X^m) in different cells of the same individual; however, after the "decision" as to which X will be inactivated has been made in a particular cell, all the clonal descendants of that cell will "abide by the decision" (that is, they will inactivate the same X). In other words, inactivation is random but fixed.

Exactly when and how X inactivation occurs has been extensively investigated, but much remains to be learned. In mouse embryos, both X's are active at the two-cell stage, but only one functions by the late blastocyst stage. X inactivation does not take place in all tissues simultaneously. What little information there is about human embryos indicates that the sex chromatin body has been detected in the implanting blastocyst at 9 to 12 days of development. It is first seen in the syncytiotrophoblast and later in the chorionic mesoderm and yolk sac. In the embryo proper, it has not been seen before the 18th day.

The mechanism of X inactivation is believed to involve DNA methylation, specifically methylation of cytosine residues to 5-methyl cytosine at certain sites in DNA. The pattern of DNA methylation produced would be stable through successive cell cycles. The altered affinity of DNA-binding proteins resulting from DNA methylation could account for the subsequent changes in gene activity. However, the initiation of inactivation and the exact method by which DNA methylation inactivates genes remain to be determined.

The proximal part of Xq appears to contain an inactivation center, around which the Barr body condenses. Evidence for the inactivation center is twofold: 1) Abnormal X's lacking the proximal part of Xq have not been observed, though other structural variants are common. 2) An abnormal X with a duplication of this segment forms an abnormal, bipartite Barr body.

Exceptions to the randomness of inactivation occur in some specialized circumstances. In females with loss of material from one X chromosome (a deletion, ring or isochromosome), the structurally abnormal X always forms the Barr body, which may then appear appropriately larger or smaller than usual. In contrast, if a female has a translocation between an X and an autosome, it is the intact X that becomes inactivated. These patterns are readily explained. In a deletion, if the intact X were to become inactivated, the functional X, lacking a chromosome segment with its component genes, would render any cell containing it inviable. In X-autosome translocations, inactivation of the translocated part of the X appears to bring about inactivation of the autosome to which it is translocated; the cell concerned would then be inviable because it would have only one functional copy of that autosome. In either case, the consequence of nonrandom inactivation is that a female with an X chromosome abnormality is, in effect, hemizygous for any allele on her active X and thus can express an X-linked disorder as a consequence of a gene on her only functional X. For this reason any female who expresses an X-linked disorder should have chromosome analysis as part of the investigation of her case.

An extreme example of nonrandom expression was shown by Nyhan and colleagues (1970) in women heterozygous both for the common A and B types of G6PD and for HPRT deficiency. Only one type of G6PD is present in their tissues; this is explained on the basis of random inactivation followed by selection against HPRT-deficient cells.

Genetic Consequences of X Inactivation

X inactivation has three principal genetic consequences: dosage compensation, variability of expression in heterozygous females and mosaicism.

DOSAGE COMPENSATION

The amount of product of X-linked genes, such as G6PD, HPRT or antihemophilic globulin (the substance deficient in classic hemophilia), is equivalent in the two sexes. For many years there was no satisfactory explanation as to how compensation for the dosage effect of the two X chromosomes of the female as compared with the male's single X was accomplished. X inactivation adequately explains this phenomenon. However, a number of problems remain. One of these is the abnormal phenotype shown by individuals with abnormal sex chromosome complements. If one and only one X is active regardless of the number present, it is difficult to understand why a 45,X or XXY individual should show any phenotypic abnormality at all, and why more than one extra X produces a progressively more abnormal phenotype.

Several hypotheses have been proposed to explain why persons with missing or extra chromosomes have abnormalities (Therman et al., 1980). There are two obvious possibilities:

1. The abnormalities may be initiated at an early embryonic stage before inactivation takes place.

2. The original X inactivation may not be regular or random in all cells with abnormal X chromosome constitutions. Cells with abnormal X chromosome numbers, being genetically unbalanced, would give rise to abnormal phenotypes.

VARIABILITY OF EXPRESSION IN HETEROZYGOUS FEMALES

Since inactivation is random, females heterozygous for X-linked genes have varying proportions of cells in which a particular allele is active. Among such females, considerable phenotypic variability is to be expected.

Clinically, the variation in expression of X-linked disorders can be extreme, ranging from the entirely normal to full expression of the defect. A carrier who exhibits an X-linked trait is called a manifesting heterozygote. Affected carriers, in whom the deleterious allele is functional in a majority of the cells, may be regarded as extreme examples of "unfavorable Lyonization." Color blindness, classical hemophilia, Christmas disease, Duchenne muscular dystrophy and many X-linked eye disorders are examples of conditions in which manifesting heterozygotes have been noted.

MOSAICISM

Women are mosaics with respect to the X chromosome; that is, they have two populations of cells, one population with one X active, the other with the alternative X active. The first direct demonstration of mosaicism at the cellular level was provided by cloning of cultured fibroblasts from a woman heterozygous for two different G6PD alleles. Two different clonal populations were demonstrated, differing as to which G6PD allele was functioning (Davidson et al., 1963). The demonstration of mosaicism in cultured fibroblasts by cloning is a useful tool for carrier identification and genetic counseling in some X-linked disorders.

Some human X-linked genes are not inactivated. Two loci definitely known not to undergo inactivation are Xg, the locus for the Xg blood group system, and STS, the locus for steroid sulfatase, which when deficient produces placental steroid sulfatase deficiency in fetal life and ichthyosis, a skin disorder, after birth. The reason why these two loci, which are close together on the distal end of Xp (and perhaps others in their vicinity), escape the inactivation of the rest of the X may have something to do with the mechanism of X-Y pairing during meiosis (Shapiro and Mohandas, 1982).

Sex Chromosome Abnormalities

INCIDENCE

The incidence of the common sex chromosome disorders in newborn surveys and in spontaneous abortions was discussed in Chapter 6. In general, the trisomic types are quite common with an incidence of about 1 in 1000 births each (that is, 1 per 1000 male births for XXY and XYY, and 1 per 1000 female births for XXX). The frequency of 45,X is much lower, about 1 in 10,000 female births. The only sex chromosome abnormality that appears in spontaneous abortions with appreciable frequency is 45,X. This is the most common cytogenetic abnormality in abortuses; as noted earlier, it accounts for 18 percent of all chromosomally abnormal abortions and is estimated to be present in 9 percent of spontaneous abortions and about 1.4 percent of all conceptions.

Mosaicism involving the X chromosomes is more common than mosaicism involving autosomes; it occurs in about 50 percent of patients with Turner syndrome and 15 percent of those with Klinefelter syndrome, whereas only about 1 percent of Down syndrome patients are mosaics. Patients with mosaic karyotypes usually have a lesser degree of abnormality.

DEVELOPMENT

Prospective collaborative studies of growth and development in children with sex chromosome abnormalities who were ascertained in newborn surveys or by prenatal chromosome analysis have been under way for several years. Information from a recent report of these studies (Stewart, 1982) is summarized in the sections that follow. Though on average they differ from 46,XX and 46,XY children in the respects noted, many children with sex chromosome anomalies are well within the normal range of size, appearance and ability. It is stressed that none of the 47,XXY, 47,XYY or 47,XXX groups is characterized by a constellation of physical features diagnostic of a "syndrome," even though they are so classified.

Clinical Aspects

KLINEFELTER SYNDROME

Figure 7–5 illustrates a patient with a 47,XXY karyotype. The condition is named after Harry Klinefelter, who described it in 1942. Because the phenotype is male but Barr bodies are present, Klinefelter syndrome was a prime candidate for chromosome analysis; its XXY chromosome complement was found shortly after human chromosome studies became possible (Jacobs and Strong, 1959).

A boy with Klinefelter syndrome is not recognizable before puberty unless

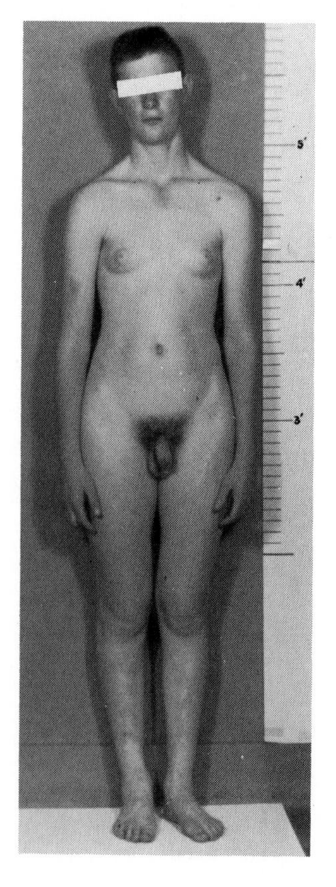

Figure 7–5. Phenotype of 47,XXY Klinefelter syndrome. Note long limbs, relatively small genitalia. Gynecomastia, present here, is not a constant feature. Courtesy of M.L. Barr.

his chromosomes are analyzed for some unrelated reason, such as participation in a newborn survey. After puberty, the chief characteristics are small testes and hyalinization of the seminiferous tubules. Usually the secondary sexual characteristics are poorly developed and gynecomastia may appear; many patients are tall and eunuchoid. The patients are almost always sterile, and sterility may be the presenting complaint.

About 15 percent of Klinefelter males are mosaics, with two or more distinct cell lines. The most common mosaic form is XY/XXY, and in these patients it is noteworthy that though the buccal smear may be chromatin-positive, testicular development and mental status may not be abnormal.

Maternal age is advanced in the XXY syndrome, and about 60 percent of the group owe their origin to either meiotic or postzygotic nondisjunction of the maternal X chromosome(s); that is, they are X^mX^mY. The remaining 40 percent are X^mX^pY, indicating nondisjunction of the X and Y in the first meiotic division of spermatogenesis.

Follow-up study of 90 XXY boys, the majority 8 to 14 years of age, has indicated that school problems are frequent, and that verbal IQ tends to be lower than normal. The boys are increasingly taller than normal, from 5 years of age on. The legs are proportionately long. Weight is also above normal but less so than height, so that the boys tend to be tall and thin. Head circumference is significantly small at birth and continues small. Bone age is below the mean in childhood but catches up almost to the mean by 12 years of age. The boys tend to be normal in performance IQ but less so in verbal IQ. They have a rather high incidence of educational problems. They may be less aggressive and active and more susceptible to social stress than their peers.

There are several variants of Klinefelter syndrome: XXYY, XXXY, XXXXY and others. As a rule the additional X's cause a correspondingly more abnormal phenotype, with a greater degree of dysmorphism, more defective sexual development and more severe mental retardation.

XYY SYNDROME

Males with a second Y chromosome have aroused great interest since they were found to be frequent among males in a maximum security prison (Jacobs et al., 1968). About 3 percent of males in prisons and mental hospitals are XYY, and among the group over 6 feet tall the proportion is much higher, over 20 percent. Since in newborn surveys the incidence is about 1 in 1000 births, most XYY males must be indistinguishable from XY males on the basis of behavior or physical appearance.

The origin of the XYY karyotype is paternal nondisjunction at the second meiotic division, which produces YY sperm. The less common XXYY and XXXYY variants, which share the features of the XYY and Klinefelter syndromes, probably also originate in the father, by a sequence of nondisjunctional events.

The relationship between XYY and aggressive, psychopathic or criminal behavior has aroused great public interest. XYY males are perhaps six times as likely to be imprisoned as XY males. Parents whose child is found, prenatally or postnatally, to be XYY are often extremely concerned about the implications, and some physicians question the advisability of disclosing such information.

The 59 XYY boys 5 to 13 years of age who have been followed from birth are tall, but on average not as tall as XXY boys, and are not heavy for their height. Their head circumference is normal. They appear to be generally normal intellectually but seem to have a high incidence of educational difficulties. Excess temper tantrums, higher levels of activity and a more negative mood are mentioned by their parents.

XXX

Trisomy X and the rare tetra-X and penta-X karyotypes are the counterparts in the female of Klinefelter syndrome in the male. Triple X females are not phenotypically abnormal. Some are first identified in infertility clinics, others in institutions for the mentally retarded, but probably many remain undiagnosed. Some have borne children, virtually all of whom have normal karyotypes.

Follow-up of 54 XXX girls aged 2 to 16 years shows that most are tall with increased height velocity beginning at 4 to 8 years of age. They tend to be underweight for their height and to be long-legged. They have reduced bone age in early childhood but not later, and significantly reduced head circumference from birth onward. Deficiencies in speech development and verbal IQ are common, school achievement is significantly below normal and the children are thought to have greater than average difficulty in interpersonal relationships.

The presence of four X chromosomes often leads to retardation in both physical and mental development, and XXXXX usually causes severe developmental retardation with multiple physical defects.

TURNER SYNDROME (X MONOSOMY)

The syndrome of sexual infantilism, short stature, webbing of the neck and cubitus valgus (reduced carrying angle at the elbow) originally described in 1938 by Turner is shown in Figure 7–6. The phenotype is female though the patients are (usually) chromatin-negative. The discrepancy suggested a chromosome abnormality, and this was confirmed by Ford and colleagues (1959), who demonstrated the 45,X karyotype.

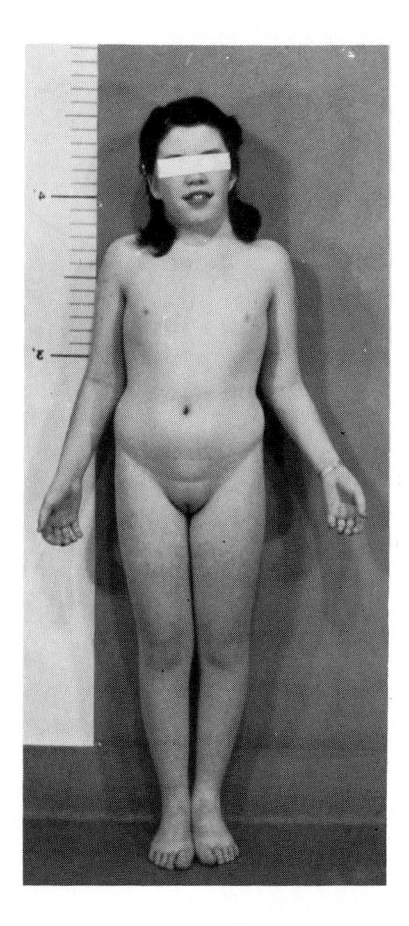

Figure 7–6. Phenotype of 45,X Turner syndrome. For details, see text. From Barr ML. Sexual dimorphism in interphase nuclei. Am J Hum Genet 1960; 12:118–127.

Other features of the phenotype include the low hairline at the nape of the neck, characteristic facial appearance, unusual dermatoglyphics with high total ridge count, wide chest with broadly spaced nipples, coarctation of the aorta and, especially in newborns, lymphedema of the feet, which together with neck webbing should alert the physician to the need for chromosome studies. The external genitalia are juvenile, and the internal sexual organs are female although the ovary is often only a streak of connective tissue; however, the streak may be arranged in the manner of ovarian stroma and ovarian follicles may be present in fetal life, though usually not postnatally. Axillary and pubic hair are usually present but sparse. Primary amenorrhea is usual, though not invariable.

The high incidence of the 45,X karyotype in spontaneous first-trimester abortions was noted earlier. It is curious that 45,X is so severe a defect prenatally, yet relatively benign after birth. Probably the chief explanation is the high proportion of mosaicism among the survivors.

Only about 60 percent of patients with the Turner syndrome have monosomy X. The remainder have a variety of karyotypes with a structural alteration of the X or mosaicism involving one or more cell lines with abnormal number or structure. The most frequent of these is 46,X,i(Xq), that is, an isochromosome for the long arm of the X. Deletions of part of Xp or Xq and ring-X chromosomes are not uncommon. In non-mosaics deletions of Xp are associated with short stature, whereas deletions of Xq are unlikely to be associated with short stature, but in general are associated with the presence of streak gonads and consequent infertility.

Mosaicism accounts for about 40 percent of all cases. About 15 percent are X/XX or X/XXX, and 10 percent or more are XX/X,i(Xq). As usual with mosaicism, the phenotype in such patients varies depending on the time of the postzygotic accident and the proportion of abnormal cell lines in different tissues. X/XY mosaicism also leads to variable phenotypic changes.

Unlike other sex chromosome aneuploids, children with Turner syndrome can often be identified at birth or before puberty by their characteristic phenotypic features. Their intelligence is normal or only slightly reduced as compared with their sibs. Many patients with Turner syndrome have defective spatial perception. As they grow up, their short stature, failure to develop secondary sexual characteristics, and infertility may create psychological problems. Therapy is not totally effective, but most patients with Turner syndrome are given anabolic steroids when 10 to 14 years of age to increase their height, and receive estrogen replacement therapy at an appropriate age to permit the development of secondary sexual characteristics, with cycling to allow menstruation.

XX MALES

Phenotypic males with an XX karyotype, who may have a frequency of about 1 in 15,000 male births, have created much interest because they appear to contradict the rule that a Y chromosome is essential to differentiation of the primitive gonad into a testis.

One of the hypotheses developed to account for XX males has been that they may simply be XX/XXY mosaics in whom the Y-containing cell line has not been identified. An argument in favor of this interpretation is that XX males typically resemble males with the Klinefelter syndrome, except that they are usually not so tall and seem to have normal intelligence. A second possible explanation is that an interchange between Xp and Yp during paternal meiosis results in XX males in whom the paternal X chromosome carries translocated male-determining Y chromosomal material.

Disorders of Sexual Development with Normal Chromosomes

In some newborn infants, assignment of sex is difficult or impossible because the genitalia are ambiguous, with anomalies that tend to make them resemble those of the opposite sex. The anomalies vary through a spectrum from a mild form of hypospadias in the male to enlarged clitoris in the female, with many intermediate states and many degrees of severity. Such problems do not necessarily involve abnormalities of the sex chromosomes, but may be due to single-gene defects or environmental causes. However, determination of the child's karyotype is an essential part of the investigation of such patients.

Ambiguous genitalia or anomalies of the genitalia may also be seen in association with other dysmorphic features in a large number of malformation syndromes. In these syndromes the total pattern of malformation may be quite specific. One such condition is the autosomal recessive Smith-Lemli-Opitz syndrome, characterized by mental retardation, abnormal facies and skeletal anomalies and by cryptorchidism and hypospadias in males (Fig. 7–7).

Homosexuality, transvestitism and other sexual psychological variations do not as a rule have their basis in either chromosomal aberrations or single-gene defects.

TRUE HERMAPHRODITISM

True hermaphroditism is very rare. A true hermaphrodite has both testicular and ovarian tissue, either as two separate organs or as a single ovotestis, not necessarily functional but histologically identifiable. The internal and external sexual organs are very variable and not in any way diagnostic. Sex hormone studies are not helpful.

The great majority of hermaphrodites are XX, some are XY and others are

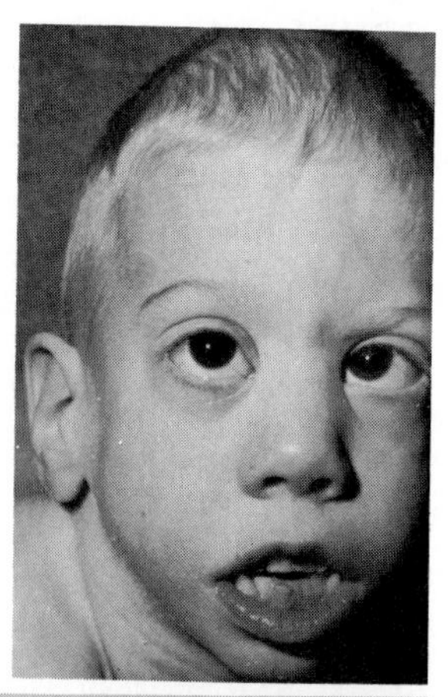

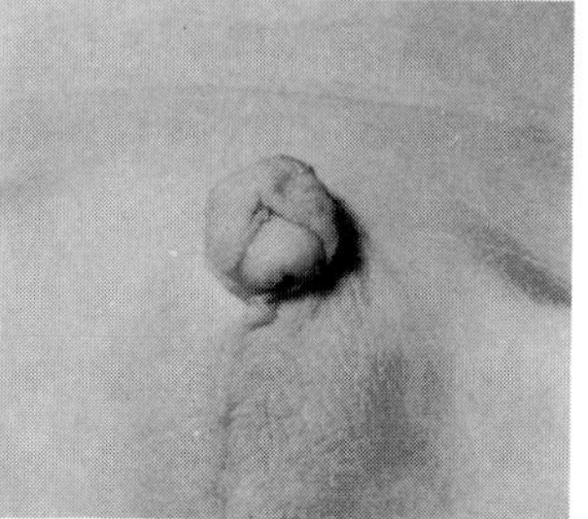

Figure 7–7. Smith-Lemli-Opitz syndrome. Facies and genitalia of a five-year-old patient. Failure to thrive and mental retardation are characteristic of this autosomal recessive syndrome. From Smith DW. Recognizable patterns of human malformation, 3rd ed. Philadelphia: W. B. Saunders Company, 1982.

not chromosomally normal but are XX/XY chimeras or mosaics. Both XX and the less common XY true hermaphroditism have occasionally been reported in sibs.

PSEUDOHERMAPHRODITISM

Pseudohermaphrodites, unlike true hermaphrodites, have gonadal tissue of only one sex. Male pseudohermaphrodites are XY and have testicular tissue (or occasionally have no detectable gonads); female pseudohermaphrodites are XX and have ovarian tissue. Both types may have ambiguous genitalia, in which the anomalies range from hypospadias or enlarged clitoris to variants in which assignment of sex is virtually impossible. The causes may be cytogenetic (usually 45,X/46,XY mosaicism), Mendelian or teratogenic.

Male Pseudohermaphroditism

Male pseudohermaphroditism has a number of different causes including dysgenesis of the gonads during embryological development, abnormalities of gonadotropins, inborn errors of testosterone biosynthesis and abnormalities of androgen target cells. These disorders are heterogeneous genetically as well as clinically. Only one will be discussed here: a syndrome of androgen insensitivity usually called testicular feminization.

Testicular feminization is an X-linked disorder in which an XY infant has female external genitalia, a blind vagina, and no uterus or uterine tubes. Testes are present either within the abdomen or in the inguinal canal, where they are sometimes mistaken for hernias. At puberty breast development occurs, though there is sparse pubic and axillary hair.

The defect in testicular feminization is end-organ unresponsiveness to androgens due to complete absence of androgen receptors in the cytosol of the appropriate target cells. The receptor protein which is specified by the normal allele at the locus (symbol *TFM*) has the role of complexing with testosterone or dihydrotestosterone. If the complex fails to form, the hormone cannot enter the nucleus, become attached to chromatin and stimulate the transcription of messenger RNAs. Ohno (1979) has called *TFM* a major sex-determining gene of man, "the master regulatory gene of the extragonadal sex-determining mechanism."

In addition to true testicular feminization, there are other, incomplete forms of androgen sensitivity. One of these, a partial form with external genital undermasculinization and pubertal subvirilization, may vary markedly in degree even within a kindred. In an even milder type the external genitalia are male and there is virilization at puberty, but the patients are infertile. Each clinical type is genetically heterogeneous, as shown by quantitative and qualitative analysis of androgen receptor-binding activity in cultured cells derived from genital skin of patients (Pinsky et al., 1984).

Female Pseudohermaphroditism

By far the most common cause of female pseudohermaphroditism is congenital adrenal hyperplasia. The incidence is about 1 in 25,000 births. Several distinct genetic and clinical forms are known, all inherited as autosomal recessives and each characterized by a block in a specific step in cortisol biosynthesis, resulting in increased secretion of ACTH (adrenocorticotropic hormone) and hyperplasia of the adrenal glands. This in turn leads to masculinization of female fetuses. Affected baby girls frequently have major anomalies of the external genitalia, often to the point that sex assignment may be impossible. The clitoris may be enlarged and the labia majora rugose and even fused. In males the same genotype produces premature virilization, but there is no difficulty in identifying the sex. In the 21-hydroxylase deficiency type of congenital adrenal hyperplasia, salt-losing crises may be very severe.

Because the gene for the 21-hydroxylase type is known to be on chromosome 6, within the HLA linkage group, there is a possibility of prenatal diagnosis by linkage analysis (see Chapter 11).

The external genitalia of a female fetus may also be masculinized if the fetal circulation contains excessive amounts of either male or female sex hormones. These hormones reach the fetal circulation from the maternal circulation, and may originate either endogenously or exogenously. Androgenic hormones may be present in excessive amounts if the mother's adrenal cortex is overactive or if she has received hormone therapy, for example, some progestins used to prevent spontaneous abortion.

A NOTE ON MANAGEMENT

The chromosomal sex is not the final indicator of the sex to which a patient with ambiguous genitalia should be assigned. Much more important factors in the decision are the presenting phenotype and the sex of rearing. If by medical or surgical intervention the patient can be allowed to lead a relatively normal though infertile life as a member of one sex, this is usually the best option. Altering the sex of rearing is usually unwise after early infancy.

The decision should be made as early as possible, both to prevent clinical problems that might arise immediately or later, and to avoid the severe psychological trauma of possible sex reversal at a later age. These disorders cause great concern and embarrassment to the family, and the concern is intensified if the disorder is one that might recur in a subsequent child. Thus these cases must be handled promptly and with tact and good judgment.

Hydatidiform Moles

Occasionally in an abnormal pregnancy the placenta is converted into a mass of tissue resembling a bunch of grapes or a hydatid cyst (Fig. 7–8). This is due to abnormal growth of the chorionic villi, in which the epithelium proliferates and the stroma undergoes cystic cavitation. Such an abnormality is called a mole. (The word mole simply means mass.) It may be complete, with no fetus or normal placenta present, or partial, with remnants of placenta and perhaps a small atrophic fetus. The two types differ in their origin.

In a complete mole, the chromosomes are all paternal in origin, the karyotype is almost always 46,XX, and with rare exceptions all the genetic markers are homozygous. The interpretation is that the mole originates when a single 23,X sperm fertilizes an egg that lacks a nucleus, and its chromosomes then double.

About half of all cases of choriocarcinoma (a malignancy of fetal, not maternal tissue) develop from hydatidiform moles. Given our present knowledge of the association of homozygosity with malignant change, for example in retinoblastoma, it may be that homozygosity for markers in hydatidiform moles has a bearing on their risk of malignant transformation.

Partial moles have been known for some time to be related to triploidy of the fetus, probably due to a double paternal contribution of chromosomes. The risk of malignancy in partial moles is still unknown.

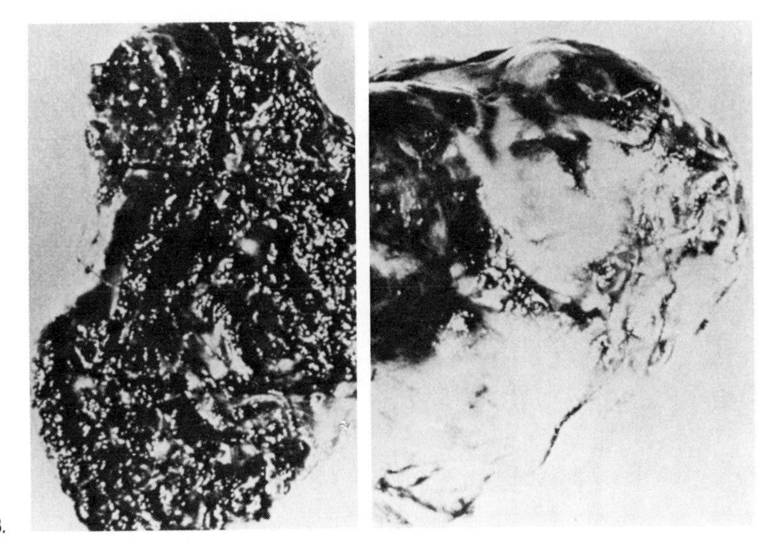

Fighure 7–8. Complete hydatidiform mole. Left, external view; right, internal view. From Stanhope CR, Stuart GCE, Curtis KL. Primary ovarian hydatidiform mole. Review of the literature and report of a case. Am J Obstet Gynecol 1983; 145:886–888.

GENERAL REFERENCES
See Chapter 6 General References also.

Fichman KR, Migeon BR, Migeon CJ. Genetic disorders of male sexual differentiation. Adv Hum Genet 1980; 10:333–371.

Gartler SM, Riggs AD. Mammalian X-chromosome inactivation. Ann Rev Genet 1983; 17:155–190.

Ohno S. Major sex-determining genes. New York: Springer-Verlag, 1979.

Pinsky L. The nosology of male pseudohermaphroditism due to androgen insensitivity. Birth Defects 1978; 14(6C):73–95.

Sandberg AA, ed. Cytogenetics of the mammalian X chromosome. Part B. X chromosome anomalies and their clinical manifestations. New York: Liss, 1983.

Simpson JL. Abnormal sexual differentiation in humans. Ann Rev Genet 1982; 16:193–224.

Stewart DR, ed. Children with sex chromosome aneuploidy: follow-up studies. Birth Defects 1982; 18(4).

PROBLEMS

1. A color-blind woman with Turner syndrome and a 45,X karyotype has a color-blind father. From which parent did she receive a chromosomally abnormal gamete?

2. If a man is color-blind and his 45,X daughter has normal color vision, can the source of the abnormal gamete be determined? Explain.

3. In a color-blind male with 47,XXY Klinefelter syndrome:
   a) What is the probable genotype?
   b) If his parents have normal color vision, what are their genotypes?
   In which parent and at which meiotic division did nondisjunction occur?

4. a) In a triple-X female, what types of gametes should one expect to find and in what proportions?
   b) If each type of gamete is equally viable and equally likely to result in a living child, what are the theoretical karyotypes and phenotypes of her progeny? (Actually, women who are XXX almost always have normal offspring only).

5. The inheritance of the Xg blood groups has been described in Chapter 4.
   a) An XXY, Xg(a+) male has an Xg(a−) mother. In which parent and at which meiotic division was the abnormal gamete formed?

b) An Xg(a+) man married to an Xg(a−) woman has an Xg(a−) son with an XXY karyotype. Which parent produced the abnormal gamete? At which meiotic division?

6. The birth incidence of XXY and XYY anomalies is roughly equal. Is this to be expected on the basis of theoretical considerations? Explain.

7. Describe in words and state whether the karyotype is balanced.
   a) 46,X,i(Xq)
   b) 46,X, −X, −21, +t(X;21) (p21; p12)

8

IMMUNOGENETICS

Immunogenetics is concerned with the genetic mechanisms underlying immunologic phenomena. To respond effectively to antigenic challenge, an individual must be able to distinguish between what is "self" and "nonself" and mount a specific defense against the nonself invader. Knowledge of the genetic background of these functions has expanded enormously in recent years; this chapter can do little more than introduce the basics of a number of topics that are relevant both to immunology and to medical genetics.

The human immune system is thought to have two chief components. One of these, involving the thymus gland, functions in cellular immunity. The other, often called the bursa system because in birds its role is performed by a separate organ called the bursa, is responsible for humoral immunity. The cell types concerned—**T cells** and **B cells**—derive from the same lymphoid stem cell population, but they differentiate along separate pathways, as shown in Figure 8–1.

The T ("thymus dependent") cells that are concerned with **cellular immunity** are the major population of peripheral blood lymphocytes, and are further differentiated into helper, suppressor and cytotoxic cell types. Their role is in defense against a variety of infections by intracellular organisms, in delayed-type skin reactions to common antigens and in rejection of grafts or malignant cells. The B ("bursa dependent") cells act in humoral immunity by synthesizing and secreting **immunoglobulins**, and are chiefly concerned with defense against extracellular invading organisms.

Response to an antigen—the **immune response**—involves several important reactions which will not be described in detail here. Initially the foreign substances, bacteria for example, are engulfed by macrophages and degraded. In this process the antigenic determinant sites of the invader are spared and are taken up by B lymphocytes. The B lymphocytes are thereby stimulated to proliferate and differentiate into **plasma cells**, the cells that secrete immunoglobulins. T cells interact with B cells to regulate antibody production.

In an individual's **primary response** to exposure to an antigen, it is a matter of days before measurable quantities of antibody are formed. The **secondary response** (a response to subsequent exposure to the same antigen) is much more rapid, more pronounced and of longer duration. This so-called anamnestic ("without forgetting") reaction happens because "memory cells" in the population of small B lymphocytes are already primed to respond vigorously to a second stimulus.

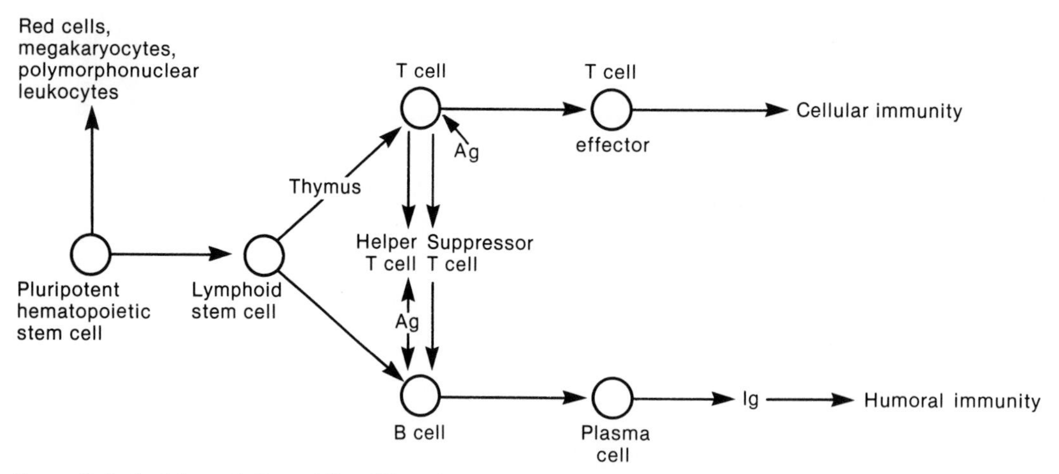

Figure 8–1. An interpretation of the differentiation of T and B cells from a common progenitor, and their interaction in the development of cellular and humoral immunity. Ag, antigen; Ig, immunoglobulin.

Immunoglobulins

Immunoglobulins constitute about 20 percent of the total plasma proteins. Most if not all immunoglobulins function as antibodies; the terms immunoglobulin and antibody are used interchangeably. The molecule is composed of two identical light chains and two identical heavy chains, held together by disulfide bonds (Fig. 8–2). There are two types of light chains, κ and λ, but only one kind of heavy chain in a single Ig class.

The five Ig classes, their percentage of the total Ig, their chain composition and their molecular weights are as follows:

Ig Class	Percent of Total	Light Chain	Heavy Chain	MW
IgG	80	κ or λ	γ (4 subclasses)	150,000
IgM	6	κ or λ	μ (2 subclasses)	900,000
IgA	13	κ or λ	α (2 subclasses)	160,000, 400,000
IgD	1	κ or λ	δ (2 subclasses)	180,000
IgE	low	κ or λ	ε	190,000

The differences in molecular weight shown above depend partly on the carbohydrate content of the molecules (which varies from 3 to 13 percent in different Ig classes), but chiefly on their unit structure. IgG, IgD and IgE are monomers, but IgM is a pentamer, composed of five basic units; IgA may have one, two or more basic units.

The various classes of immunoglobulins have different functions. IgG is largely responsible for combating infections. It is easily transported across the placenta to give passive immunity to the fetus, and makes its appearance late in the primary response. IgM precedes IgG in the immune response, is more effective than IgG in fixing complement and forms the rheumatoid factor in autoimmune disease. IgA's main role is in the defense of body surfaces and protection against microorganisms in the gut. So far the function of IgD is unclear, but it is thought to function as a receptor for antigen on the B cell membrane and thus as a trigger for B cell proliferation and differentiation. IgE is responsible for allergic reactions and the release of histamine; it may also combat intestinal parasites.

The immunoglobulin chains are folded into regions of homology called

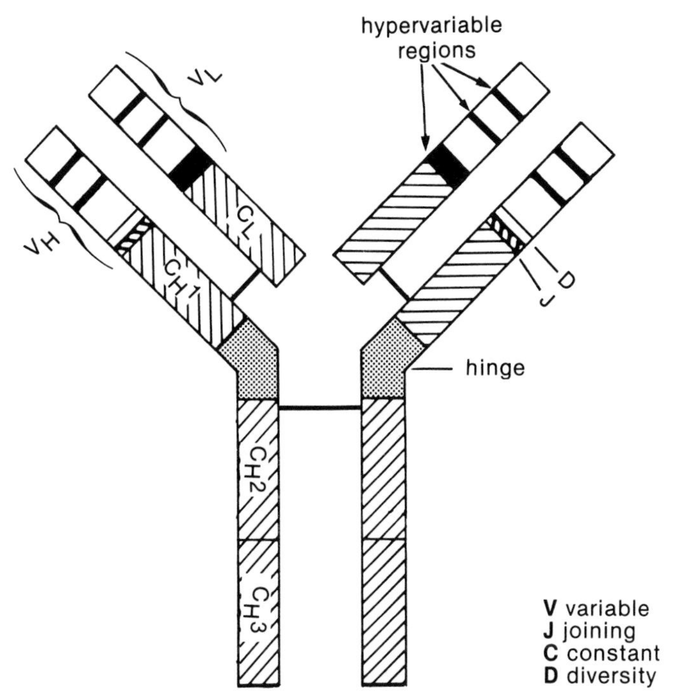

Figure 8–2. A representation of the immunoglobulin molecule. See text for details. Based on Leder P, Sci Amer 1982; 246:102–115, 1982.

domains, two in the light chains and four or five in the heavy chains. The domains of the light chains are designated the **variable** (V) and **constant** (C) regions, the V regions being extremely heterogeneous in amino acid sequence whereas the C regions of a single immunoglobulin class are alike. The heavy chains have a V region followed by three or four C regions.

The **antibody-combining site** (that is, the antibody specificity) of the Ig molecule is composed of parts of the variable region of both the light and the heavy chain, in particular the three hypervariable (HV) regions of each, where the amino acid sequence is especially changeable.

THE GENETIC BASIS OF ANTIBODY DIVERSITY

An essential and unique property of the immune response is its ability to generate an enormous number of different types of antibody molecules, differing in amino acid sequence; there may be up to 10^8 different kinds of antibodies. This great diversity appears to have evolved to protect vertebrates from a vast variety of environmental infections and toxic agents, as well as from malignant cells originating within the body. How so much specific genetic information could be encoded in the genome has been the major problem in immunogenetics. One of the first accomplishments of DNA technology has been to solve this problem, at least in broad outline.

Three separate **gene families** code for the immunoglobulin chains: the **H** (heavy chain) family on chromosome 14, the κ (kappa light chain) family on chromosome 2, and the λ (lambda light chain) family on chromosome 22. As an

CHROMOSOME 14

Figure 8–3. A scheme of the arrangement of the genes on chromosome 14 that encode an immunoglobulin heavy chain. See text for details.

exception to the familiar "one gene, one polypeptide chain" rule, the variable and constant regions of each chain are coded by separate genes which are brought into proximity before being transcribed into messenger RNA (Fig. 8–3). For example, in the biosynthesis of a heavy chain during the differentiation of a B lymphocyte into an antibody-secreting plasma cell, one of 100 to 300 V region heavy chain (V_H) genes is joined to one member of a small cluster of heavy chain C region (C_H) genes. Similarly, one of 100 to 300 V region light chain (V_L) genes is linked to the C region of one of the two types of light chain genes (C_L), each of which is present in a single copy on its chromosome.

Between the V_L and C_L regions, just upstream of the C_L gene, there is a series of four **J** (for joining) "minigenes," only one of which will be used in joining the V_L and C_L regions. In heavy chain formation, an additional **D** (for diversity) segment, coded by one of 12 or more D minigenes, is situated between the V_H and J_HC_H segments.

The multiplicity of genes in the germ line that can be combined to code for an immunoglobulin molecule allows for a very large number of antibodies differing in amino acid sequence. Even more diversity is allowed for by the fact that the VJ or VDJ combinations may occur at slightly different sites and that one or more nucleotides may be inserted, apparently only into the heavy chain gene, between V_H and D_H or between D_H and J_H. Moreover, somatic mutation is frequent in the V region, thus further increasing antibody diversity.

Within immunoglobulin genes the functional domains are coded by DNA sequences separated by introns. For example, in C_H the four domains (C_H1, C_H2, C_H3 and the hinge region, as shown in Figure 8–3) are individually coded by exons, which are separated by introns that are spliced out during messenger RNA maturation. This demonstrates a general principle of exon-intron arrangement within genes.

Allelic exclusion is a unique feature of light chain production. Although the cell has two sets of κ genes and two of λ genes, it synthesizes only one type of light chain.

HYBRIDOMAS AND MONOCLONAL ANTIBODIES

The study of the basis of antibody diversity at the molecular genetic level has been aided by the use of hybridomas, which are made by fusing mutant mouse myeloma cells that will no longer make antibody to normal spleen cells from mice immunized with a particular antigen (see Cell Hybridization, Chapter 10). Spleen cells cannot grow continuously in culture, but myeloma cells will grow in a suitable culture medium. When the two cell types are mixed in appropriate conditions, some cells fuse to form hybrids or hybridomas, which will grow in long-term culture, whereas the parent cells die. A hybridoma secretes a single antibody of a single immunoglobulin class, specific for the antigen originally used to immunize the mouse. The ability to obtain large quantities of a pure (**monoclonal**) antibody has led to fruitful advances in our understanding of antibodies and their genetic determinants, and has also improved our understanding of differentiation antigens and histocompatibility antigens.

Human lymphoblastoid cell lines (described further in Chapter 10) have been fused to mouse myeloma cells to form hybridomas that secrete human immunoglobulins. It was through the use of these hybridomas that mapping of the human genes for the κ and λ light chains and the μ, γ and α heavy chains was accomplished.

Histocompatibility

In the earliest attempts at transplantation, the grafted tissue survived for only a few days; the only exception was that grafts between monozygotic twins were usually accepted. In other words, grafts from genetically different sources are subject to the same reactions against "nonself" as are invading organisms or malignant cells. Today progress has been made in understanding the genetic basis of graft rejection and means for circumventing it. Organ transplantation has become an important area of medicine.

The genetic system primarily responsible for distinguishing between self and nonself is the **major histocompatibility complex** (MHC). The ABO blood group system is also important in transplantation; as a general rule, grafts are rejected unless donor and host are matched with respect to both systems.

GRAFT TERMINOLOGY

There is a standard nomenclature for different types of grafts. An **autograft** is a graft of the host's own tissue. An **isograft** (isogeneic graft) is a graft between genetically identical individuals, that is, MZ twins, members of an inbred strain, or the F_1 of a cross between two inbred strains. Both autografts and isografts are normally accepted. An **allograft** (allogeneic graft) is a graft between two genetically different members of a species, and is rejected unless special precautions are taken. A **xenograft** is a graft between members of different species (for example, from ape to man); this type of graft is rejected more rapidly than an allograft. Human grafts, except for autografts or those between MZ twins, are allografts.

Autografts of Cultured Epithelium

An autograft is the tissue of choice for transplantation, but tissue for autografts is often unavailable or not present in sufficient quantity. Recently the development of a culture technique for generating large sheets of epithelial tissue from small skin biopsies has proved to be a successful method of obtaining skin autografts. Skin biopsies 2 cm^2 in size can be expanded in surface area by a factor of 10,000 in less than a month. In severely burned children, grafts cultured from a child's own skin have been used to generate permanent epidermis over much of the body surface (Gallico et al, 1984).

THE MAJOR HISTOCOMPATIBILITY COMPLEX

The **HLA** (human leucocyte antigen) complex is the major histocompatibility complex in man. It is homologous to the H–2 locus of the mouse, the system in which the principles of histocompatibility were originally revealed.

The HLA complex (Fig. 8–4) has been localized to a specific region on chromosome 6p. It contains genes for histocompatibility antigens and for complement components, and presumably also contains *Ir* (immune response) genes,

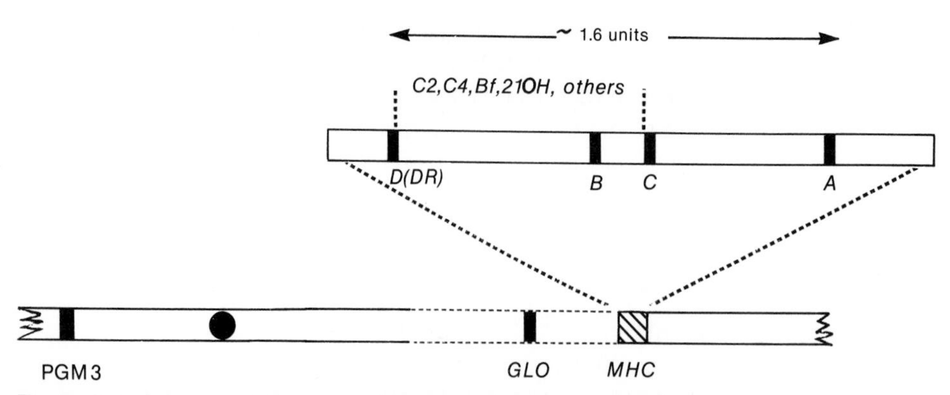

Figure 8–4. Scheme of the major histocompatibility complex *(MHC)* on the short arm of chromosome 6. The relative positions of HLA-A, HLA-B, HLA-C and HLA-D *(DR)*, and the approximate loci for the second and fourth components of complement *(C2, C4)*, the properdin factor B *(Bf)* and 21-hydroxylase (deficient in a form of congenital adrenal hyperplasia) are shown. Centromere, black circle. *GLO*, glyoxalase locus. PGM3, a phosphoglucomutase locus.

which have been identified in the mouse. The loci for at least one polymorphic enzyme, glyoxalase (GLO), and two genetic disorders, the 21–hydroxylase deficiency form of congenital adrenal hyperplasia and oligopontocerebellar ataxia, also lie within or near the complex.

The five known HLA loci are designated HLA-A, HLA-B, HLA-C, HLA-D and HLA-DR ("D-related"). As shown in Table 8–1, each has numerous alleles (currently, at least 20 for HLA-A, over 40 for HLA-B, and 8 or more for each of the other three). The "complotype" of alleles for four complement components mentioned later in this chapter may also be considered as part of the complex. The set of HLA genes on one chromosome constitutes a **haplotype**; thus an individual has two haplotypes, each with five determinants. HLA genes are so closely linked that they move as a unit (except for rare recombinants, which originate by crossing over with a frequency of 1 to 2 percent). Thus there is one chance in four that two sibs will have identical HLA haplotypes (Fig. 8–5).

The great number of alleles at each of the five HLA loci allows for an enormous variety of HLA haplotypes. Because the HLA complex is so highly polymorphic, it is the most useful single genetic system for population studies and paternity testing.

Table 8–1. EXAMPLES OF DEFINED ANTIGENS OF THE HLA SYSTEM

HLA–A	HLA–B	HLA–C	HLA–D	HLA–DR
At least 20, including:	Over 40, including:			
A1	B5	8, designated Cw1–Cw8	12, designated Dw1–Dw12	10, designated DR1–DR10
A2	B7			
A3	B8			
A9	B12			
A10	B13			
A11	B14			
A28	B15			
Others	B17			
	B18			
	B27			
	B37			
	B40			
	Others			

Data from Kostyu DD, Reisner EG. Human leukocyte and platelet antigens and antibodies. In Williams WJ et al., eds. Hematology, 3rd ed. New York: McGraw-Hill, 1983.
w = "workshop," that is, a specificity that is less thoroughly understood.

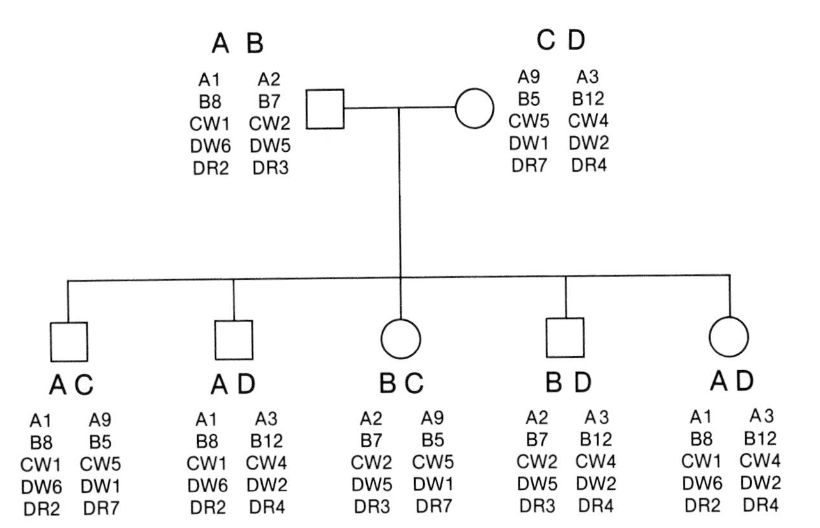

Figure 8–5. The inheritance of HLA haplotypes. A haplotype consists of five loci, each with different antigenic specificities. A haplotype is transmitted as a unit (indicated here by large letters); thus there is a one-in-four probability that two sibs will match exactly.

Immunologists classify the products of the genes in the HLA complex as follows:

Class I products: HLA-A, B and C antigens
Class II products: D and DR antigens (B-cell antigens)
Class III products: complement components (C2, C4, properdin factor B)

Linkage Disequilibrium

Ordinarily one expects that alleles at linked loci will be in equilibrium; that is, that the proportions of combinations of alleles will be the product of their population frequencies. For example, HLA-A1 has a frequency of 0.17 and HLA-B8 has a frequency of 0.11. If the two are in linkage equilibrium, the expected frequency of the combination A1 B8 is $0.17 \times 0.11 = 0.019$, which is far below the observed frequency of 0.088. This is an example of linkage disequilibrium, which is a striking characteristic of certain HLA alleles. It has been found that linkage disequilibrium exists over the complete distance from HLA-A to GLO, a distance of five or more recombination units.

The cause of linkage disequilibrium has been the subject of much discussion but is still unclear. It could be due to natural selection for certain combinations, or simply to genetically different populations too recently mixed for equilibrium to have been achieved. One theory is that linkage disequilibrium results from selection, not for specific combinations of HLA alleles, but for other closely linked loci, such as the "immune response" genes that are known in the mouse and presumed to exist in man as well. (For further discussion, see Chapter 11.)

HLA AND TRANSPLANTATION

The principles of transplantation genetics have been uncovered largely through experimental work on inbred mice. Members of an inbred strain, some of which have been developed by brother-sister mating for well over 100 generations, are virtually identical (isogenic) and consequently do not reject tissue from animals of the same strain.

Tissue from a mouse of an inbred strain can, with rare exceptions, be successfully implanted into the F_1 of a cross between a mouse of the donor strain and one of another inbred strain. An inbred mouse is homozygous at the H–2 loci and all other loci (except for rare mutations), so its progeny receive one copy of each of its genes. Thus an F_1 hybrid does not regard tissue from either parent as foreign and does not mount an immune attack against it. However, since the tissue of the hybrid has antigens from each parent it will not be accepted by either parent strain.

Cross-Matching Tests

Matching tests are performed in vitro to minimize the chances of rejection of the graft by the recipient. There are two general types of tests:

1. **Mixed lymphocyte culture (MLC) test.** The principle of this test is that foreign antigens have a mitogenic (mitosis-stimulating) effect on lymphocytes. Since not all histocompatibility differences can yet be identified by tissue typing, the MLC test is used to determine directly the presence and degree of antigenic differences between donor and host.
2. **Leucocyte typing by lymphocytotoxicity testing.** This test depends on the ability of living lymphocytes to remain impervious to certain dyes such as trypan blue that stain dead lymphocytes. An antiserum to an antigen borne by a sample of lymphocytes will, in the presence of complement, kill the cells. Antisera are prepared from women who have been pregnant or from recipients of blood transfusions. Lymphocytes of the host and each of the prospective donors are incubated in a battery of antisera, after the addition of complement. Stain is then added and the number of dead lymphocytes determined. If host and prospective donor lymphocytes react differently to any antiserum, they differ in one or more antigens.

Graft Rejection

Both B cells and T cells are involved in graft rejection, the relative roles of immunoglobulins and cellular immunity varying according to the circumstances. The primary response to an allograft chiefly involves the activation of cellular immunity by T lymphocytes, which attack foreign cells both directly and by the production of extracellular lymphokines. The graft may at first appear to be accepted, but during this time sensitization develops; within 10 to 20 days the graft dies and is sloughed.

After rejection of one implant the host rejects a second graft from the same source even more rapidly, by means of a secondary response brought about by the primed antigen-sensitive cells (memory cells) of the T cell series.

Acceptance of allografts can be prolonged by **immunosuppression**, that is, by treatment with agents that will suppress the immune response.

GRAFT-VERSUS-HOST REACTION

Not only does the host react against antigenically different donor cells, but the donor cells, if they are immunocompetent, also react against the antigens of the host. This is the graft-versus-host (GVH) reaction. For many years the GVH reaction prevented marrow transplantation because the donor tissue, containing immunocompetent cells, reacted to the antigens of the host with serious and often fatal consequences. Improvements in techniques of donor selection and immunosuppression now allow marrow grafts to be used therapeutically in certain immune disorders, in aplastic anemias and even in leukemias after destruction of the marrow cells of the host by radiation.

If a GVH reaction occurs in a young animal, it may not be fatal but will inhibit growth and render the host weak and sickly; this is **runt disease**.

DEVELOPMENT OF IMMUNE COMPETENCE

Tissue antigens form throughout fetal development, but the immune response does not develop until shortly before birth. By this time the self antigens that might cause an immune response are normally present, so do not elicit an immune response. Tolerance to one's own antigens is **immunological homeostasis**. Indeed, a fetus or in some species a newborn will accept cells from a genetically different donor and allow them to persist and proliferate. The host then has lifelong tolerance to tissue from that donor. This phenomenon, which is known as **acquired tolerance**, might eventually be exploited as a means of overcoming problems of graft rejection.

A few cases are known of human DZ twins who have exchanged hematopoietic stem cells in utero, and thus have a population of blood cells derived from the co-twin as well as their own. Such a person, known as a **blood-group chimera,** will accept tissue from the co-twin.

An exception to immunological homeostasis is **autoimmunity**, in which antibodies are formed against one's own antigens.

THE FETUS AS ALLOGRAFT

A human mother rejects tissue from her child, and an inbred female mouse rejects tissue from her offspring unless the father is of the same strain, though in both cases rejection may take longer than would otherwise be expected. However, the trophoblast (fetal tissue), which certainly acts as a graft on the mother, is not rejected. The mechanism that prevents rejection of the fetus is still unknown.

Fetal blood cells enter the maternal circulation, allowing the mother an opportunity to make antibodies or T effector cells, or both, directed against fetal tissue. Some IgG antibodies cross the placental barrier from mother to fetus, but they do not appear to harm the fetus except in cases of blood group incompatibility (described in Chapter 9); some give the newborn **passive immunity** to infections against which the mother has circulating antibodies.

The depression of the maternal immune response to fetal tissue may be caused by blocking factors, particularly chorionic gonadotropin, which has been shown to inhibit the mixed lymphocyte reaction. Alpha-fetoprotein, a glycoprotein present in large amounts in the fetus and placenta (and in excessive amounts if the fetus has a neural tube defect or certain other malformations), may also inhibit the immune response.

It may also be that trophoblast cells have relatively weak tissue antigens. Fetal antigens appear to develop very slowly as the products of conception grow; this slow development may allow gradual reduction of the function of the maternal immune system for these particular antigens.

Complement

After formation of an antigen-antibody complex, the inactivation of foreign material requires the collaboration of a group of serum proteins known collectively as complement. The complement system is the primary humoral mediator of antigen-antibody reactions. Its range of activities includes direct lysis of target cells, opsonization (facilitation of the destruction of foreign material, such as bacteria) and activation of other components of the immune response, for example, macrophages.

The complement system has numerous components that are activated sequentially as a "cascade," and is controlled by a number of regulatory proteins. The process will not be described in detail here. Among the components, several are polymorphic: C3 (the third component), C6, C7, the two subunits of C8, Bf and the two alpha chains of C4. Genes at the C4 locus determine the Rodgers (Rg) antigens and C4B locus genes determine the Chido (Ch) antigens; in other words, Rodgers and Chido are antigens on C4 molecules in plasma and become attached to red cells only when complement is activated.

Four complement loci—Bf, C2, C4A and C4B—are very closely linked to one another, and form part of the major histocompatibility complex (MHC) on chromosome 6 (see Figure 8–4). The set of four complement genes present on a single chromosome form a **complotype**. C3 is a well-known marker on chromosome 19, in close linkage with the Lewis, secretor, Lutheran and myotonic dystrophy loci.

There are a number of genetic disorders associated with deficiency of specific complement components. Perhaps the best known is deficiency of the inhibitor of the activated C1 component, which causes the autosomal dominant disorder hereditary angioneurotic edema.

Genetic Disorders of the Immune System

The disorders of the immune system involve genetic defects of several kinds that impair defense mechanisms against infection. These mechanisms include (1) disorders of functions of phagocytes, which normally recognize, ingest and destroy foreign particles; (2) disorders of cellular or humoral immunity, and (3) deficiencies of specific complement components, which have been mentioned briefly and will not be discussed further. The following examples will indicate the variety of etiology of immunodeficiency disorders.

Chronic granulomatous disease is an X-linked disorder of phagocytes in which phagocytosis of microorganisms proceeds normally but bactericidal activity is deficient. The molecular basis (or bases) is not yet known. A peculiar feature of many but not all cases is an association with mutation at the Xk locus, which normally specifies a precursor of the Kell blood group antigens; it is not clear whether the association is functional or results from deletion or other disorganization of two closely linked genes.

Severe combined immunodeficiency disease (SCID) is the most severe of the immunodeficiency diseases. Both cellular and humoral immunity are profoundly defective. Genetically SCID is heterogeneous; most cases are X-linked (and are described as Swiss-type agammaglobulinemia), but some are autosomal recessive. Among the latter is the type with inherited absence of the enzyme adenosine deaminase (ADA), the first molecular defect to be recognized in any immunodeficiency disorder (Giblett et al, 1972). About half of all cases of autosomal recessive SCID are thought to involve ADA deficiency. Deficiency of nucleoside phosphorylase (NP) or inosine phosphorylase, other enzymes in the pathway of purine metabolism, can also cause autosomal recessive forms of SCID. In NP deficiency, T cell function is severely impaired, but B cell function seems unaffected.

Bruton-type agammaglobulinemia is the classical example of B cell deficiency. Onset is typically at 6 to 12 months of age. Immunoglobulins are virtually absent, and the affected children are highly susceptible to pneumococcal and streptococcal infections, though not to viral infections. The B cells and plasma cells are absent, and the tonsils are very small. There are nearly normal numbers of circulating lymphocytes, but these are T cells; the thymus is normal, and grafts are rejected. The condition is X-linked.

Several disorders involving other organ systems have immunodeficiency as an associated manifestation. **Ataxia-telangiectasia** (AT), previously mentioned as a chromosome breakage syndrome, is an autosomal recessive condition marked by the development of cerebellar ataxia and of telangiectasia, which are most readily seen in the sclerae. Other features of AT include severe immunodeficiency, chromosomal instability and a notable increase in the risk of malignancy, which apparently applies to otherwise unaffected heterozygotes as well as to homozygotes. **Wiskott-Aldrich syndrome** is X-linked, characterized by thrombocytopenia, eczema and susceptibility to many infections, and associated with complex immune deficiencies, both cellular and humoral. At least three distinct syndromes of short-limbed dwarfism with immunodeficiency exist.

A noteworthy feature of immune deficiency diseases as a group is that they include a large number of X-linked forms, only a few of which are mentioned here. It has been proposed that a set of related X-linked genes may determine the sequential differentiation from primitive stem cells to the functional elements of the immune system, and that different mutations result in the different X-linked immune disorders (Hirschhorn and Hirschhorn, 1983). Molecular analysis may soon help to clarify the current incomplete picture.

In immune deficiency diseases autoimmune phenomena are unusually frequent, showing that defective self-recognition can underlie autoimmunity, as one would expect. Many associations of autoimmune disorders with specific HLA types have also been recognized (see next section).

HLA and Disease Associations

Ankylosing spondylitis is a joint disease, chiefly affecting the sacroiliac joints, in which inflammation causes ossification of the ligaments, eventually leading to fusion (ankylosis). In 90 percent of persons with this disease the HLA antigen B27 is present, though in the general population the frequency of B27 is only 8 percent. This is one of the most striking disease associations known.

In recent years there have been hundreds of reports of an increased frequency of one or more of the HLA antigens in association with a specific disease. Not all the reports have been reliable, but an appreciable number have been confirmed. The relative risk (defined as the ratio of the risk of developing the disease in those with the antigen to the risk in those without it) is about a hundredfold in ankylosing spondylitis, but is much lower in most other conditions showing positive associations.

The associations vary in specificity as well as in strength, some diseases showing association with more than one antigen. In general, the kinds of disorders for which positive associations exist show familial clustering, but do not have a clearly Mendelian pattern of inheritance. Most are regarded as immune-related. Some of the well-known associations are shown in Table 8–2.

DIABETES MELLITUS

The genetic basis of diabetes mellitus is still unclear, even though the disease is known to run in families and is very common, with a prevalence of about 1 percent in European populations.

The two major forms of diabetes mellitus, the insulin-dependent juvenile-onset form (IDDM) and the non-insulin-dependent maturity-onset form (NIDDM), appear to be genetically different. Family studies suggest that each type "breeds true," and studies in monozygotic twin pairs show that only about 50 percent

Table 8–2. EXAMPLES OF ASSOCIATIONS BETWEEN HLA ANTIGENS AND DISEASE

Disease	Antigen		Frequency of Antigen (%) Patients	Controls
Ankylosing spondylitis	B27		90	8
Celiac disease	DR3(B8)		96	27
Diabetes mellitus,	DR3, B8		50	21
insulin-dependent	DR4, B15		38	13
Graves disease	DR3(B8)		53	18
Multiple sclerosis	DR2(B7)		55	23
Myasthenia gravis	DR3(B8)		30	17
Psoriasis	Cw6	Caucasian	50	23
		Japanese	53	7
Reiter syndrome	B27		80	9
Systemic lupus erythematosus	DR3(B8)		70	28

Chiefly from Kostyu DD, Reisner EG. Human leukocyte and platelet antigens and antibodies. In Williams WJ et al., eds. Hematology, 3rd ed. New York: McGraw-Hill, 1983.

are concordant with respect to IDDM, whereas close to 100 percent are concordant for NIDDM. It is also clear that only the IDDM type shows HLA associations. The HLA antigens B8, B15 and D3, DR3, D4 and DR4 are all associated with an increased risk for IDDM, and the combination of B8 and B15 or DR3 and DR4 puts the individual at substantial extra risk.

Recurrence risk figures for genetic counseling in IDDM and NIDDM are unreliable, and will remain so until the genetic heterogeneity of diabetes is much better understood. The rare form called "maturity onset diabetes of young people" (MODY) is clearly autosomal dominant. There are also numerous genetic syndromes in which diabetes occurs for which accurate recurrence risks can be determined from the pattern of inheritance. For IDDM and NIDDM, empiric risk figures must suffice at present. Typical risks estimated by Zavala and colleagues (1979) for two common situations are as follows:

1 parent with IDDM—Risk to child 3.5 percent

Neither parent affected, one child with IDDM—Risk to 2nd child 4.4 percent

Genetic counseling in diabetes must also take into account the special hazards to offspring of diabetic mothers. Infants born to diabetic mothers are large, obese and bloated, and have higher than average perinatal mortality. The risk of major congenital malformation is approximately tripled overall, to about 6 to 9 percent. The increase is chiefly accounted for by the risk of cardiac and cardiovascular problems and of neural tube defects. A rare kind of malformation, the so-called caudal regression syndrome of sacral agenesis and lower limb defects, seems to occur chiefly in infants of diabetic mothers but has a low recurrence risk.

Recent reports indicate that good diabetic control during very early pregnancy is likely to reduce the risk of major congenital malformations in infants of diabetic mothers. A minor hemoglobin constituent known as HbA1c represents normal HbA that has had a glucose added to the amino-terminal valine of the β chain during red cell circulation. It therefore provides a retrospective measure of the adequacy of glucose control in the previous 4 to 8 weeks. The level of HbA1c in the first trimester is significantly higher in mothers who later deliver infants with major anomalies than in those who deliver normal infants (Miller et al, 1981). Elevated HbA1c in the first trimester is an indication of problems that can in some cases be identified by ultrasonography later in the pregnancy.

Table 8–3. EXAMPLES OF AUTOIMMUNE DISEASES

	Antibodies
Systemic	
Rheumatoid arthritis	Antibody to IgG
Scleroderma	Antibody to chromosomal centromeres
Systemic lupus erythematosus	Antibodies to DNA and many other autoantibodies
Organ-Specific	
Addison disease	Antibody to adrenal cells
Diabetes mellitus, juvenile insulin-dependent*	Antibody to islet cells
Goodpasture syndrome	Antibody to kidney basement membrane
Hashimoto thyroiditis	Antibody to thyroglobulin
Myasthenia gravis	Antibody to acetylcholine receptor
Pernicious anemia	Antibody to vitamin B_{12} binding site of intrinsic factor
Autoimmune hemolytic anemia	Antibody to red blood cells
Idiopathic thrombocytopenic purpura	Antibody to platelets

*Diabetes mellitus is heterogeneous and the extent of the role of autoimmunity in its etiology is still undetermined.

Autoimmune Diseases

Autoimmunity, as the name implies, is immune response to self-antigens and thus represents a major exception to the rule of immunological tolerance to self. Normally a state of immunological homeostasis exists, but this can be disturbed on either side of the antigen-antibody relationship; an antigen not previously present may elicit antibody formation, or an antibody capable of self-sensitization may be formed. Clearly, self-tolerance is not absolute.

Initiation of the autoimmune process can happen through a variety of mechanisms, including genetic defects in various components of the immune system itself. The chief genes that confer susceptibility or resistance to autoimmune diseases appear to be HLA alleles, but genes for deficiency of certain complement components and immunoglobulin genes also appear to be associated with particular autoimmune diseases.

Table 8–3 lists some of the many known autoimmune diseases. As a group these disorders seem to share a weak degree of association with particular DR alleles within the HLA complex. Note that some are systemic, others organ-specific. Most are diseases of later life. Patients suffering from one autoimmune disease may simultaneously develop another type, for example, patients with Hashimoto thyroiditis may develop pernicious anemia, and 70 percent of patients with autoimmune hemolytic anemia develop other autoimmune diseases.

Some autoimmune disorders, the so-called collagen diseases, produce distinctive pathological changes in connective tissue.

GENERAL REFERENCES

There are numerous excellent references to the topics discussed in Chapter 8. The following are samples and sources of additional references.

Amos DB, Kostyu DD. HLA—a central immunological agency of man. Adv Hum Genet 1980; 10:137–208.

Barrett JT. Textbook of immunology: an introduction to immunochemistry and immunobiology, 4th ed. St. Louis: C. V. Mosby, 1983.

Ellison JW, Hood LE, Human antibody genes: evolutionary and medical genetic perspectives. Adv Hum Genet 1983; 13:113–147.
Leder P. The genetics of antibody diversity. Sci Am 1982; 246:102–115.
Snell GD, Dausset J, Nathenson S. Histocompatibility. New York: Academic Press, 1976.
Stites DP, Stobo JD, Fudenberg HH, Wells JV, eds. Basic and clinical immunology, 4th ed. Los Altos, California: Lange Medical Publications, 1982.
Williams WJ, Beutler E, Erslev AJ, Lichtman MA, eds. Hematology, 3rd ed. New York: McGraw-Hill, 1983.

PROBLEMS

1. Arrange the following relatives of a transplant recipient in order of their suitability as donors: sib, father, MZ twin, unrelated person, cousin, mother, DZ twin.

2. Table 8–2 gives examples of associations between HLA antigens and disease. The relative risk of a disease in a person with a given antigen is ad/bc, where a is the number of patients with a given antigen, b is the number of controls with that antigen, c is the number of patients without it and d is the number of controls without it. Calculate the relative risk in the following situations:
   a) Diabetes mellitus and B8
      $a = 48, b = 590, c = 59, d = 1900$
   b) Ankylosing spondylitis and B27
      $a = 57, b = 160, c = 6, d = 1837$
   c) Celiac disease and DR3
      $a = 30, b = 275, c = 4, d = 750$

9

BLOOD GROUPS AND SOME OTHER POLYMORPHISMS IN BLOOD

Numerous polymorphisms are known in the components of human blood, especially in the antigens of red blood cells. Because of their ready classification into different phenotypes, simple mode of inheritance and different frequencies in different populations, red cell antigens (blood groups) are useful genetic markers in family and population studies and in linkage analysis. Some, especially the ABO and Rh blood group systems, have clinical importance. Although all the genetic polymorphisms of blood, including the HLA system (discussed in the previous chapter), might appropriately be called blood groups, in general usage the term is restricted to the red cell antigens.

The usefulness of any genetic trait as a genetic marker depends on the following characteristics:

1. A simple and unequivocal pattern of inheritance. Most of the blood groups, enzyme variants, serum protein variants and DNA polymorphisms are determined by systems of codominant alleles; that is, in heterozygotes the products of both genes are detectable, permitting the genotype to be inferred directly from the phenotype. (There are important exceptions such as the *O* gene of the ABO blood group system, in which the gene product is enzymatically inactive.)

2. Accurate identification of each phenotype.

3. A relatively high frequency of each of the common alleles at the locus. If the rarer of a pair of alleles is encountered in only a small percentage of a population, the locus is not likely to be very useful for family studies and linkage analysis. In two-allele systems equal frequency of the two alleles is the ideal, since this situation provides the best opportunity for segregation to be observed at the locus. The presence of several or many fairly common alleles, as in the HLA system, is even more useful. Recently, a method of evaluating the polymorphism information content (PIC) of any marker locus has been developed (see Chapter 11).

4. Absence of effect of environmental factors, age, interaction with other genes or other variables on the expression of the trait.

The study of the blood group systems has made many important contributions to human genetics. In man, multiple allelism at a locus was first demonstrated by the ABO blood group genes. The first four autosomal linkages to be found all involved blood groups. The X-linked blood group system Xg is a standard point of reference on the X chromosome. Xg has also been useful in investigation of X inactivation and sex chromosome aneuploidy. Population genetics has made extensive use of the blood groups. Interaction of nonallelic genes is demonstrated by the *ABO-H-secretor-Lewis* relationship and by other biosynthetic pathways less thoroughly explored so far. The long debate over whether the antigens of the Rh system are determined by different closely linked alleles or are different antigenic expressions of a single allele illustrates the difficulty of deciding between these possibilities, and has provided insight into the nature of genes, although in the absence of molecular analysis the problem is still unsolved.

Blood Groups

The human blood groups are important in blood transfusion, tissue transplantation and hemolytic disease of the newborn. Whether any particular blood group system is clinically important depends chiefly on whether antibodies of the system can shorten the survival of red cells in vivo (in a fetus or in a transfused patient), and how common such antibodies are. Some blood group antigens are important to the structure and function of cell membranes, and some antigens are known to be affected by some diseases.

Blood groups are particularly important in transfusion therapy. For all patients, the blood donors must be ABO-compatible. For a patient previously immunized by pregnancy or transfusion, the laboratory must select donors whose red cells are most likely to survive after transfusion to that patient. All Rh-negative women of childbearing age, whether immunized to Rh or not, must receive only Rh-negative blood.

In tissue transplantation, ABO compatibility of donor and recipient is essential to graft survival.

Hemolytic disease of the newborn occurs when antibodies formed by a pregnant woman in response to antigens of her fetus or of a blood donor (usually, though not always, antigens of the Rh system) are transferred via the placenta to the fetus (see later in this chapter). At one time hemolytic disease of the newborn ranked among the first 10 causes of perinatal death, and led to physical or mental handicap in surviving children who were inadequately treated. The discovery in 1963—by Clarke and his associates in England and Freda and his associates in the United States—of a means of suppressing the formation of anti-Rh during pregnancy was a significant medical advance.

TERMINOLOGY

Although the terminology used for blood groups and blood group genes is somewhat inconsistent, it is so firmly established that it has remained unchanged even though in recent years many other marker symbols have been revised and standardized to accommodate the rapidly increasing amount of information and the requirements of computerization. Three chief kinds of notation are used for alleles in a blood group system:

1. Letter sequences (*A* and *B*, *M* and *N*)
2. Large and small letters (*S* and *s*, *K* and *k*, *C* and *c*). Note that here the use of a lower-case letter does not imply that the gene is recessive.
3. Alleles designated by a symbol with a superscript (Lu^a and Lu^b, Fy^a and Fy^b).

Some blood group systems are named for the person in whom the antibody was first recognized (for example, Duffy and Kidd), and some part of this name is used to designate the gene locus concerned (Fy for Duffy, Jk for baby J. Kidd).

Table 9–1. SOME HUMAN BLOOD GROUP SYSTEMS

System	Year of Discovery	Comments
ABO	1900	Major clinical importance
MNSs	1927	Useful marker; little clinical importance
P	1927	
Secretor	1930	Determines presence of ABH antigens in secretions
Rh	1940	Major clinical importance
Lutheran	1945	Lutheran-secretor was the first autosomal linkage known
Kell	1946	Clinical importance
Lewis	1946	Related to ABO; see text
Duffy	1950	First to be assigned to a chromosome
Kidd	1951	Clinical importance
Diego	1955	Oriental marker
Cartwright	1956	
Auberger	1961	
Xg	1962	X-linked
Dombrock	1965	
Colton	1967	
Sid	1967	
Scianna	1974	
Complement C4	1967 1976	After activation of complement, red cells carry C4 antigens (Chido and Rodgers)

Data chiefly from Race and Sanger, 1975.

The Lutheran system, on the other hand, was named for the donor of the provoking antigen.

Most of the well-known blood group systems and the honorary blood group system secretor are listed in Table 9–1. The first example of the antibody defining each system was found as follows: naturally occurring in healthy subjects (ABO, Lewis); in immunized animals (MN, P, Rh); in the mothers of infants with hemolytic disease of the newborn (Rh, Kell, Kidd, Diego); or during cross-matching tests (Duffy). The list does not include several systems whose independent rank is uncertain at present, and other antigens that are very common (public) or very rare (private). It is hard to know whether a public or private antigen belongs to some known blood group system or is actually part of a new system, but in either case its frequency detracts from its usefulness as a genetic marker because so few families show segregation at the locus concerned.

THE ABO SYSTEM

The first notable event in blood group history was the discovery of the ABO blood groups by Karl Landsteiner and his pupils at the University of Vienna. They found that human blood can be assigned to one of four types according to the presence of two antigens, A and B, on the red cells and two corresponding antibodies, anti-A and anti-B, in the plasma.

There are four major phenotypes: O, A, B and AB. The reaction of the red cells of each type with anti-A (−A) and anti-B (−B) is as follows:

Red cell phenotype	Reaction with −*A*	−*B*
O	−	−
A	+	−
B	−	+
AB	+	+

− no agglutination
+ agglutination

Table 9–2. THE ABO BLOOD GROUPS

Blood Group (Phenotype)	Genotype	Antigens on Red Cells	Antibodies in Serum
O	*OO*	Neither	anti-A, anti-B
A	*AA* *AO*	A	anti-B
B	*BB* *BO*	B	anti-A
AB	*AB*	A and B	Neither

Type A subjects have antigen A on their red cells, type B subjects have antigen B, type AB subjects have both A and B, and type O subjects have neither. The *A*, *B*, and *O* genes are alleles at a locus on chromosome 9, as described in Chapter 4.

A remarkable feature of the ABO groups not shared by other blood group systems is the reciprocal relationship, in an individual, between the antigens present on the red cells and the antibodies in the serum (Table 9–2). (Serum and plasma are interchangeable terms in this connection. Serum is often more satisfactory than plasma for carrying out laboratory tests.) When the red cells lack A, the serum contains anti-A; when the cells lack B, the serum contains anti-B. The reason for this reciprocal relationship is uncertain, but formation of anti-A and anti-B is thought to be a response to A and B antigens occurring naturally in the environment (for example, in bacteria).

In the ABO blood group system there are compatible and incompatible combinations, a compatible combination being one in which the red cells being tested (or the red cells of a donor) do not carry an A or B antigen corresponding to the antibodies in the test (or recipient) serum (Table 9–3). Antibodies in the donor's plasma are not usually taken into account in transfusion, presumably because they are greatly diluted in the recipient's circulation. Although theoretically there are universal donors (group O) and universal recipients (group AB), a patient is given blood of his own ABO group, except in emergencies. The regular presence of anti-A and anti-B explains the failure of many of the early attempts to transfuse blood, since these antibodies can cause immediate destruction of ABO-incompatible cells.

Subtypes of A and B have been recognized. The most important is the separation of type A into A_1 and A_2, with a corresponding separation of type AB into A_1B and A_2B. In agglutination tests with anti-A or anti-A_1, the red cells of individuals of genotypes A^1A^1, A^1A^2 and A^1O cannot be distinguished. About 85 percent of type A bloods are A_1. Other variants of both A and B have been identified but they are rare.

Table 9–3. AGGLUTINATION REACTIONS WITHIN THE ABO BLOOD GROUP SYSTEM

Recipient: *Blood Group*	Recipient: *Antibodies (Serum or Plasma)*	Blood Group of Donor's Red Cells: *O*	*A*	*B*	*AB*
O	anti-A, anti-B	−	+	+	+
A	anti-B	−	−	+	+
B	anti-A	−	+	−	+
AB	Neither	−	−	−	−

+ agglutination.
− no agglutination.

Frequency of the ABO Groups and Genes

Because the relative proportions of the O, A, B and AB types differ in different populations, frequency figures are valid only for the population from which they are derived. The following figures will serve as an illustration:

	Approximate Frequency		
Blood type	*W. European*	*African*	*Oriental*
O	0.46	0.50	0.30
A	0.42	0.29	0.35
B	0.09	0.17	0.23
AB	0.03	0.04	0.13

Note that a high frequency of blood type B is characteristic of Asian populations. Surprisingly, though American Indians are of Asian origin, almost all lack the *B* gene and most also lacked *A* before the influx of Europeans to this continent.

ABO and Disease Associations

In Chapter 8 some of the disease associations of genes at the HLA locus were listed, and the theoretical reasons for the associations were briefly discussed. Similar but weaker associations have been found for the ABO blood groups (Table 9–4). Theoretically, if a gene is to be maintained in a population at a frequency higher than can be explained by recurrent mutation, it must confer some selective advantage. However, the genes *A*, *B* and *O* are all maintained at quite high frequencies even though there is no obvious advantage conferred by any one genotype (Reed, 1969). Although the relation of the ABO groups to disease is well established, it may not be of great genetic import because the diseases concerned usually affect people in middle or later life, after the peak reproductive period.

The first strong evidence for an association of a particular blood group with a disease was produced by Aird et al (1953), who described an excess of group A among patients with gastric cancer, an observation that has been repeatedly confirmed. An even closer association exists between duodenal ulcer, type O and nonsecretion of ABH antigens.

Since the level of Factor VIII is slightly higher in type A than in type O subjects (Preston and Barr, 1964), the reported association of type O with duodenal ulcer may be a consequence of the greater risk of hemorrhage in duodenal ulcer patients of type O, who are more often hospitalized because of

Table 9–4. EXAMPLES OF SIGNIFICANT ASSOCIATIONS BETWEEN ABO BLOOD GROUPS AND DISEASE

Disease	Comparison	Relative Risk
Gastric cancer	A:O	1.2
Tumors of salivary glands:		
malignant	A:O	1.6
nonmalignant	A:O	2.0
Cancer of cervix	A:O	1.1
Duodenal ulcer	O:A	1.3
Gastric ulcer	O:A	1.2
Rheumatic diseases	Non-O:O	1.2
Diabetes mellitus	Non-O:O	1.1
Ischemic heart disease	Non-O:O	1.2

Summarized from Vogel, 1970.

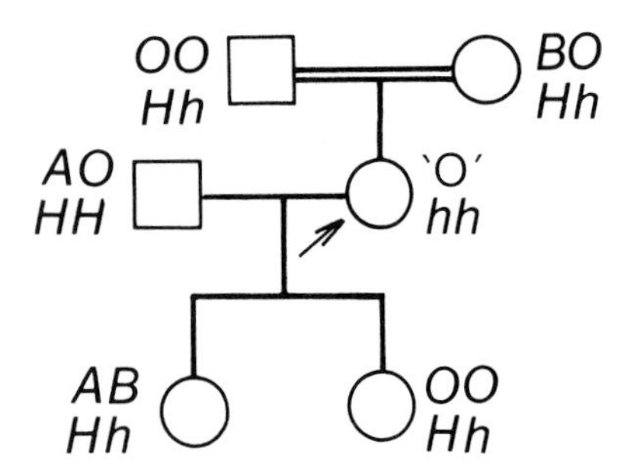

Figure 9–1. Pedigree of O_h phenotype (genotype *hh*) described in the text. (From Levine et al., Blood 1955; 10:1100–1102.)

bleeding and therefore more likely to be ascertained. In contrast, type A patients are thought to be at increased risk of thrombosis because of an increased clotting tendency. This observation, if confirmed, could have a bearing on the risk of thrombosis in type A women using oral contraceptives. Oral contraceptives reduce the level of antithrombin III and thus increase the probability of thrombosis.

H antigen

H antigen is the substrate from which the A and B antigens are made by the action of the *A* and *B* genes. The *O* gene is thought to be silent; that is, it has no active product, and thus type O cells carry unaltered H antigen. Anti-H is found in the serum of those whose red cells lack H, that is, some type A_1 subjects and all subjects of the O_h phenotype.

The O_h Phenotype

In the rare O_h phenotype the red cells and secretions lack the antigens A, B and H; the serum contains anti-A, anti-B and anti-H. Figure 9–1 is a pedigree showing the O_h phenotype. In this family a woman whose red cells were typed as O produced an AB child. Family studies led to the conclusion that the woman had a *B* gene but could not form the B antigen. (It is noteworthy that her parents were consanguineous). Most persons are homozygous or heterozygous for a gene *H*, required for the development of the H antigen from a precursor. Persons of the O_h phenotype are homozygous for a silent gene *h*. When H is not formed, the enzymes determined by *A* and *B* genes have no substrate on which to act, so that *hh* persons cannot make A or B antigen even if they have the *A* or *B* gene. This rare phenotype was first identified in Indians living in Bombay but has also been found elsewhere; thus its original name, Bombay phenotype, is no longer appropriate.

Secretion of ABH Antigens

The ABH antigens of the ABO blood group system occur in almost all cells of the body, and in most people are also present in secretions such as saliva.

Table 9–5. ABH ANTIGENS IN SECRETORS AND NONSECRETORS

Blood Group	Antigens of Secretors		Antigens of nonsecretors	
	Red Cells	*Saliva*	*Red Cells*	*Saliva*
O	H	H	H	—
A	A and H	A and H	A and H	—
B	B and H	B and H	B and H	—
AB	A, B and H	A, B and H	A, B and H	—

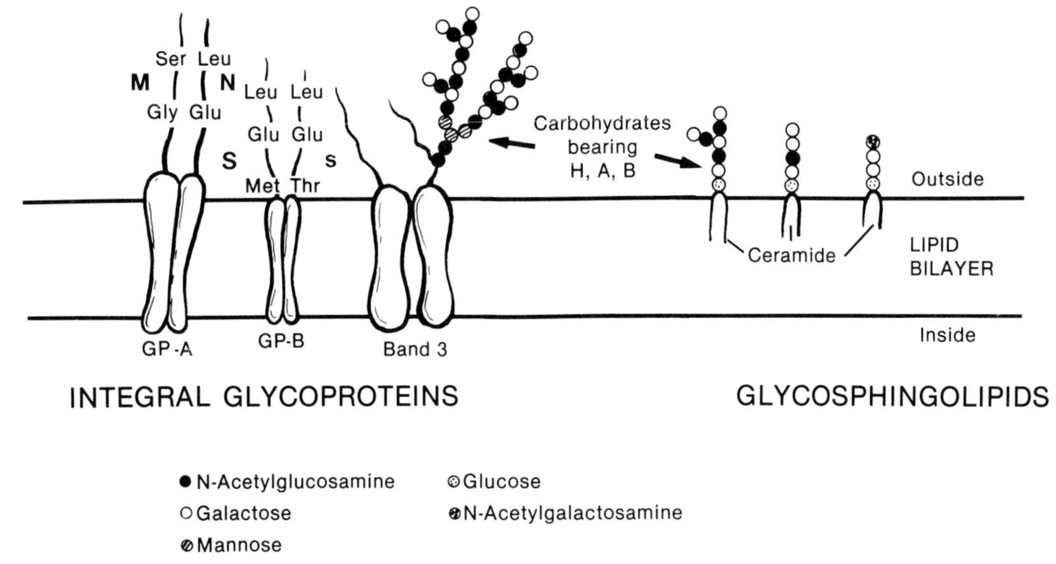

Figure 9–2. Diagram of the red cell membrane, showing the molecules that bear the antigens of the ABO and MNSs systems. The specificities H, A and B are determined by different terminal sugars, not shown here. The glycophorin molecules GP-A and GP-B also bear carbohydrate chains, but these are omitted in order to show amino acid substitutions—two differences between M and N, and one difference between S and s. Redrawn with permission from Giblett ER. Erythrocyte antigens and antibodies. In Williams WJ, Beutler E, Erslev AJ, Lichtman MA. Hematology, 3rd ed. New York: McGraw-Hill, 1983.

The ability to secrete ABH antigens is determined by a *Secretor* gene, *Se*; its allele, *se*, has no known function. Secretors of ABH are either *SeSe* or *Sese*; nonsecretors (about 22 percent of Caucasians) are *sese*. The ABH antigens on the red cells and in the saliva of secretors and nonsecretors are shown in Table 9–5. The term secretor refers only to the ability to secrete ABH antigens; it does not refer to Lewis antigens.

Genetic Pathways of Synthesis of ABH Antigens

The biosynthesis of the ABH antigens has been unraveled chiefly by the work of Morgan and Watkins; one of their many papers on the subject (Watkins, 1980) is cited in our references. Their studies were made with water-soluble A, B and H antigens that could be isolated in large amounts from ovarian cyst fluids; red cell ABH antigens and secreted ABH antigens are thought to be synthesized in a similar manner.

In basic structure the A, B and H antigens are macromolecules (more precisely, a family of closely related macromolecules) with a core to which many sugar chains are attached (Fig. 9–2). The antigenic specificity is conferred by the terminal sugars. The transferases (enzymes) that add the sugars to the core are presumably the direct products of the *H*, *A* and *B* genes. The steps may be summarized as follows:

Sugar Transferases

Gene	*Transferase*	*Acceptor*	*Sugar Added**	*Antigen*
H	H-transferase	Precursor	L-fucose	H
h	—	—	—	Unchanged precursor
A	A-transferase	H	GalNAc	A
B	B-transferase	H	D-Gal	B
O	—	—	—	Unchanged H

*Abbreviations: GalNAc, N-acetylgalactosamine; D-Gal, D-galactose

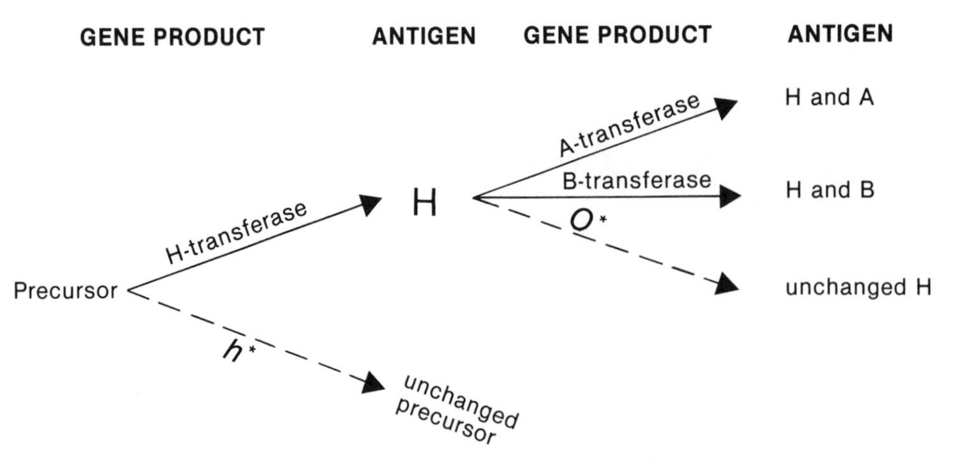

Figure 9–3. Diagram of the pathways of biosynthesis of H, A and B antigens (asterisks). Genes *h* and *O* have no detectable effect. Based on the work of Morgan and Watkins and of Ceppellini (see Watkins, 1980).

The relationship between the blood group genes, their direct products and the red cell antigens is shown schematically in Figure 9–3.

Landsteiner's discovery of the ABO blood groups in 1900 and Garrod's work on inborn errors of metabolism, dating from about 1902, both coincided closely in time with the rediscovery of Mendel's laws. For many years the two kinds of genetic variation seemed to be quite different; blood groups involved antigen differences that were found in normal individuals, whereas inborn errors were abnormalities caused by rare enzyme variants. We now see that, in both, the basic mechanism is a gene mutation causing synthesis of a variant protein. The difference in frequency is a consequence of the relative biological fitness of the normal and mutant forms. This topic is discussed in a later chapter (Chapter 15).

THE LEWIS SYSTEM AND ITS RELATIONSHIP TO ABO AND SECRETOR

In the Lewis blood group system there are two antibodies, anti-Le^a and anti-Le^b, that define two antigens, Le^a and Le^b. In most whites the red cells are either Le(a+b−) or Le(a−b+). The phenotype Le(a−b−) is quite rare in whites but relatively common in blacks from West Africa. The Le(a+b+) phenotype is common in white infants but very rare in white adults.

The Lewis antigens are not primarily red cell antigens, but are antigens of the body fluids that are only secondarily taken up from plasma by red cells. The presence of Le^a and Le^b antigens in the plasma and saliva depends on a pair of allelic genes *Le* and *le*. The complicated interactions of the Lewis system with the ABO and secretor systems are shown in Table 9–6. Secretors (*SeSe* or *Sese*) with an *Le* gene secrete ABH, Le^a and Le^b; their red cells are Le(a−b+) or, occasionally, Le(a+b+). Nonsecretors (*sese*) with an *Le* gene secrete Le^a but not Le^b or ABH; their red cells are Le(a+b−). The relatively rare people who lack *Le* (that is, are *lele*) may or may not secrete ABH, depending on whether they have an *Se* gene, but they never secrete Le^a or Le^b. Since their serum lacks Le^a and Le^b, their red cells also lack Le^a and Le^b; that is, they are Le(a−b−).

THE MN SYSTEM

After the discovery of the ABO blood group system in 1900, no other blood group antigens were found until 1927, when Landsteiner and Levine injected

Table 9–6. ANTIGENS PRODUCED BY INTERACTION OF *H, SECRETOR* AND *LEWIS* GENES

Genotype			Antigens on Red Cells			Antigens in Secretions			Approximate Frequency (Whites)
H	*Se*	*Le*	*H*	*Le*[a]	*Le*[b]	*H*	*Le*[a]	*Le*[b]	
HH or *Hh*	*SeSe* or *Sese*	*LeLe* or *Lele*	+	−	+	+	+	+	0.69
"	*sese*	*LeLe* or *Lele*	+	+	−	−	+	−	0.26
"	*SeSe* or *Sese*	*lele*	+	−	−	+	−	−	0.05
"	*sese*	*lele*	+	−	−	−	−	−	
hh	*SeSe* or *Sese*	*LeLe* or *Lele*	−	+	−	−	+	−	rare
hh	*sese*	*lele*	−	−	−	−	−	−	

+ present.
− absent.
The A and B antigens, determined by the *A* and *B* genes, are not shown.

human blood into rabbits and found that the immune serum formed by the rabbits could be used to distinguish between different human red cell samples. The antigens thus recognized were called M and N.

As originally described, the MN groups were models of genetic simplicity. They appeared to depend on a pair of codominant alleles, *M* and *N*, roughly equal in frequency, which produced three genotypes, *MM, MN* and *NN*, and three corresponding phenotypes, M, MN and N.

Blood Group (Phenotype)	Genotype	Reaction with *Anti-M*	Reaction with *Anti-N*	Approximate Frequency (European)
M	*MM*	+	−	0.28
MN	*MN*	+	+	0.50
N	*NN*	−	+	0.22

The Ss subdivisions of the MN groups were not discovered until some 20 years later, when the first example of anti-S was recognized. Combinations of MN and Ss are inherited as units, that is, *MS, Ms, NS* and *Ns*. The different antigenic specificities in the MN system result, not from terminal sugars as in the ABO system, but from differences in the amino acid sequence of the glycophorin molecules of the red cell membrane. In glycophorin A (GP-A), M and N differ in two amino acids; in glycophorin B (GP-B), there is a single amino acid difference between S and s (see Fig. 9–2). *MS, NS, Ms* and *Ns* are inherited as units because the genes for glycophorins A and B are very closely linked. Rare phenotypes of the MNSs system result from absence or changes in either GP-A or GP-B.

The MN blood group system has little importance in blood transfusion or in maternal-fetal incompatibility. Its major significance in medical genetics is that the relative frequencies of the genes and codominant pattern of their inheritance are useful in solving identification problems.

THE Rh SYSTEM

The Rh system, which is genetically complex, ranks with the ABO system in clinical importance because of its role in hemolytic disease of the newborn and in transfusion. The name comes from rhesus monkeys, which were used in the experiments that led to the discovery of the system. At the simplest level of explanation, the population is divided into Rh-positive individuals, who are either homozygous or heterozygous for a gene that specifies an antigen D, and Rh-negative individuals, who lack D.

Table 9–7. MAJOR ALLELES OF THE Rh BLOOD GROUP SYSTEM

Allele	Antigenic Determinants	Allele Frequencies *W. European*	*African*	*Oriental*
R^1	D, C, e	0.45	0.10	0.55
r	c, e	0.37	0.15	0.10
R^2	D, c, E	0.14	0.10	0.35
R^0	D, c, e	0.02	0.60	low
r''	c, E	0.01	low	low
r'	C, e	0.01	low	low
R^z	D, C, E	low	low	low
r^y	C, E	low	low	low

Data from Giblett ER, 1983.

The Rh locus (or series of loci) is on chromosome 1. There is still controversy over the precise genetic interpretation of the system, but until the molecular basis of the antigenic differences becomes clear it can be envisaged as a single but complex locus, with multiple alleles that determine the amino acid sequence of a component of the red cell membrane. A single allele specifies a polypeptide with multiple antigens, perhaps representing amino acid differences at more than one site in a single chain.

According to the Fisher-Race terminology used here, there are five Rh antigens called D, C, E, c and e. Of these antigens, C and c are antithetical, as are E and e. Since the counterpart of D is not known, it is represented by d.

Eight Rh alleles, the antigens produced and the approximate frequencies of the alleles are shown in Table 9–7. The only common allele associated with absence of D is r, with a frequency of 0.37 in white North Americans. (The allele frequency can be calculated from the frequency of Rh-negative (*rr*) individuals; if the frequency of *rr* is 0.14, the frequency of $r = \sqrt{0.14} = 0.37$.)

It is speculated that the eight common Rh alleles might have arisen from R^0 (the most common African allele) by mutation followed by recombination within the locus, as shown in Figure 9–4.

There are a number of interesting Rh variants. One is the Rh-null phenotype, in which the red cells lack all Rh antigens. Rh-null individuals have a form of hemolytic anemia, a consequence that suggests that the Rh antigens are an essential part of the normal red cell membrane.

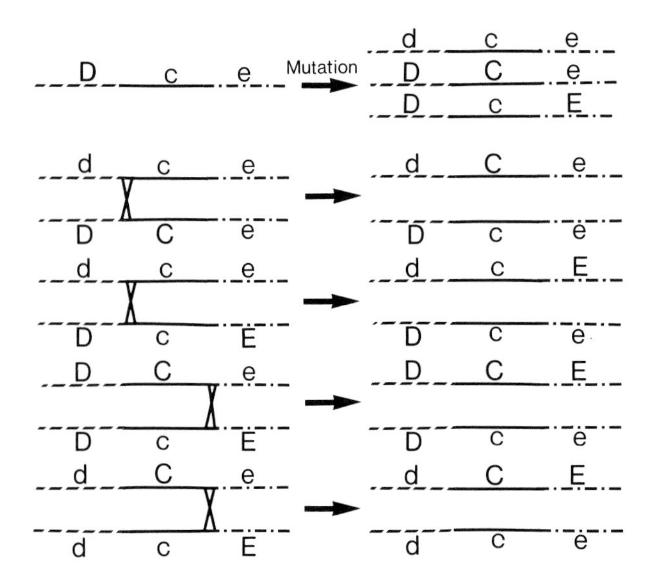

Figure 9–4. Possible origin of the Rh gene complexes by mutation and recombination. In this scheme, to generate the very rare *CdE* complex one of the two gene complexes undergoing recombination must itself be a recombinant.

Clinically the chief significance of the Rh system is that persons lacking D can readily form anti-Rh after exposure to Rh-positive red cells. When Rh-negative girls and women of childbearing age require transfusions, they must be given only Rh-negative blood. In pregnant women, the risk of immunization by fetal red cells can be minimized by giving injections of Rh immune globulin during and after pregnancy.

Hemolytic Disease of the Newborn

The discovery of the Rh system and its role in hemolytic disease of the newborn (HDN) has been a major contribution of genetics to medicine. In HDN, the life span of the fetal red cells is shortened by the action of antibodies formed by the mother against antigens of the fetus. The disease begins in utero and continues up to three months after birth, as long as some maternal antibody remains in the infant. Thus, though its basis is a genetically determined antigenic difference between mother and child, HDN is an *acquired* hemolytic anemia and must be clearly distinguished from hereditary anemias such as those caused by G6PD deficiency or hereditary spherocytosis.

There are two main types of HDN: one due to Rh incompatibility, when the mother is Rh-negative and the fetus is Rh-positive, and the other due to ABO incompatibility, when the mother is O and the fetus A or B. Most cases of HDN recognized clinically are due to Rh incompatibility; ABO incompatibility is difficult to diagnose but tends to be mild and to require no treatment. HDN is only rarely caused by incompatibility in other blood group systems, such as Kell or Duffy. In a white population, HDN occurs approximately once in 100 births.

MECHANISM OF Rh HEMOLYTIC DISEASE

Normally during pregnancy small amounts of fetal blood cross the placental barrier and reach the maternal blood stream. When this happens in an Rh-negative mother with an Rh-positive fetus, the Rh-positive fetal cells may stimulate the formation of anti-Rh, which is then transferred to the fetal circulation, where it attaches to the red cell membrane (Fig. 9–5).

In the fetus, red cells that are heavily coated with anti-Rh are rapidly removed from the circulation. The fetus becomes anemic and responds by releasing large numbers of erythroblasts (nucleated immature red cells) into the blood (thus accounting for the name erythroblastosis fetalis sometimes used for this disease). Hydrops, a consequence of the anemia, may cause intrauterine death.

After birth, the rapid destruction of red cells produces a large amount of bilirubin, which may cause jaundice during the first day of life. If hyperbilirubinemia is not promptly prevented by replacement transfusion, in which the infant's cells are replaced by Rh-negative cells, deposition of unconjugated bilirubin in the brain (kernicterus) may produce cerebral damage, which few infants survive. Those who do survive may have high-frequency deafness, mental retardation or the athetoid type of cerebral palsy.

FACTORS INFLUENCING THE DEVELOPMENT OF Rh HEMOLYTIC DISEASE

In theory HDN can develop whenever the fetus has inherited from the father an antigen that the mother lacks, but the number of genetic opportunities for HDN is far in excess of the number of cases that actually develop. The probability

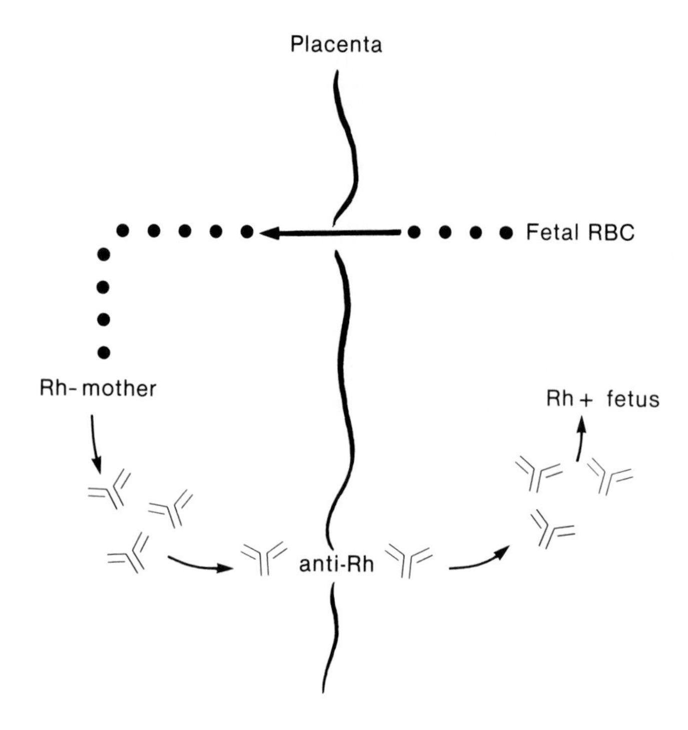

Figure 9–5. Diagram of the immunization of an Rh-negative woman by cells from her Rh-positive fetus. The woman responds by forming anti-Rh, which passes to the fetal blood stream and may cause hemolysis of fetal red cells.

that an Rh-negative mother will produce an infant with HDN due to Rh incompatibility depends on the following considerations:

1. If the father as well as the mother is Rh-negative, the fetus cannot be Rh-positive.

2. The father may be heterozygous *(Dd)* and transmit to the fetus *d* rather than *D*. Calculating from Hardy-Weinberg considerations, one can estimate that even when her mate is Rh-positive, the chance that an Rh-negative woman is carrying an Rh-positive fetus is only 60 percent.

3. The fetus may be protected because the mother is O and the fetus non-O (that is, A or B). The protection afforded by incompatibility on the ABO blood group system has long been recognized. If the mother is type O she has naturally occurring anti-A and anti-B, and these antibodies will destroy any A or B fetal red cells in the maternal circulation before the cells have an opportunity to stimulate anti-Rh. The probability that any pregnant woman is O and has a non-O fetus is about 15 percent. (The frequency of type O times the frequency of non-O alleles $= 0.46 \times 0.32 = 0.15$.)

4. Usually an Rh-negative woman does not become immunized until she has had at least one Rh-incompatible pregnancy or transfusion. In fact, with modern treatment she may never become immunized, since primary immunization to the Rh antigen on fetal cells can now be prevented in nearly all cases by treatment of the Rh-negative mother of an Rh-positive fetus with Rh immune globulin, prepared from the plasma of people who have become sensitized by transfusion or pregnancy. Because abortion and amniocentesis for prenatal diagnosis have been incriminated as a means of Rh sensitization, nonimmunized Rh-negative women who undergo abortion or amniocentesis should receive Rh immune globulin as a prophylactic.

5. There is individual variation in the ability of a pregnant woman to develop an immune response to the red cells of an Rh-positive fetus.

THE LUTHERAN SYSTEM

The particular genetic interest of the Lutheran blood group system is that it provided the first example of autosomal linkage and crossing over in man, the Lutheran-secretor linkage; these two loci are now part of a large cluster of linked loci on chromosome 19. There are a large number of genes in the Lutheran system, of which the best known are Lu^a and Lu^b. The gene Lu^a is much less common (0.04) than Lu^b (0.96).

THE KELL SYSTEM

The Kell system was originally described in terms of two phenotypes, K-positive and K-negative. Like other blood group systems it has become more complicated with many antigens, one of which (Js^a) is common in blacks but rare in other populations. Like the Rh system, the Kell system may be implicated in hemolytic disease of the newborn, when a K-negative mother forms anti-K in response to the red cells of a K-positive fetus. Protection from hemolytic disease due to anti-K is afforded if the mother and fetus are ABO-incompatible.

A precursor of the Kell antigens is determined by the X-linked gene *Xk*. When the Kx antigen determined by the normal *Xk* gene is missing, the so-called McLeod phenotype results. Patients with the McLeod phenotype have red cell acanthocytosis—a burr-like appearance of the red cells, with spiny projections—and a hemolytic anemia. These defects suggest that Kx is a structural component of the normal red cell membrane.

THE DUFFY SYSTEM

At the Duffy (Fy) locus on chromosome 1 there are three alleles of importance, Fy^a, Fy^b and a silent allele *Fy* that is almost entirely restricted to Africans.

Genotype	Phenotype	Approximate Frequency *Whites*	*Africans*
Fy^aFy^a Fy^aFy	Fy(a+b−)	0.20	0.11
Fy^aFy^b	Fy(a+b+)	0.46	0.01
Fy^bFy^b Fy^bFy	Fy(a−b+)	0.33	0.20
FyFy	Fy(a−b−)	low	0.68

The Fy(a−b−) phenotype protects red cells against invasion by the malarial parasite *Plasmodium vivax*, possibly because the antigens Fy^a and Fy^b function as the red cell receptors for the parasite. The selective advantage of the Fy(a−b−) phenotype in malaria-endemic areas may account for its high frequency in some of these areas, especially in West Africa.

OTHER BLOOD GROUP SYSTEMS

Several other independent blood group systems are listed in Table 9–1, but will not be discussed here. One of these, Xg, is of little clinical importance but of great genetic interest as one of the few known loci on the X chromosome that do not undergo X-inactivation.

Red Cell Enzyme Polymorphisms

Red cell enzymes, like red cell antigens, show extensive polymorphism. An important example, the glucose-6-phosphate dehydrogenase (G6PD) polymorphism, has been described earlier. One additional example, red cell acid phosphatase (ACP), is described here because it shows quantitative variation arising from a simple three-allele system and may be a model for other systems in which there is quantitative variation.

The three main *ACP* alleles together give rise to six relatively common phenotypes. (Other less common variants are not discussed here.) In the general population the range of activity is wide and the distribution unimodal, but the individual genotypes lead to differences in ACP activity (Fig. 9–6). It seems that each allele is responsible for a certain average activity (ACP^A, 62 units; ACP^B, 94 units; and ACP^C, 118 units). The distribution of red cell acid phosphatase activity in the general population and in individuals of each phenotype shows that the apparently unimodal curve is actually a composite of the range of activities of the various phenotypes.

Plasma Protein Polymorphisms

There are over 100 proteins in plasma, and genetic polymorphism has been found in most of those that have been studied in depth. By electrophoresis in a supporting matrix of starch, acrylamide or agarose gel, followed by protein staining, the proteins of a plasma sample can be separated in an electrically charged field. The process (electrophoresis) reveals a pattern of bands in which the different proteins assume characteristic positions determined chiefly by their

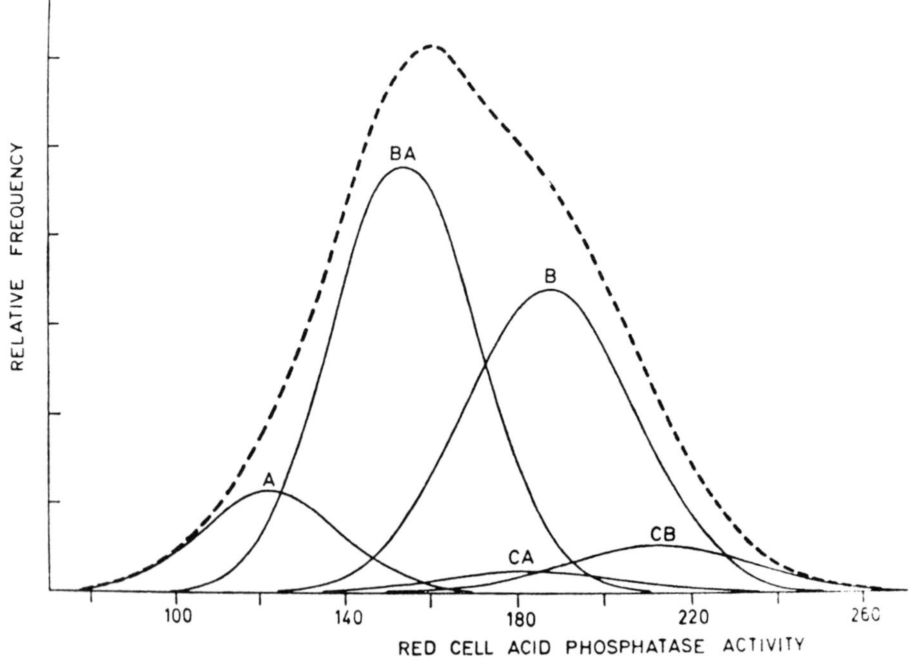

Figure 9–6. Distribution of red cell acid phosphatase activities in the general population (broken line) and in the separate phenotypes (solid lines). The apparent continuous distribution is the sum of several separate distributions. From Harris H. The principles of human biochemical genetics, 3rd ed. Amsterdam: Elsevier/North-Holland Biomedical Press, 1980, by permission.

net electrical charge. When the usual band pattern has been determined, variants can be detected by differences in the pattern. Alteration in the position of a protein band occurs when a mutation has produced a polypeptide that differs in net electrical charge from the usual form. Only those variants that have a different charge can be detected electrophoretically, but other techniques, especially isoelectric focusing, can reveal substitutions involving amino acids with similar charges.

Two well-known polymorphic plasma proteins, haptoglobin and α_1-antitrypsin, are discussed in this section. Haptoglobin is of importance as an example of genetic change in evolution; α_1-antitrypsin is clinically important.

HAPTOGLOBIN

Haptoglobins are α globulins with the property of binding hemoglobin and thus of conserving iron released from red cell destruction. The molecule contains two α and two β chains. The α and β chains are synthesized as a single polypeptide that is later cleaved; in other words, α and β messenger RNA is spliced from the same gene (Maeda et al, 1984).

There are three common variants representing polymorphism in the α chains, as well as several rare types. Two of the common variants, Hp^{1F} (fast) and Hp^{1S} (slow), differ in a single amino acid. The third variant, Hp^2, is quite different and resembles a fusion of Hp^{1F} and Hp^{1S} that produces a chain almost twice as long as the Hp^1 chain. Smithies and his colleagues proposed in 1962 that Hp^2 originated by nonhomologous crossing over in a Hp^{1F}/Hp^{1S} heterozygote (Fig. 9–7). Nonhomologous crossing over resulting in partial gene duplication is a rare event but could be of major evolutionary significance, since it permits formation of new genes without actual mutational change in amino acid sequence, as long as the reading frame of the new gene is not affected by the crossover.

Recently DNA sequencing has confirmed that the intragenic duplication within the Hp^2 gene results from a nonhomologous and probably random crossover, most likely in a Hp^{1F}/Hp^{1S} heterozygote, and has shown that the crossover took place within different introns of the two Hp^1 genes (Maeda et al, 1984).

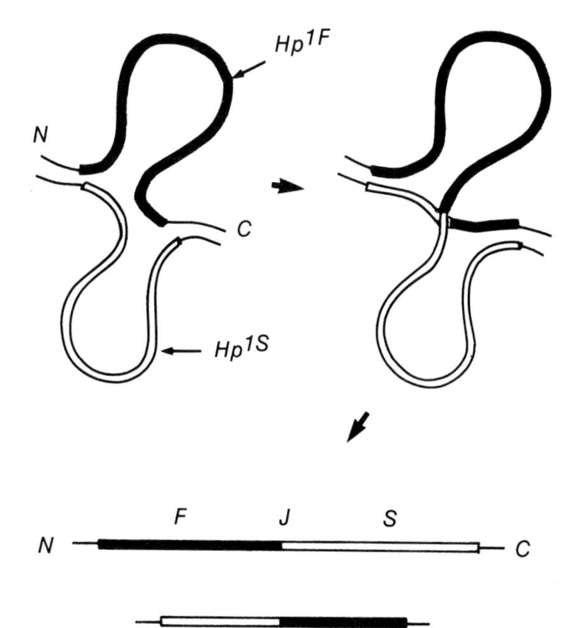

Figure 9–7. Possible origin of the gene Hp^2 by nonhomologous crossing over in an Hp^{1F}/Hp^{1S} heterozygote. Crossing over results in a new gene containing long segments of the Hp^{1F} and Hp^{1S} genes, joined at point J. A corresponding deletion product is also formed. From Smithies O, Connell GE, Dixon GH. Chromosomal rearrangements and the evolution of haptoglobin genes. Nature 1962; 196:232–236.

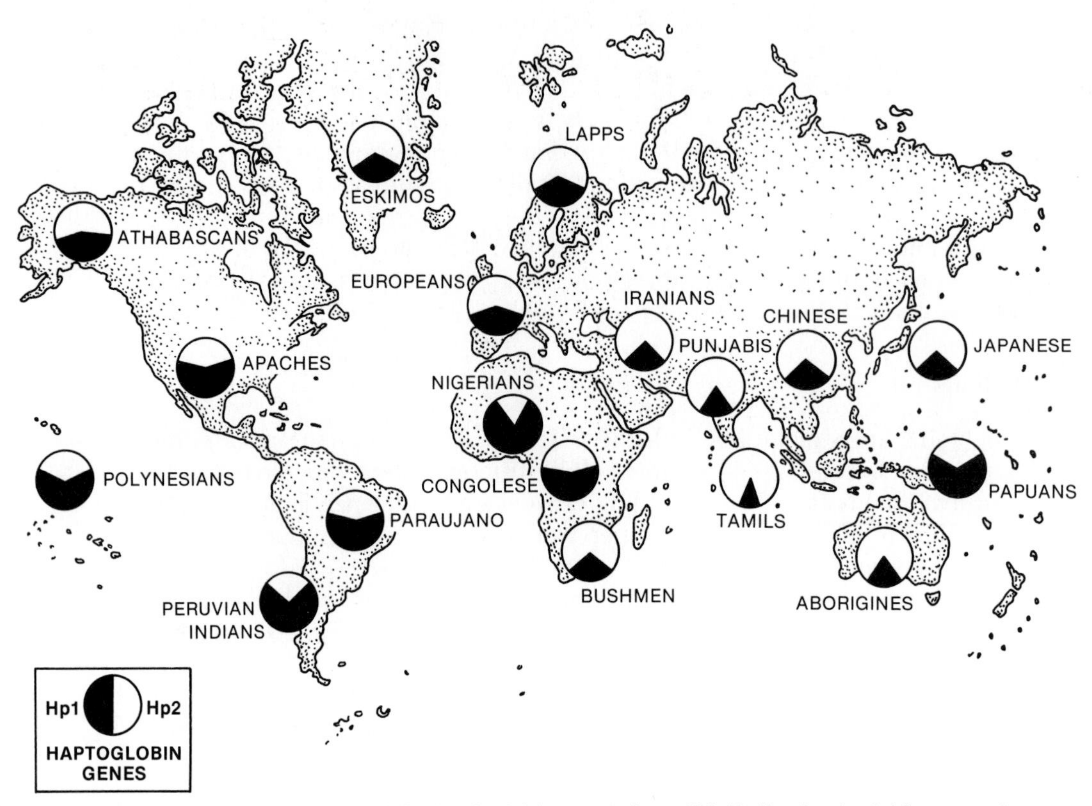

Figure 9–8. Geographic distribution of the haptoglobin genes. From Kirk RL. The haptoglobin groups in man. Basel: Karger, 1968; Harris H. The principles of human biochemical genetics, 3rd ed. Amsterdam: Elsevier/North-Holland Biomedical Press, 1980, by permission.

The Hp^2 allele, which occurs only in man, seems to be in the process of replacing Hp^{1F} and Hp^{1S} (Fig. 9–8), as if it confers some selective advantage. It is useless to speculate about what the advantage of Hp^2 might be when the physiological role of haptoglobin is poorly understood and its total or near-total absence (quite common in black populations) does not seem to be harmful.

ALPHA$_1$-ANTITRYPSIN

Alpha$_1$-antitrypsin is a serum protein that inhibits several proteolytic enzymes, such as trypsin, chymotrypsin and pancreatic elastase. Its locus, known as the Pi (protease inhibitor) locus, is highly polymorphic, with several relatively common alleles and 25 or so rare alleles. The alleles give rise to variants of the enzyme, differing in enzymatic activity. The identification is achieved by electrophoresis in one direction followed by crossed electrophoresis in agarose containing α_1-antitrypsin antibodies or, more commonly, by isoelectric focusing (Fig. 9–9).

Here we will mention only two alleles, a common allele Pi^M(M) and a rare allele Pi^z (*Z*). Individuals who are *ZZ* have α_1-antitrypsin deficiency, which is associated with a high risk of obstructive lung disease in early adult life, particularly in smokers. Liver disease in childhood may also be a consequence of the *ZZ* genotype; a minority of *ZZ* children have a clinically recognizable liver abnormality within the first few months of life, but many (about two-thirds) of these soon recover and develop normal liver function. The relatively favorable prognosis for *ZZ* individuals is a fact to be considered in genetic counseling and prenatal diagnosis. In heterozygotes, there is a suggestion of predisposition to a

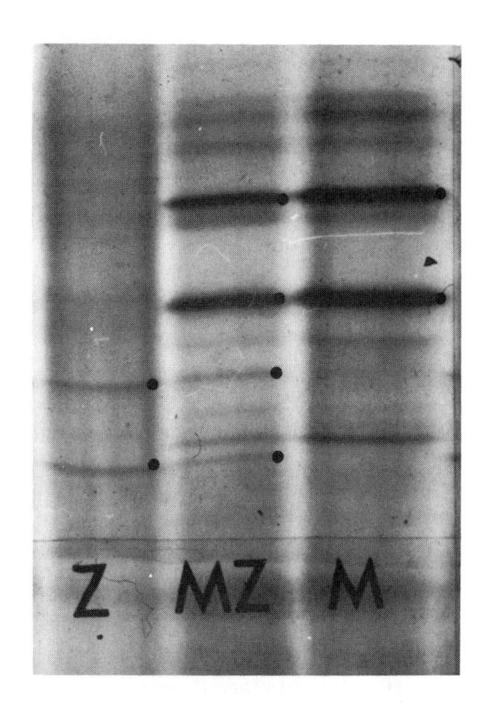

Figure 9–9. Genetic variants (Pi types) of α_1-antitrypsin, as separated by isoelectric focusing in acrylamide gel. Dots indicate major bands for each type. Deficient type is Pi type Z (genotype *ZZ*). Photograph courtesy of D. W. Cox.

number of disorders, such as rheumatoid arthritis, that seem to involve the immune system.

The Significance of Polymorphism

Genetic polymorphism has been known in human genetics since the discovery of the ABO blood groups by Landsteiner in 1900, and is now recognized as a very common if not ubiquitous phenomenon, typical of DNA and proteins in general.

A polymorphism as originally understood referred to the presence of two or more alleles at a locus, each with an appreciable frequency. By general agreement, a locus is considered to be polymorphic if its most common allele has a frequency of not more than 0.99, so that at least 2 percent of the population are heterozygotes. At some polymorphic loci there are alleles that cause disease in homozygotes; for example, as described above, *ZZ* individuals are deficient in α_1-antitrypsin and may have lung disease.

When Harris and his colleagues began to survey the human enzymes that could be examined by electrophoresis, an unexpectedly large number of variants were detected. At many loci two relatively common alleles were found, each with an appreciable frequency, as well as one or more rare alleles that could be accounted for by recurrent mutation.

In what proportion of his gene loci is a given individual likely to be heterozygous? In man, a recent estimate of about 6 percent is probably too low, since improvements in technique allow additional polymorphisms to be recognized. The number of possible combinations of genes for protein variants is so large that it is highly unlikely that any two persons (except, of course, monozygotic twins) have exactly the same set.

The extent of polymorphism has led to speculation about its significance in terms of evolution. The classical or selectionist view is that natural selection, acting upon mutations, is the primary factor in maintaining polymorphism. In some cases, a system with two common alleles is being maintained as a balanced polymorphism, with heterozygote advantage (see Chapter 15). In other cases, we may simply be observing an intermediate stage in the replacement of one major allele by another, as seems to be the case for the haptoglobins. Other special mechanisms could account for polymorphisms. For example, two different homozygotes might each be best fitted for a particular environment; a cline with high frequencies of both alleles might be observed between them. The major point is that selection in the Darwinian sense is operating.

The opposing view, first proposed by Kimura, is that natural selection need not be the essential process in maintaining polymorphism, since random genetic drift alone could account for the presence of two common alleles at a locus. One of the strong arguments in this neutralist or non-Darwinian view is that molecular evolution has proceeded at a more or less fixed rate for many proteins in widely different species over long periods of time, with such regularity that the rate of amino acid substitution can be used as a measure of the evolutionary time sequence. In other words, in the neutralist view there is no need to postulate that one allele is preferred or that the heterozygote is at an advantage. Neutral mutations could become fixed by chance.

The debate has had a stimulating effect on the study of the extent and significance of polymorphism. One observation, which supports the selectionist view, is that in the majority of enzyme polymorphisms each genotype produces a different level of activity; this variation would provide a basis for selection. It is likely that both selection and random genetic drift are significant in evolution.

DNA POLYMORPHISMS

As described earlier, the term polymorphism is now also applied to variants of DNA sequence demonstrable by the use of restriction enzymes (restriction fragment length polymorphisms, or RFLPs). A polymorphism revealed by variation in restriction fragment length is analogous to polymorphism shown by biochemical variation in enzymes or other proteins, since both derive from relatively frequent alterations in DNA sequences. However, RFLPs occur in non-coding DNA segments, whereas polymorphic loci in the original sense are structural loci; that is, their genes encode proteins. The high frequency of RFLPs suggests that the non-coding sequences of the genome are under much less selective pressure than are the structural genes.

GENERAL REFERENCES

Bowman BH, Kurosky A. Haptoglobin: the evolutionary product of duplication, unequal crossing over and point mutation. Adv Hum Genet 1982; 12:189–261.

Fagerhol MK, Cox DW. The Pi polymorphism: genetic, biochemical and clinical aspects of human α_1-antitrypsin. Adv Hum Genet 1981; 11:1–62.

Giblett ER. Erythrocyte antigens and antibodies. In Williams WJ, Beutler E, Erslev AJ, Lichtman MA, eds. Hematology, 3rd ed. New York: McGraw-Hill, 1983.

Harris H. The principles of human biochemical genetics, 3rd ed. New York: Elsevier/North-Holland, 1980.

Mollison PL. Blood transfusion in clinical medicine, 7th ed. Oxford: Blackwell, 1983.

Race RR, Sanger R. Blood groups in man, 6th ed. Oxford: Blackwell, 1975.

Stebbins GL, Ayala FJ. The evolution of Darwinism. Sci Am 1985; 253:72–82.

Watkins WM. Biochemistry and genetics of the ABO, Lewis, and P blood group systems. Adv Hum Genet 1980; 10:1–136.

PROBLEMS

1. A woman of blood group AB has an AB child.
   a) What are the possible blood groups of the father?
   b) What proportion of men are ruled out as the father of this child?

2. A woman of blood group B,N has a child of blood group O,MN. She states that a certain man of blood group A,M is the father of the child.
   a) Can the man be excluded as the father on the blood-group evidence?
   b) Which of the following men *could* be excluded as the father?
      1) Man of blood group B,N
      2) Man of blood group AB,MN
      3) Man of blood group O,M
      4) Man of blood group A,N
   c) What genes *must* the child have received from the father?

3. Five pairs of parents claim four kidnapped children. Assign as many as possible of the following children to the right parents.

Children	Parents
A	B × B
A	A × A
B	A × O
AB	O × O
	A × AB

4. Review Figure 9–2.
   a) Name two blood group antigens in which the specificity is determined by enzymes that add specific sugars.
   b) Name two blood group antigens in which the specificity is determined by amino acid sequences in the protein chains of the antigens.

10 SOMATIC CELL GENETICS

Somatic cell genetics, in broad terms, is the study of genetic organization, expression, and regulation in cultured cells of somatic origin. As an approach to problems in human and medical genetics it has been remarkably successful in recent years, particularly by virtue of its contribution to gene mapping. It is likely to continue to be widely used because of its promise for solving the enigma of genetic regulation.

Originally, somatic cell genetics was seen as a way of investigating genetic disorders, either Mendelian or chromosomal, in long-term culture rather than in living patients. This approach has many obvious advantages. If a cell line from a patient with a rare disorder or unusual karyotype can be established when the patient is available, it can be frozen in liquid nitrogen and studied later at any convenient time. Established cell lines can be revived after having been frozen in liquid nitrogen for long periods. They can be exchanged with other laboratories or stored in a central depository for use by many different investigators. Cell lines can also be used for a variety of experimental procedures that might raise both practical and ethical problems if they were attempted with living patients.

However, a number of pitfalls in the use of somatic cell genetics await the unwary investigator. Many factors can influence biochemical observations in cultured cells, and must be controlled in genetic experiments. For example, the extent of genetic variation in the human population is known to be so large that probably no two people (except identical twins) have exactly the same genotype, and failure to take the underlying variability into account can lead to misinterpretation. Because many biochemical and morphological changes occur with age, samples must be matched for donor age and culture age as closely as possible. Matching for sex is also necessary. Fibroblast cultures from different sites do not necessarily yield identical cell strains. In particular, fibroblast cultures derived from foreskin biopsies, which are frequently used as controls since they are readily available because of circumcision, are derived from genital skin and may not be suitable controls for biopsies from nongenital sources. Considerable heterogeneity exists within a culture in cell type, cell viability and enzyme levels, even when the culture is clonal (that is, derived from a single cell); this must be taken into account if diagnostic errors are to be avoided. Clones derived from female carriers of X-linked biochemical disorders are of two distinct populations, depending on which X is the active one in any specific clone, and therefore may differ considerably. Nutritional factors and infection, particularly mycoplasma infection, can also alter biochemical findings.

The usefulness of cultured somatic cells for research in medical genetics depends upon three main factors:

1. The feasibility of growing cells in long-term culture.
2. The possibility of identifying and characterizing genetic differences in cultured cells.
3. The possibility of performing some type of genetic analysis with the cells.

Much of the success of somatic cell genetics has come about because each of these prerequisites has been achieved. In particular, the transfer of genetic material between cells by somatic cell hybridization or other methods (described later in the chapter) has enormously expanded the range of basic and clinical information that can be derived from such studies.

Cell Cultures

Techniques for cell culture have taken a long time to develop and to become widely used, partly because of the demanding nutritional requirements of cultured cells and partly because successful maintenance of a culture for a long period requires elaborate precautions to avoid contamination. The nutritional requirements include the use of a semidefined medium with 10 or 15 percent fetal or newborn calf serum, which contains as yet undefined growth factors, as well as glucose, amino acids, vitamins, minerals, a buffering system and NaCl. Prevention of contamination requires vigilance and constant testing.

Only a few types of cells can be cultured successfully; highly differentiated tissues usually deteriorate very rapidly in culture, though recently there has been success in culturing epithelial cells derived from such sources as skin, breast tissue and sweat glands. The kinds of cells most frequently used in cultures are the following:

Peripheral Lymphocytes in Short-Term Culture. These cells do not meet the requirement of long-term survival, as they persist in culture for only 72 hours or so and are never available in the quantities needed for most biochemical assays, but they are used extensively for chromosome analysis and in immunogenetics.

Fibroblasts. Fibroblasts are the most useful cell type for genetic studies. They are cultured from small explants of skin or other tissue, set up in culture vessels in nutrient media and maintained under closely controlled environmental conditions. The cells, which have a characteristic spindle shape (Fig. 10–1), grow in monolayers and must be subcultured at frequent intervals because overcrowding leads to contact inhibition and cessation of growth. Usually they have normal karyotypes, though clones of cells with abnormal karyotypes arise in culture with a fairly high frequency, possibly reflecting loss of rigid control over the stability of the karyotype in culture. Though fibroblast cultures have a fairly long life span, they undergo senescence after 10 to 100 generations in culture, the duration of survival being longer in cultures set up from a young donor than in cultures from an older person.

Permanent Lines of Transformed Cells. Some cell lines undergo malignant transformation in culture, either spontaneously or experimentally by viral transformation; other cell lines have been established directly from malignant tissue. These lines do not have the property of contact inhibition, do not undergo senescence and do not have stable karyotypes. A well-known example is the HeLa cell line, established many years ago from a patient with cervical carcinoma. The HeLa cell line is heteroploid, with 63 to 65 chromosomes in its various sublines or even within a line. The growth properties of HeLa cells are so exceptional that some years ago it was found that cell lines in many laboratories had become contaminated by HeLa cells in the course of laboratory procedures and that the HeLa cells had overgrown and replaced the original lines.

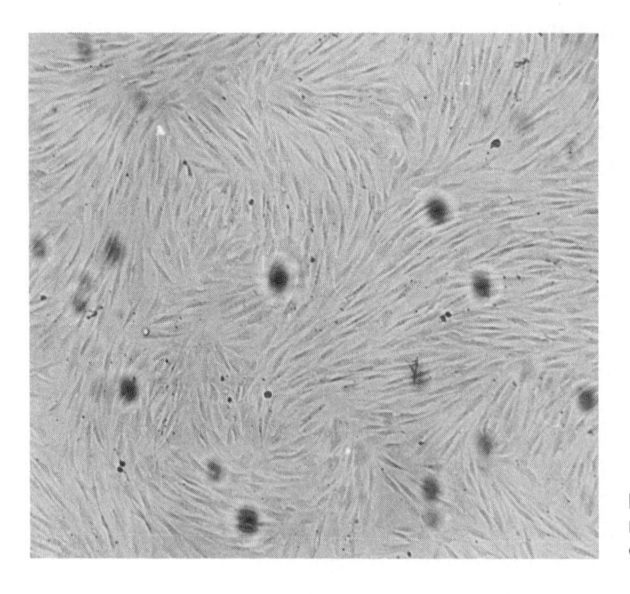

Figure 10–1. Fibroblast culture under low-power magnification. The cells are spindle-shaped and grow in a monolayer.

Lymphoblastoid Cell Lines. Though cultures of peripheral blood lymphocytes do not ordinarily persist in culture, they can be induced to do so if they are transformed by Epstein-Barr virus, the cause of infectious mononucleosis. Lymphoblasts have several advantages: they are easier to manipulate experimentally than fibroblasts because they grow in small clumps in suspension rather than in monolayers (Fig. 10–2), and they do not undergo senescence. A disadvantage is that the karyotype is somewhat unstable, and specific types of karyotypic alteration develop within the cultures over a period of time.

Amniocyte Cultures. Samples of amniotic fluid obtained from a pregnant mother by amniocentesis contain cells derived from the fetal skin and mucous

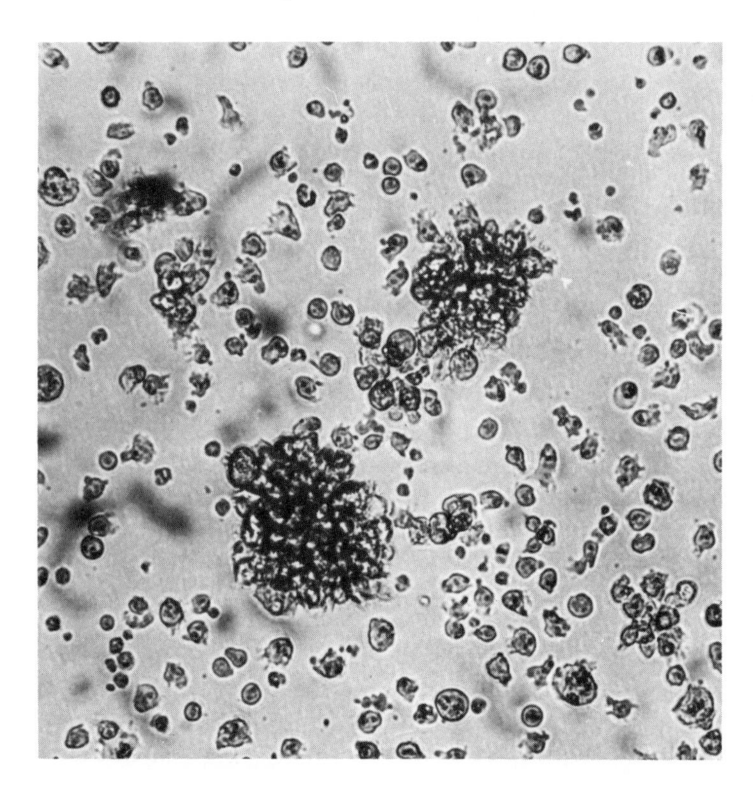

Figure 10–2. Lymphoblastoid cells in culture, under high-power magnification. The cells grow in suspension and typically form small clumps.

membranes and from the inner surface of the amnion. These cells of fetal origin include a small fraction that have at least some growth potential in culture, though as a rule amniocyte cultures do not grow vigorously or for many generations. The fetal cells may be epithelial cells or fibroblasts. They are used for determining the fetal karyotype, for analysis of enzymes or other genetic markers and increasingly for DNA analysis in prenatal diagnosis.

Phenotypic Expression in Cultured Cells

Medical cytogenetics rests on the principle that the karyotype of a cultured cell is, with rare exceptions, characteristic of the person from whom it was derived. Fortunately, this is also true of many gene products; many enzymes, structural proteins and surface antigens are expressed in cultured cells. In addition, at the DNA level, if a probe has been made for a specific gene or DNA segment, the probe can be used to identify the matching gene or segment in cultured cells, allowing a direct approach without the requirement of expression of the gene. The ability to bypass the gene product and analyze the DNA directly has allowed mapping of many genes whose products are still completely unknown, such as the genes for Huntington chorea and Duchenne muscular dystrophy.

Somatic Cell Hybridization

Somatic cell hybridization has already been mentioned as a productive new genetic technique. In human genetics the term implies the fusion of human cells in culture with cultured cells from another species, usually the mouse (Fig. 10–3). When human and mouse cells are cultured together, some of the cells form

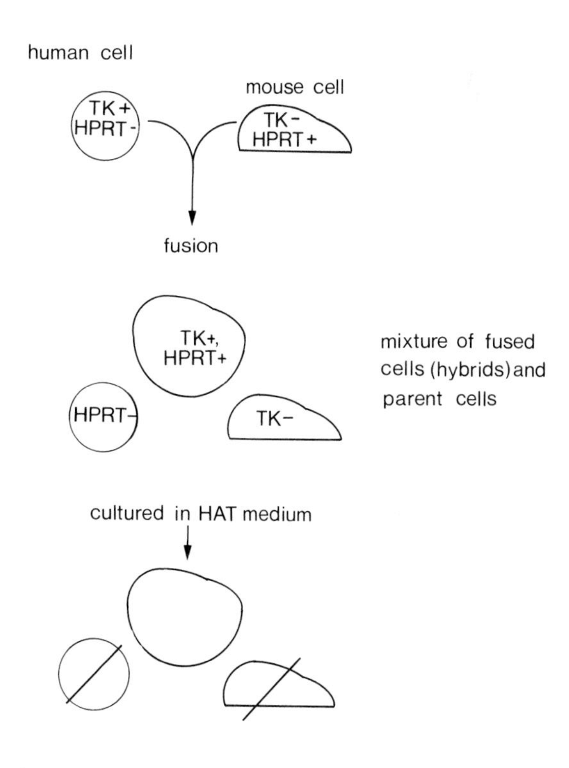

Figure 10–3. Fusion of human and mouse cells in vitro, with selection of fused cells by the HAT technique. See text for discussion.

heterokaryons (fused cells with separate nuclei). The nuclei may then fuse to form true hybrid cells. The hybrid cells continue to proliferate in culture, but the number of human chromosomes they contain is progressively reduced. The fact that it is the human chromosomes, not the mouse ones, that are preferentially lost is the basis of the usefulness of somatic cell hybrids for human linkage studies.

When hybrid cells have formed, the next problem is how to select the fused cells from the background of other cells (unfused cells, or fused cells of identical instead of different origin) in which they are growing. In the earliest experiments only hybrid cells with growth advantages over the parental cells could be isolated, but many methods of selection for hybrid cells have since been devised. The first such selective method is the so-called HAT technique, named because it makes use of hypoxanthine, aminopterin and thymidine. Purine analogues such as 8-azaguanine (8AZA) are toxic when converted to nucleotides by the action of the enzyme hypoxanthine-guanine phosphoribosyl transferase (HPRT). Cells lacking HPRT (HPRT−) are resistant to 8AZA. Cells from patients with Lesch-Nyhan syndrome (HPRT deficiency, an X-linked disorder) are naturally HPRT−, and mutants with the same deficiency arise occasionally in culture. Either of these can be selected from a mixed culture because they resist 8AZA and survive, whereas HPRT+ cells are killed. Similarly, thymidine kinase deficient (TK−) cells are resistant to bromodeoxyuridine (BUdR), an analogue of thymidine. The drug aminopterin inhibits the synthesis of purine and pyrimidines. Thus, cells that lack either HPRT or TK fail to grow if aminopterin is present in the culture. If an HPRT− human cell line and a TK− mouse cell line are fused and plated in HAT medium, both the parent lines are killed and only the fused cells, which have mouse HPRT and human TK, are capable of survival (Fig. 10–3).

When hybrid cells have been selected by their survival, they may be grown as individual clones for further study. For linkage studies, the clones may be subcloned until only a few different human chromosomes, or ideally only one, remains.

Use of Somatic Cell Hybrids in Gene Mapping

The greatest contribution of somatic cell hybridization has been in linkage analysis. A human gene can be assigned to a specific chromosome when its phenotype is present when that chromosome is retained, and lost when it is lost. The following example (Creagan and Ruddle, 1977) is a simplified illustration. Clones of somatic cell hybrids are selected in which there are certain combinations of human chromosomes, as shown. The three clones are examined for the expression of a given phenotype. If the phenotype is expressed in all three clones, its locus is on chromosome 1; if it is expressed in clones A and B but not C, its locus is on chromosome 2; and so on.

	Chromosomes							
	1	2	3	4	5	6	7	8
Clone A	+	+	+	+	−	−	−	−
Clone B	+	+	−	−	+	+	−	−
Clone C	+	−	+	−	+	−	+	−

Panels of hybrid clones covering all 24 human chromosomes (22 autosomes, X and Y) are available, so it is usually unnecessary to make new hybrids to look for linkages. Instead, the existing panels can be used to find the chromosomal locus of any gene product that remains unmapped. The method can be used for any gene that is expressed in hybrid cells, or for any gene or DNA sequence that can be identified by means of a DNA probe. It has been most useful in identifying the loci of human enzymes that can be distinguished electrophoretically from the homologous mouse enzymes. Enzyme markers coded by genes on each arm of each human chromosome are known and can be used to facilitate speedy mapping.

Regional localization of genes can be determined by making hybrids that contain structural abnormalities of chromosomes, especially translocations or deletions, and correlating the absence of a particular band with the loss of a gene product or identifiable DNA sequence.

Linkage is discussed further in Chapter 11.

Other Methods of Transfer of Genetic Information

Though somatic cell hybridization is the most widely used method of transferring genes, several other methods are also utilized, for example:

1. Microcell-mediated gene transfer. Under certain experimental conditions, chromosomes of dividing cells form microcells containing only one or a few chromosomes, which can be used as donors in hybridization experiments.

2. Chromosome-mediated gene transfer. Whole chromosomes, in purified preparations made by the use of a fluorescence-activated cell sorter, are taken up by endocytosis, then broken down into smaller fragments.

3. Transfer of purified DNA fragments, either by direct microinjection into the recipient cell using microcapillary pipettes or by a technique requiring calcium phosphate precipitation of DNA and endocytosis of the DNA by the recipient cell. In both cases the transferred DNA becomes incorporated into one or more chromosomes of the recipient.

Thus several experimental approaches are available for combining information from different genetic sources.

Complementation

Complementation analysis as a method of detecting genetic heterogeneity has been mentioned in earlier chapters. A **complementation test** determines whether two mutations are in the same gene. The original use of complementation in a human cell culture experiment involved co-cultivation of different mutant cell lines from patients with mucopolysaccharidoses in the same culture vessel, to learn whether the two cross-corrected. This was the test used by Neufeld and colleagues (1970) to analyze the mucopolysaccharidoses. Fibroblasts from patients with Hurler syndrome and Hunter syndrome, which are phenotypically very similar, cross-corrected; therefore, the defects were not allelic. This was to be expected, since Hunter syndrome is X-linked and Hurler syndrome autosomal recessive. However, it was surprising to find that fibroblasts from patients with two phenotypically distinct conditions, Hurler syndrome and Scheie syndrome, failed to cross-correct, indicating that the same protein was affected in both disorders; in other words, the two conditions were allelic.

Complementation can also be sought in heterokaryons made by fusion of cells from two different deficient lines. Such experiments were used to demonstrate nonallelic complementation in the propionic acidemias, a heterogeneous group of disorders of organic acid metabolism in which there is deficiency of propionyl-CoA carboxylase (Gravel et al., 1977). Heterokaryons were also used to demonstrate interallelic complementation (complementation between alleles at a single locus) in argininosuccinic acid lyase (ASAL) deficiency, a disorder of the urea cycle with significant clinical heterogeneity (McInnes et al., 1984).

Though co-cultivation or fusion of cells to form heterokaryons has been adequate for some complementation experiments, others require true hybrids that will proliferate in culture. For example, cells from patients with Fanconi anemia have an increased frequency of chromosome breaks. Demonstration of complementation of this phenotype between cells from different patients required the isolation of true hybrid lines (Duckworth-Rysiecki et al., 1985).

Mutation in Cell Cultures

Studies of mutagenesis in somatic cell cultures are subject to some technical difficulties, but the very large number of cells in a culture and the availability of various selective systems to isolate mutant cells so that they can be cloned for later analysis has allowed some progress to be made. The mutagenic potential of exposure to X-rays, non-ionizing radiation and a number of chemical agents has been established, and mutation frequency in relation to dose has been measured. In the future, it should be possible to use cell cultures to study the possible mutagenicity of environmental pollutants, one of today's most challenging and controversial problems in public health.

Mutations occurring in somatic cell cultures are an important source of material for genetic analysis. This can be demonstrated by an example. Familial hypercholesterolemia, as described in Chapter 5, is an autosomal dominant disorder in which the enzyme HMG CoA reductase plays an important role as the rate-limiting enzyme that controls de novo cholesterol synthesis. It is regulated in its turn by the presence of free cholesterol, liberated from LDL in the lysosomes (see Figure 5–14), which exerts negative feedback control over its transcription. The structural gene for HMG CoA reductase is a so-called housekeeping gene which is active at a low rate in many tissues; many such enzymes are subject to negative feedback control.

HMG CoA reductase has been cloned by a method that exploited a mutant arising in cell culture (Reynolds et al., 1984). Chinese hamster ovary (CHO) cells were exposed to compactin, a competitive inhibitor of reductase. This selective procedure yielded a mutant line that had a 15-fold amplification of the HMG CoA reductase gene and produced more than 100 times the normal amount of gene product. Recombinant DNA technology allowed cloning and characterization of the gene, which was then found to have certain differences from other known genes that may be relevant to its role in negative feedback regulation.

Studies of Differentiated Cell Function

The expression of differentiated function in cell culture is currently an active area of research, since it provides a way of obtaining answers to the still mysterious question of gene regulation.

Although the "housekeeping genes" responsible for cell survival are regularly expressed in cell hybrids, genes coding for differentiated cell function often remain unexpressed. However, there are exceptions. Some liver cell functions are expressed in hybrids in which one parent is a rodent hepatoma (a permanent line derived from a liver tumor), the other a relatively undifferentiated fibroblast or epithelial line; for example, human serum albumin is made by mouse hepatoma/human leucocyte hybrids though not by either parent (Darlington et al., 1974). In hybrids between the Syrian hamster melanoma line, which normally produces melanin, and a nonpigmented cell line, pigment is not synthesized; here the loss of pigment production is thought to depend on a diffusible suppressor, produced by the hybrid as well as by the non-pigmented parent type.

Cells of central nervous system origin, muscle cells and hemoglobin-producing cells are examples of other cell types that express their differentiated function in appropriate hybrids. Many oncogenes have also been identified by their phenotype in cultured cells (see Chapter 6). It is thought that oncogenes (or rather, proto-oncogenes) are required for normal differentiation in particular tissues, and that mutation of such a gene may interfere with normal differentiation and produce a specific type of malignancy.

Conclusion

When somatic cell genetics first became possible, it was seen primarily as a way to analyze cellular phenotypes directly. Since then it has become a technique that is even more valuable as a means of allowing combination of DNA of different origins, thus allowing analysis of chromosomes, protein markers and DNA markers, and integration of information derived from the three lines of evidence. Much of the last decade's progress in genetic mapping both at the protein level and at the molecular level would have been impossible if somatic cell hybridization had not provided a way of isolating a few chromosomes, a single chromosome or even a part of a chromosome against a defined background of rodent DNA.

GENERAL REFERENCES

Buchwald M. Use of cultured human cells for biochemical analysis. Clin Biochem 1984; 17:143–150.
Morrow J. Eukaryotic cell genetics. New York: Academic Press, 1983.
Puck TT, Kao FT. Somatic cell genetics and its application to medicine. Ann Rev Genet 1982; 16:225–271.
Ruddle FH. A new era in mammalian gene mapping: somatic cell genetics and recombinant DNA methodologies. Nature 1981; 294:115–120.
Shows TB, Sakaguchi AY, Naylor SL. Mapping the human genome, cloned genes, DNA polymorphisms, and inherited diseases. Adv Hum Genet 1982; 12:341–352.

PROBLEMS

1. A certain metabolic disorder is expressed in cultured fibroblasts. Cell cultures from six different patients are grown in all possible paired combinations, with the following results:

	A	B	C	D	E	F
A	−					
B	+	−				
C	−	+	−			
D	+	+	+	−		
E	+	−	+	+	−	
F	−	+	−	+	+	−

+ represents complementation in culture; − represents failure to complement.

a) How many complementation groups are there?
b) Assign each patient to a complementation group.
c) What do these results tell you about the genetics of the condition?

2. A series of mouse-human hybrid clones have the following human chromosomes:

Clone I 4, 7, 8, 10, 20, X

Clone II 1, 3, 7, 21, X

Clone III 9, 13, 19, 21, X

Clone IV 1, 4, 10, 19, X

Clone V 2, 3, 13, 20, X

Human enzyme A is present in clone V but absent in the other four clones. Human enzyme B is present in all five clones. Human enzyme C is present in clones III and IV but not in the other clones. All five clones lack human enzyme D. What conclusions can be drawn about the chromosomal location of these four enzymes?

3. Match the following:

A

—Not a characteristic of lymphoblastoid cultures
—Ability of different genetic defects to cross-correct
—A hybrid cell that secretes a single antibody of a single class
—Used to select HPRT deficient cells in culture
—A cell with two separate nuclei, formed by fusion of two separate cells
—A fusion cell in which the separate nuclei have fused and mitosis occurs
—Characteristic of fibroblast cells in culture
—Used to select thymidine kinase deficient cells in culture
—Permanent line of transformed cells
—A malignant tumor of plasma cells, usually of bone marrow origin

B

1) Microcell
2) Myeloma
3) Senescence
4) True hybrid cell
5) 8-azaguanine
6) Heterokaryon
7) Complementation
8) Hybridoma
9) Bromodeoxyuridine
10) Contact inhibition
11) HeLa cell line

11

LINKAGE AND MAPPING

Genes close together on the same chromosome tend to be inherited together, and are said to be **linked** or in linkage. Linkage must not be confused with association, which refers to the presence together of two or more characteristics with a frequency greater than expected by chance—such as the pleiotropic association of multiple effects of a single gene or the HLA disease associations described earlier—and has no implication of physical proximity of genes on a chromosome.

Linkage has in the past been a peripheral topic in medical genetics, regarded as an important genetic concept but of little medical significance. This is no longer the case. Spectacular progress has been made in recent years in delineation of the human gene map; for some conditions linkage analysis can already be used for carrier detection, identification of presymptomatic cases of certain disorders, and prenatal diagnosis. For these reasons, linkage has moved front and center as a topic about which physicians need to become knowledgeable.

According to Mendel's law of independent assortment, genes that are not allelic assort independently of one another. Linkage is a major exception to this law; if two loci are linked—that is, if they are on the same chromosome and not too far apart—alleles at those loci do not assort independently but are transmitted together to the same gamete more than 50 percent of the time. It is only when the loci are on different chromosomes, or far apart on the same chromosome, that the law of independent assortment holds true. Genes on the same chromosome are described as **syntenic** ("on the same thread"), whether or not they are close enough together to show linkage.

The physical basis of recombination is the exchange of chromosomal material (crossing over) that occurs between chromatids of homologous chromosomes in prophase of meiosis I, and is evidenced by chiasmata which are seen holding bivalents together at that stage.

Classical family study of traits with Mendelian inheritance used to be the only source of information about human linkage and still is a major mapping tool, but until other methods were invented, progress in mapping human genes was disappointingly slow. Apart from X linkage, which is easy to identify by its pedigree pattern, only three linkages came to light before about 1968; the linkages of the Lutheran blood group and the secretor trait, Rh and elliptocytosis, and ABO and the nail-patella syndrome. Though these pairs of loci were known to be linked, none of them had been mapped, that is, assigned to a specific chromosomal location. The first to be mapped was the Duffy blood group locus, in 1968. Since then, the rate of progress of gene mapping has been transformed by means of new technologies: chromosome banding, isozyme analysis, somatic cell hybridization, recombinant DNA technology and the development of computer programs to cope with the mass of information that has accumulated. By 1983 over 800 human genes, as well as a number of restriction fragment length polymorphisms, had been mapped. Delineation of the human gene map is a major medical achievement, in McKusick's words "a triumph of human anatomy analogous to the discovery that the kidney is the site of urine production and that the heart pumps blood."

Linkage Phase

In a double heterozygote, there are two possible arrangements of the two pairs of linked genes, as shown in Figure 11–1. The genes on the same chromosome are said to be in **coupling**, while those on opposite chromosomes of the pair are in **repulsion**. The arrangement in a particular genotype is called the **linkage phase**. In Figure 11–1, the linkage phase is shown as *AB/ab* on the left, and as *Ab/aB* on the right. In *AB/ab* individuals, *A* and *B* are in coupling on one chromosome, *a* and *b* on the homologous chromosome; in *Ab/aB* individuals, *A* is in coupling with *b*, and *a* with *B*.

If the linkage phase is known, the recognition and measurement of linkage is more efficient than when it is unknown. Later in this chapter an example is given of the measurement of linkage in phase-known and phase-unknown pedigrees.

LINKAGE EQUILIBRIUM AND DISEQUILIBRIUM

Normally, linked loci should be in equilibrium; that is, the relative proportions of the possible combinations should be determined only by the population frequencies of the alleles of the loci. As an example, if the gene frequencies are:

A 0.90	*B* 0.60
a 0.10	*b* 0.40

the corresponding frequencies of the combinations are expected to be:

AB 0.54	*aB* 0.06
Ab 0.36	*ab* 0.04

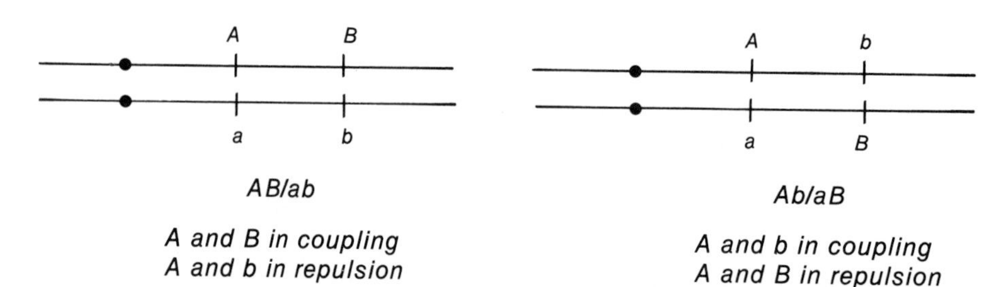

Figure 11–1. Possible linkage phases of genes *A* and *a*, *B* and *b*.

If linkage equilibrium is not present, there are several possible explanations, including the following:

1. Population admixture or mutation may have occurred too recently for equilibrium to have become re-established.

2. There may be a chromosome inversion involving the linked loci. Recombination is known to be suppressed within an inversion, because recombination results in duplication-deficiency chromatids that usually do not result in viable zygotes (see Chapter 6). There also seems to be an actual reduction in the frequency of recombination both within and just outside an inverted segment.

3. Natural selection may favor particular combinations of alleles at the linked loci. Linkage disequilibrium is particularly obvious for alleles at some of the HLA loci, and this phenomenon is thought to underlie the associations of HLA and disease (see Chapter 8).

Detection and Measurement of Linkage in Family Studies

Mendel's first law states that alleles segregate. As a corollary, recombination is proof that two genes are at different loci. Unless they are seen to undergo recombination, they may be alleles at a single locus.

The classical way to detect linkage is by observing the simultaneous transmission of nonallelic genes through successive generations of families. Not every family is informative for linkage. The primary requirement is that at least one parent must be heterozygous at both loci, so that recombination between the loci can be recognized. The most informative loci are those that are highly polymorphic, such as the HLA locus, at which as a rule each parent has two different haplotypes, neither of which is present in the other parent. Families with a large number of children and families with three or more generations available for analysis are more informative than smaller families. Three-generation families are more useful than two-generation ones because, as explained in further detail later, they may allow the linkage phase in the doubly heterozygous parent to be determined. If the linkage phase in the parents is unknown and only one child can be examined, that family is not informative for linkage.

Suppose we are interested in measuring linkage between two loci: locus A with alleles *A* and *a* and locus B with alleles *B* and *b*. There are two possible linkage phases, as shown in Figure 11–1.

If the linkage phase of genes *A* and *B* in the doubly heterozygous parent is known, the expected genotypes of the offspring are easy to predict. When exceptions occur, the cause is recombination between the A and B loci, and the exceptional offspring are recombinants.

Parents	AB/ab × ab/ab	Ab/aB × ab/ab
Expected offspring (nonrecombinants)	1/2 AB/ab, 1/2 ab/ab	1/2 ab/ab, 1/2 aB/ab
Exceptional offspring (recombinants)	Ab/ab, aB/ab	AB/ab, ab/ab

However, if the linkage phase is unknown, a twop-generation pedigree does not show which of the offspring are the nonrecombinants and which are the recombinants.

Figure 11–2 shows crossing over in a double heterozygote in whom the linkage phase is known to be *AB/ab*, resulting in the formation of four types of gametes: two parental types, *AB* and *ab*, and two recombinants, *Ab* and *aB*. In Figure 11–3, II–5 is a recombinant. (Note that the term recombinant refers to either the gamete or the individual resulting from it.)

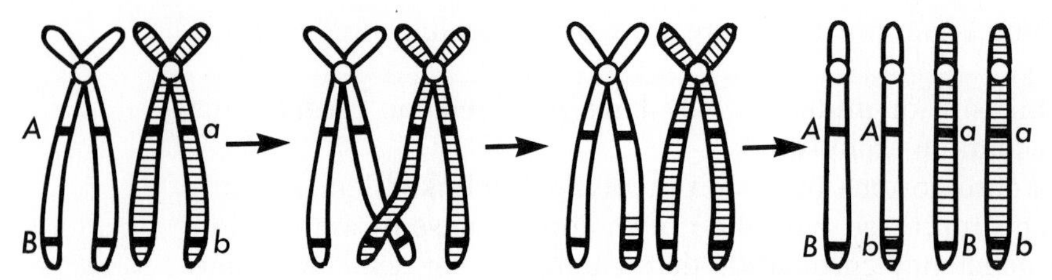

Figure 11–2. Recombination at meiosis in an *AB/ab* double heterozygote. Four types of gametes are formed: two nonrecombinant types, *AB* and *ab*, and two recombinants, *Ab* and *aB*.

RECOMBINATION FREQUENCY

A single two-generation family is rarely, if ever, large enough to give definitive evidence of linkage, but in a series of families the following observation might be made:

Parents	*AB/ab* × *ab/ab*
Offspring	45% *AB/ab* 5% *Ab/ab* 5% *aB/ab* 45% *ab/ab*

Thus among the offspring 90 percent are nonrecombinants, with the parental combinations of alleles, and 10 percent are recombinants. The recombination frequency (θ) is 10 percent or 0.10, and genes *A* and *B* (or, more specifically, the A and B loci) are said to be 10 centimorgans or map units apart.

A centimorgan (cM), or map unit, equals 1 percent recombination (θ = 0.01). The centimorgan was named in honor of Thomas Hunt Morgan, the Nobel prizewinning geneticist who discovered genetic linkage in 1911. It is estimated that on the average 1 cM is approximately 1000 kb long. However, the genetic map of the chromosome is not identical to the physical map; recombination occurs more frequently toward the ends of the chromosomes than at the centromeres, where there appears to be little recombination. In the human, crossing over is more frequent in females than in males.

We can now begin to construct a linkage map for the A and B loci.

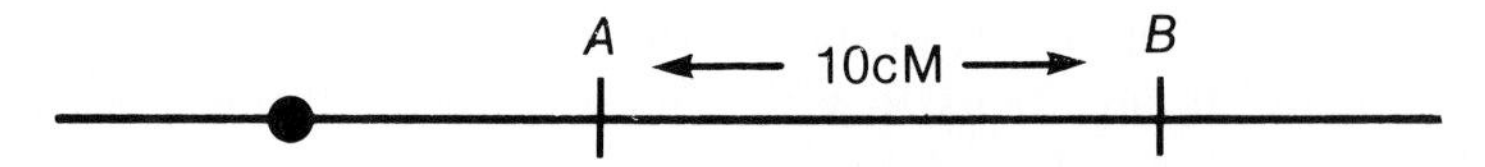

A third linked locus C can be added to this linkage map if its distance from the A and B loci can be measured. Consider the following findings:

A–C θ = .13

B–C θ = .05

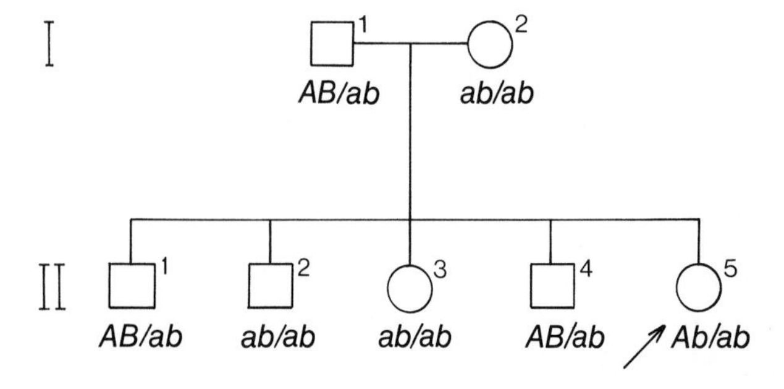

Figure 11–3. A pedigree showing a recombinant (arrow).

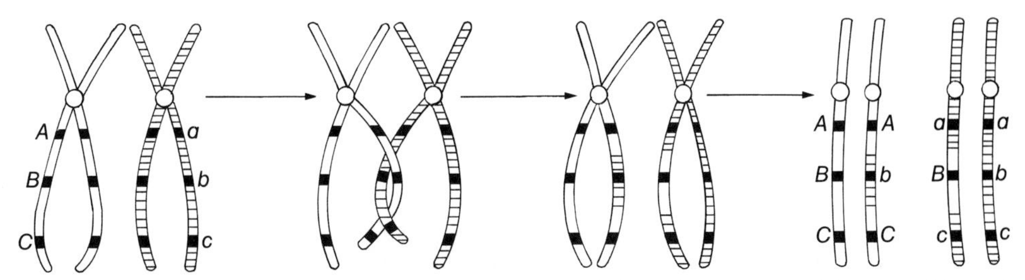

Figure 11–4. A double crossover between the A and C loci does not result in recombination between the loci. When an intermediate locus B is included, the crossover is recognizable.

The order of the genes must then be A–B–C, and C can be mapped as follows:

A ← 10cM → B ← 5cM → C

Note that the A–C distance as measured by recombination frequency is less than the sum of the A–B and B–C distances. The reason is that double crossovers (shown in Fig. 11–4) do not result in recombination between A and C, and lead to an underestimate of the distance between them.

X LINKAGE

For linkage analysis in X-linked pedigrees, the mother's father's genotype is particularly important because, as Figure 11–5 shows, it can demonstrate the linkage phase in the mother. Since there can be no recombination between X-linked genes in the male, and the mother always receives her father's only X chromosome, any X-linked marker in her genotype but not on her father's X must have come from her mother (unless, of course, it is a new mutation). If such a marker is close to the gene for an X-linked disorder known to be carried

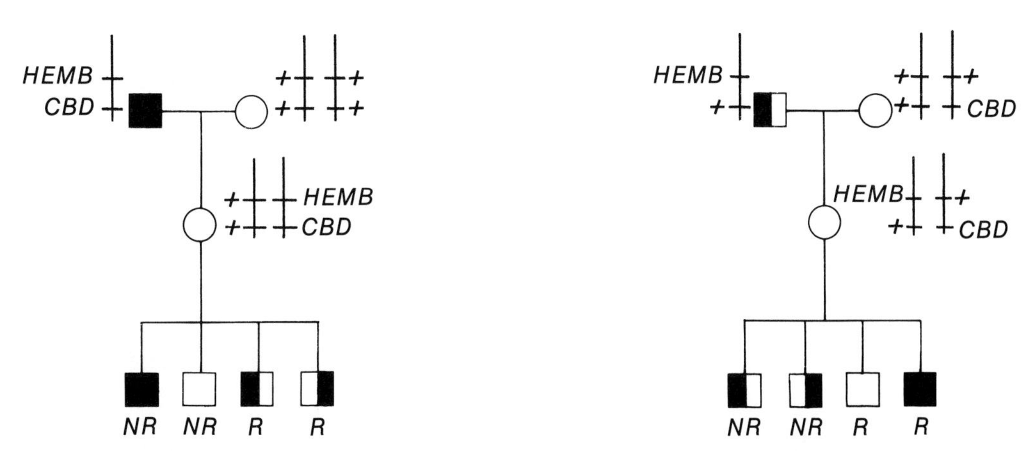

Figure 11–5. In X linkage, the maternal grandfather's phenotype can show the linkage phase in his daughter. On left, the maternal grandfather's X chromosome has genes for hemophilia B *(HEMB) and deutan color blindness (CBD)*. Thus his daughter has these two genes on one X and their normal alleles (+) on the other. Sons who have both conditions or neither are nonrecombinants (NR); those with one or the other are recombinants (R). On right, the mother has *HEMB* on one X, *CBD* on the other. Sons with one condition or the other are nonrecombinants; those with both or neither are recombinants.

Table 11–1. LOD SCORES FOR FAMILIES OF 1–7 CHILDREN, PHASE-KNOWN AND PHASE-UNKNOWN

Children Tested		Recombination Fraction																
Number	*Distribution*	*0.00*	*0.01*	*0.02*	*0.03*	*0.04*	*0.05*	*0.075*	*0.10*	*0.125*	*0.15*	*0.175*	*0.20*	*0.25*	*0.30*	*0.35*	*0.40*	*0.45*
1	1 NR:0 R	0.301	0.297	0.292	0.288	0.283	0.279	0.267	0.255	0.243	0.230	0.217	0.204	0.176	0.146	0.114	0.079	0.041
	0 NR:1 R	−∞	−1.699	−1.398	−1.222	−1.097	−1.000	−0.824	−0.699	−0.602	−0.523	−0.456	−0.398	−0.301	−0.222	−0.155	−0.097	−0.046
2	2 NR:0 R	0.602	0.593	0.585	0.576	0.567	0.558	0.534	0.511	0.486	0.460	0.435	0.408	0.352	0.292	0.228	0.158	0.082
	1 NR:1 R	−∞	−1.402	−1.106	−0.934	−0.814	−0.721	−0.557	−0.444	−0.359	−0.293	−0.238	−0.194	−0.125	−0.076	−0.041	−0.018	−0.005
	0 NR:2 R	−∞	−3.398	−2.796	−2.444	−2.194	−2.000	−1.648	−1.398	−1.204	−1.046	−0.912	−0.796	−0.602	−0.444	−0.310	0.194	−0.092
	Z 2:0	0.301	0.292	0.284	0.275	0.266	0.258	0.236	0.215	0.194	0.173	0.153	0.134	0.097	0.064	0.037	0.017	0.004
	Z 1:1	−∞	−1.402	−1.106	−0.934	−0.814	−0.721	−0.557	−0.444	−0.359	−0.292	−0.238	−0.194	−0.125	−0.076	−0.041	−0.018	−0.004
3	3 NR:0 R	0.903	0.890	0.877	0.863	0.850	0.837	0.802	0.765	0.729	0.690	0.652	0.612	0.528	0.438	0.342	0.237	0.123
	2 NR:1 R	−∞	−1.106	−0.813	−0.646	−0.530	−0.442	−0.290	−0.189	−0.116	−0.063	−0.021	0.010	0.051	0.070	0.073	0.061	0.036
	1 NR:2 R	−∞	−3.101	−2.504	−2.156	−1.911	−1.721	−1.381	−1.143	−0.961	−0.816	−0.694	−0.592	−0.426	−0.298	−0.196	−0.115	−0.051
	0 NR:3 R	−∞	−5.097	−4.194	−3.666	−3.291	−3.000	−2.472	−2.097	−1.806	−1.569	−1.368	−1.194	−0.903	−0.666	−0.465	−0.291	−0.138
	Z 3:0	0.602	0.589	0.576	0.562	0.549	0.533	0.501	0.465	0.429	0.393	0.356	0.318	0.243	0.170	0.104	0.049	0.013
	Z 2:1	−∞	−1.402	−1.106	−0.934	−0.814	−0.721	−0.557	−0.444	−0.359	−0.292	−0.238	−0.194	−0.125	−0.076	−0.041	−0.018	−0.004
4	4 NR:0 R	1.204	1.187	1.169	1.151	1.133	1.116	1.069	1.020	0.972	0.920	0.870	0.816	0.704	0.584	0.456	0.316	0.164
	3 NR:1 R	−∞	−0.809	−0.521	−0.358	−0.247	−0.163	−0.022	0.066	0.127	0.167	0.197	0.214	0.227	0.216	0.187	0.140	0.077
	2 NR:2 R	−∞	−2.805	−2.211	−1.868	−1.627	−1.442	−1.113	−0.888	−0.718	−0.586	−0.477	−0.388	−0.250	−0.152	−0.082	−0.036	−0.010
	1 NR:3 R	−∞	−4.800	−3.902	−3.378	−3.007	−2.721	−2.205	−1.842	−1.563	−1.339	−1.150	−0.990	−0.727	−0.520	−0.351	−0.212	−0.097
	0 NR:4 R	−∞	−6.796	−5.592	−4.887	−4.388	−4.000	−3.296	−2.796	−2.408	−2.092	−1.824	−1.592	−1.204	−0.888	−0.620	−0.388	−0.184
	Z 4:0	0.903	0.886	0.868	0.850	0.832	0.814	0.768	0.720	0.671	0.621	0.570	0.517	0.409	0.298	0.190	0.094	0.025
	Z 3:1	−∞	−1.110	−0.822	−0.659	−0.547	−0.464	−0.321	−0.229	−0.165	−0.119	−0.085	−0.060	−0.028	−0.011	−0.003	−0.001	−0.000
	Z 2:2	−∞	−2.805	−2.211	−1.868	−1.627	−1.442	−1.113	−0.887	−0.718	−0.585	−0.477	−0.388	−0.250	−0.151	−0.082	−0.035	−0.009
5	5 NR:0 R	1.505	1.483	1.461	1.439	1.417	1.395	1.336	1.275	1.215	1.150	1.087	1.020	0.880	0.730	0.570	0.395	0.205
	4 NR:1 R	−∞	−0.512	−1.229	−1.071	0.036	0.116	0.245	0.321	0.370	0.397	0.414	0.418	0.403	0.362	0.301	0.219	0.118
	3 NR:2 R	−∞	−2.508	−1.919	−1.580	−1.344	−1.163	−0.846	−0.633	−0.475	−0.356	−0.259	−0.184	−0.074	−0.006	0.032	0.043	0.031
	2 NR:3 R	−∞	−4.504	−3.609	−3.090	−2.724	−2.442	−1.937	−1.587	−1.320	−1.109	−0.933	−0.786	−0.551	−0.374	−0.237	−0.133	−0.056
	1 NR:4 R	−∞	−6.499	−5.300	−4.600	−3.028	−3.721	−3.028	−2.541	−2.165	−1.862	−1.606	−1.388	−1.028	−0.742	−0.506	−0.309	−0.143
	0 NR:5 R	−∞	−8.495	−6.990	−6.109	−5.485	−5.000	−4.120	−3.495	−3.010	−2.615	−2.280	−1.990	−1.505	−1.110	−0.775	−0.485	−0.230

	Z 5:0	1.204	1.182	1.160	1.138	1.115	1.093	1.035	0.975	0.914	0.851	0.787	0.720	0.581	0.436	0.288	0.149	0.042
	Z 4:1	−∞	−0.813	−0.530	−0.372	−0.265	−0.186	−0.056	0.022	0.070	0.099	0.117	0.124	0.118	0.095	0.063	0.031	0.008
	Z 3:2	−∞	−2.805	−2.211	−1.868	−1.627	−1.442	−1.113	−0.887	−0.718	−0.585	−0.477	−0.388	−0.250	−0.151	−0.082	−0.035	−0.009
6	6 NR:0 R	1.806	1.780	1.754	1.727	1.700	1.674	1.603	1.530	1.458	1.380	1.305	1.224	1.056	0.876	0.684	0.474	0.246
	5 NR:1 R	−∞	−0.216	0.063	0.217	0.320	0.395	0.511	0.576	0.613	0.627	0.631	0.622	0.579	0.508	0.415	0.298	0.159
	4 NR:2 R	−∞	−2.211	−1.627	−1.292	−1.061	−0.884	−0.579	−0.378	0.232	−0.126	−0.042	0.020	0.102	0.140	0.146	0.122	0.072
	3 NR:3 R	−∞	−4.207	−3.317	−2.802	−2.441	−2.163	−1.670	−1.332	−1.077	−0.879	−0.715	−0.582	−0.375	−0.228	−0.123	−0.054	−0.015
	2 NR:4 R	−∞	−6.203	−5.007	−4.312	−3.821	−3.442	−2.761	−2.286	−1.922	−1.632	−1.389	−1.184	−0.852	−0.596	−0.392	−0.230	−0.102
	1 NR:5 R	−∞	−8.198	−6.697	−5.821	−5.201	−4.721	−3.852	−3.240	−2.767	−2.385	−2.062	−1.786	−1.329	−0.964	−0.661	−0.406	−0.189
	0 NR:6 R	−∞	−10.194	−8.388	−7.331	−6.581	−6.000	−4.943	−4.194	−3.612	−3.138	−2.736	−2.388	−1.806	−1.332	−0.930	−0.582	−0.276
	Z 6:0	1.505	1.479	1.453	1.426	1.399	1.371	1.302	1.231	1.157	1.082	1.004	0.924	0.756	0.578	0.393	0.211	0.061
	Z 5:1	−∞	−0.517	−0.238	−0.084	0.019	0.093	0.211	0.276	0.312	0.329	0.331	0.323	0.284	0.222	0.149	0.076	0.021
	Z 4:2	−∞	−2.512	−1.928	−1.593	−1.361	−1.185	−0.877	−0.673	−0.524	−0.412	−0.324	−0.254	−0.153	−0.087	−0.044	−0.018	−0.004
	Z 3:3	−∞	−4.207	−3.317	−2.802	−2.441	−2.164	−1.670	−1.331	−1.077	−0.877	−0.715	−0.582	−0.375	−0.227	−0.123	−0.053	−0.013
7	7 NR:0 R	2.107	2.077	2.046	2.015	1.983	1.953	1.870	1.785	1.701	1.610	1.522	1.428	1.232	1.022	0.798	0.553	0.287
	6 NR:1 R	−∞	0.081	0.356	0.505	0.603	0.674	0.779	0.831	0.833	0.857	0.849	0.826	0.755	0.654	0.529	0.377	0.200
	5 NR:2 R	−∞	−1.915	−1.335	−1.005	−0.777	−0.605	−0.312	−0.123	0.011	0.104	0.175	0.224	0.278	0.286	0.260	0.201	0.113
	4 NR:3 R	−∞	−3.910	−3.025	−2.514	−2.158	−1.884	−1.403	−1.077	−0.834	−0.649	−0.498	−0.378	−0.199	−0.082	−0.009	0.025	0.026
	3 NR:4 R	−∞	−5.906	−4.715	−4.024	−3.538	−3.163	−2.494	−2.031	−1.679	−1.402	−1.171	−0.980	−0.676	−0.450	−0.278	−0.151	−0.061
	2 NR:5 R	−∞	−7.902	−6.405	−5.534	−4.918	−4.442	−3.585	−2.985	−2.524	−2.155	−1.845	−1.582	−1.153	−0.818	−0.547	−0.327	−0.148
	1 NR:6 R	−∞	−9.897	−8.095	−7.043	−6.298	−5.721	−4.676	−3.939	−3.369	−2.908	−2.518	−2.184	−1.630	−1.186	−0.816	−0.503	−0.235
	0 NR:7 R	−∞	−11.893	−9.786	−8.553	−7.678	−7.000	−5.767	−4.893	−4.214	−3.661	−3.192	−2.786	−2.107	−1.554	−1.085	−0.679	−0.322
	Z 7:0	1.806	1.776	1.745	1.714	1.682	1.650	1.569	1.486	1.400	1.312	1.221	1.128	0.932	0.723	0.502	0.278	0.084
	Z 6:1	−∞	−0.220	0.055	0.204	0.302	0.371	0.478	0.532	0.555	0.559	0.548	0.526	0.456	0.360	0.247	0.131	0.037
	Z 5:2	−∞	−2.216	−1.636	−1.306	−1.078	−0.907	−0.613	−0.422	−0.289	−0.192	−0.121	−0.070	−0.007	0.019	0.022	0.014	0.004
	Z 4:3	−∞	−4.207	−3.317	−2.802	−2.441	−2.164	−1.670	−1.331	−1.077	−0.877	−0.715	−0.582	−0.375	−0.227	−0.123	−0.053	−0.013

Sources: Smith et al., 1961; Smith 1968.

by her mother, the probability that she herself carries the disease gene is high. An application of this principle to genetic counseling for Duchenne muscular dystrophy is given later in this chapter.

LOD SCORES

Often in human linkage studies information can be obtained for only two generations. The lod score method, originated by Morton (1955), allows extraction of linkage information even if the linkage phase is unknown, and combination of information from phase-known and phase-unknown families. The term lod is simply an abbreviation of "logarithm of the odds." The principle is to set up a series of theoretical recombination frequencies, $\theta = 0.0$ (absolute linkage), 0.01, 0.02, 0.05, etc., and for each theoretical value to calculate the relative odds Z for a particular pedigree on the basis of that recombination fraction, as compared with the likelihood of there being no linkage ($\theta = 0.50$).

$$Z(\theta) = \log_{10} \frac{(\text{Probability of family for } \theta = 0.01 \text{ etc.})}{(\text{Probability of family for } \theta = 0.50)}$$

As the ratio is expressed as a logarithm, information from different pedigrees can be combined by simple addition. A lod score of 3 (equivalent to a 1000:1 probability in favor of linkage) is arbitrarily accepted as proof of linkage, and a score of -2 or even -1 is taken to rule out linkage.

Virtually all the early studies of human linkage used blood groups, so it is no coincidence that one of the most lucid descriptions of the use of lod scores appears in Race and Sanger's *Blood Groups in Man* (see General References).

Examples of lod scores are given in Table 11–1, for a range of recombination frequencies and for families of up to seven children. When the phase is known, the children can be scored as nonrecombinants (NR) or recombinants (R). When the phase is unknown, it is necessary to calculate the probability of obtaining the observed numbers of nonrecombinant and recombinant offspring separately for both the possible linkage phases in the doubly heterozygous parent; the average is taken as the Z score.

In some situations the doubly heterozygous parent's genotype must be derived from one of the offspring. For such cases, a correction e is used. As the e scores are often unnecessary and in any case make little difference to the Z scores, they are omitted from Table 11–1. The original source (Morton, 1955) or general references to this chapter should be consulted for a more extensive discussion.

Because of the higher recombination frequency in females than in males, it is necessary to separate the data for families in which the father is doubly heterozygous from those in which the mother is doubly heterozygous; otherwise a loose linkage can be missed.

Lod scores for a number of individual families can be added, and the totals graphed. The values of θ at which the relative probability is greatest is accepted as the best estimate of the recombination fraction.

MEASUREMENT OF LINKAGE IN PHASE-UNKNOWN AND PHASE-KNOWN PEDIGREES

The examples shown so far have all been ones in which the linkage phase of the genes in the doubly heterozygous parent is known. Linkage analysis is

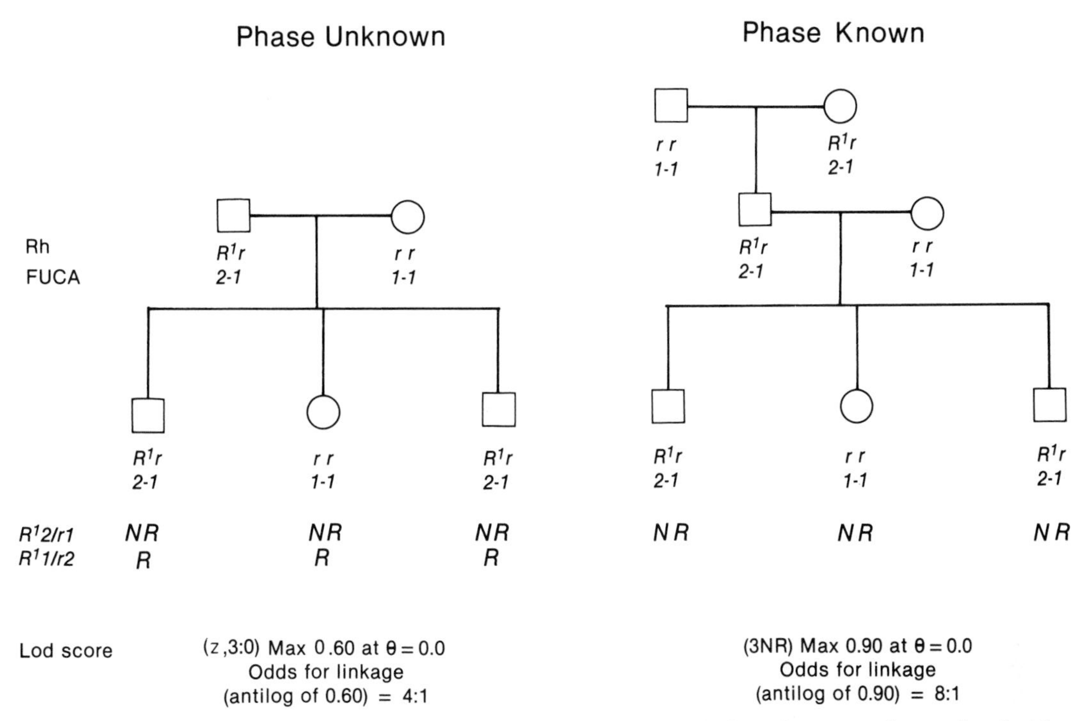

Figure 11–6. Comparison of linkage information in phase-unknown and phase-known pedigrees. See text for discussion.

much more informative in phase-known than in phase-unknown pedigrees. Figure 11–6 demonstrates this point in a family segregating at two loci known to be very closely linked $\theta = 0.0$) on chromosome 1: the Rh blood group locus and the locus for α-L-fucosidase. (The Rh blood group system is described in Chapter 9. Fucosidase is a polymorphic enzyme involved in glycoprotein metabolism; there are two normal alleles Fu^1 and Fu^2 and an abnormal silent allele Fu^0 which when homozygous causes the disease fucosidosis.)

Phase-unknown pedigree: Here the father is R^1r and Fu^1Fu^2, but we do not know whether his phase is R^1Fu^1/rFu^2 or R^1Fu^2/rFU^1. Two children have received R^1 and Fu^2 from him, and one has received r and Fu^1. If he is R^1Fu^2/rFu^1, all three children are nonrecombinants, but if he is R^1Fu^1/rFu^2, all are recombinants. Table 11–1 shows that when the count is 3:0, the following are the lod scores for different values of θ:

θ	0.00	0.01	0.02	0.03	0.04	etc.
Z 3:0	0.602	0.589	0.576	0.562	0.549	etc.

The maximum lod score is 0.602 at $\theta = 0.00$. The probability of linkage is the antilog of 0.602 = 4:1.

Phase-known pedigree: Information about the grandparents is now available and we see that the father received R^1 and Fu^2 from his father, r and Fu^1 from his mother; he is R^1Fu^2/rFu^1. All three children are nonrecombinants. Table 11–1 gives the following lod scores for different values of θ:

θ	0.00	0.01	0.02	0.03	0.04	etc.
3 NR:0 R	0.903	0.890	0.877	0.863	0.850	etc.

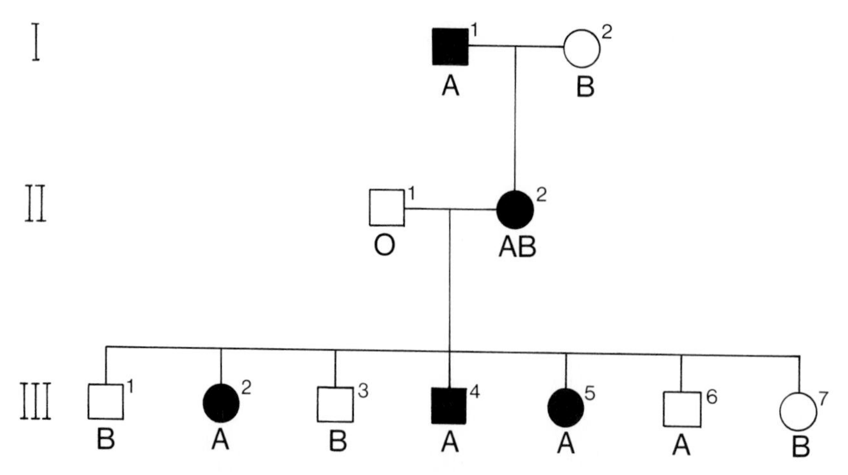

Figure 11–7. The ABO/nail-patella syndrome linkage in a three-generation family. Black symbols: nail-patella syndrome. Letters: ABO blood groups.

The maximum lod score is now 0.903 at $\theta = 0.00$; the probability of linkage is the antilog of $0.903 = 8:1$.

In our example, the family size is small (3 offspring). The larger the family, the more advantageous it is to know the linkage phase.

Figure 11–7 is another example of a three-generation pedigree, giving data for the nail-patella syndrome and the ABO blood groups. Nail-patella syndrome is an autosomal dominant disorder with nail dysplasia, hypoplastic patellae, other bony abnormalities and sometimes nephropathy. Is this a phase-known or phase-unknown pedigree? Which children, if any, are recombinants? (Answers in back of text.)

USEFULNESS OF POLYMORPHISMS

It is intuitively obvious that the more polymorphic a locus is, the more useful it is for detection of linkage. In Figure 11–8 the probability that a linkage between locus A (at which *A* codes for a dominant trait) and locus B will be much more readily identified if there are several reasonably common alleles at the B locus (shown as B^1, B^2, B^3, B^4) than if B is not highly polymorphic.

In the example shown on the left in Figure 11–8, each of the offspring is informative. However, if both the affected and the unaffected parent were B^1B^2,

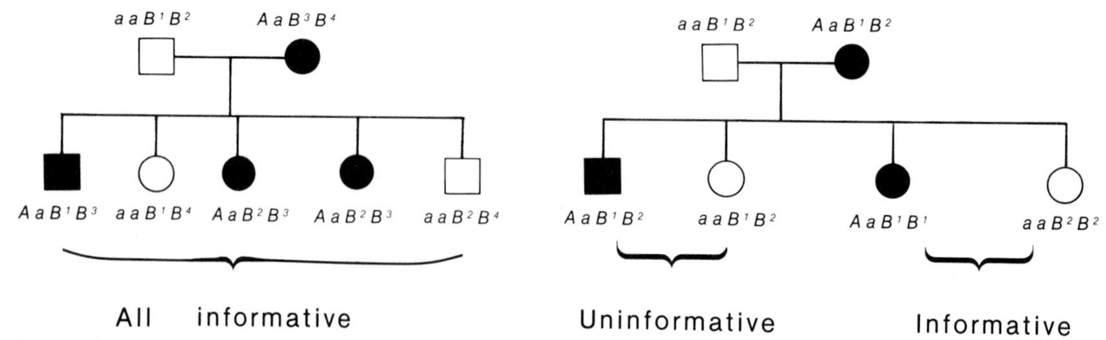

Figure 11–8. Linkage is more likely to be recognized at more highly polymorphic loci. In this figure, if there are four different alleles at the B locus, all offspring are informative; if there are only two alleles and both parents are B^1B^2, only half the offspring are informative.

Table 11–2. PIC OF NINE CODOMINANT HLA AND BLOOD GROUP MARKERS (Estimates Based on Caucasian Gene Frequencies)*

Locus	PIC	Number of Alleles
HLA-A	0.976	12
HLA-B	0.976	16
MNSs	0.638	4
Rh	0.580	6
Jk (Kidd)	0.375	2
P	0.375	2
Fy (Duffy)	0.368	2
K (Kell)	0.084	2
Lu (Lutheran)	0.058	2

*Data from Botstein et al., 1980.

as shown on the right, the probability that any one offspring would be informative is only 0.5.

The usefulness of any genetic marker can be expressed as its PIC (polymorphism information content). The PIC of a marker locus is the sum of the frequency of each mating type multiplied by the probability that an offspring of that mating type will be informative. Table 11–2 shows PICs for some of the more common markers. Note that in general the loci with the greatest number of alleles (here, HLA-A and HLA-B) have the highest PICs. Note also that among the two-allele systems, those in which the two alleles are close to equal in frequency are more informative than those in which one allele is rare; for example, Lutheran, in which the frequency of the rare *Lu*[a] is only 0.04, is at the bottom of the list in PIC and thus in usefulness. Students who wish to have a more complete introduction to PIC values are advised to consult the original reference (Botstein et al., 1980).

Gene Mapping

So far this account has been restricted to linkage—what it means and how it is measured—but the delineation of the human gene map goes further. Mapping is the actual assignment of genes to specific chromosomal locations. In addition to family studies, several additional methods, not all of which are mentioned here, are used for the actual assignment of genes to specific chromosomes. Mapping utilizes family studies, chromosome analysis, somatic cell studies and molecular technology, often in combination.

The methods of somatic cell genetics, especially somatic cell hybridization, have been particularly useful. In Chapter 10 mention was made that, in somatic cell hybrids, the presence of a particular gene from a donor, or the product of such a gene, can be correlated with a specific donor chromosome or even with part of a chromosome. Hybrid cell lines carrying deleted chromosomes or translocated segments rather than single whole chromosomes have helped to narrow down the localization of genes to specific chromosomal regions.

In situ hybridization of DNA probes to genes on chromosomes is a new technique which has recently been developed to the point of being sensitive enough to detect single-copy genes.

Trisomy mapping (by observation of three different alleles or a proportionate increase in gene product in a trisomic individual) has also been useful occasionally.

The use of recombinant DNA techniques has substantially increased the amount of the genome that can be covered for mapping purposes. Given 15 six-child families in which an autosomal dominant disease is segregating, the probability of mapping the disease gene is 8 percent if only the classical markers are used, but 44 percent if the mapped RFLPs are added (Skolnick et al., 1984).

Several diseases have already been shown to be linked to RFLPs. The first such linkage to be identified, a polymorphism near the β globin locus in sickle cell anemia (Kan and Dozy, 1978), was described in Chapter 5. Other examples are given later in this chapter.

THE CURRENT STATUS OF THE HUMAN GENE MAP

Since 1973, conferences have been held at two-year intervals to update the gene map in the light of current findings. By 1983, 824 genes had been mapped, 276 more than were mapped by 1981. Of these, 226 were mapped with the aid of recombinant DNA probes. Many more genes and anonymous DNA fragments were mapped within the next two years. Though this is still a small percentage of the total number of human genes, the accelerating pace of discovery suggests that the task of mapping the human genome may be complete within a few years. The more complete the map becomes, the easier it will be to map additional gene loci.

So far, relatively few of the mapped genes are disease genes. McKusick's map of the "morbid anatomy" of the human genome (Figure 11–9) shows about 75 well-known mapped disorders. Identification of polymorphic markers (loci of enzymes, red cell or white cell antigens, or RFLPs) tightly linked to any of these loci will allow presymptomatic diagnosis, carrier identification, genetic counseling and prenatal diagnosis to be given for a whole range of disorders for which little can be offered at present. Meanwhile, the number of disorders that are mapped and therefore potentially open to genetic management is rapidly increasing.

Clinical Applications of RFLPs

RFLPs are polymorphisms of DNA fragments that can be detected by means of electrophoresis in agarose gels of the fragments produced by specific DNA restriction enzymes (see Chapter 3). They are an important class of genetic markers because they are inherited as simple Mendelian codominants and in many cases are highly polymorphic.

Several significant diseases are already known to be linked to RFLPs. If the pace of discovery continues, it should soon be possible to show linkage of almost any disease locus to an RFLP, map it and use the information in genetic medicine.

HUNTINGTON DISEASE

As described earlier (see Chapter 4), Huntington disease (HD) is autosomal dominant but its genetics is complicated by late onset and somewhat variable onset age. A number of attempts have been made to find a genetic marker linked to the Huntington locus, but linkage studies with the usual polymorphic protein markers were uniformly negative. It is therefore of great interest that a very closely linked DNA polymorphism has been found (Gusella et al., 1983).

The family in which the linkage was first identified is an 8000-member Venezuelan group in which there are numerous cases of Huntington disease, all inherited from a common ancestor. An "anonymous" DNA fragment from chromosome 4, known as G8, reveals variation at two sites when digested with

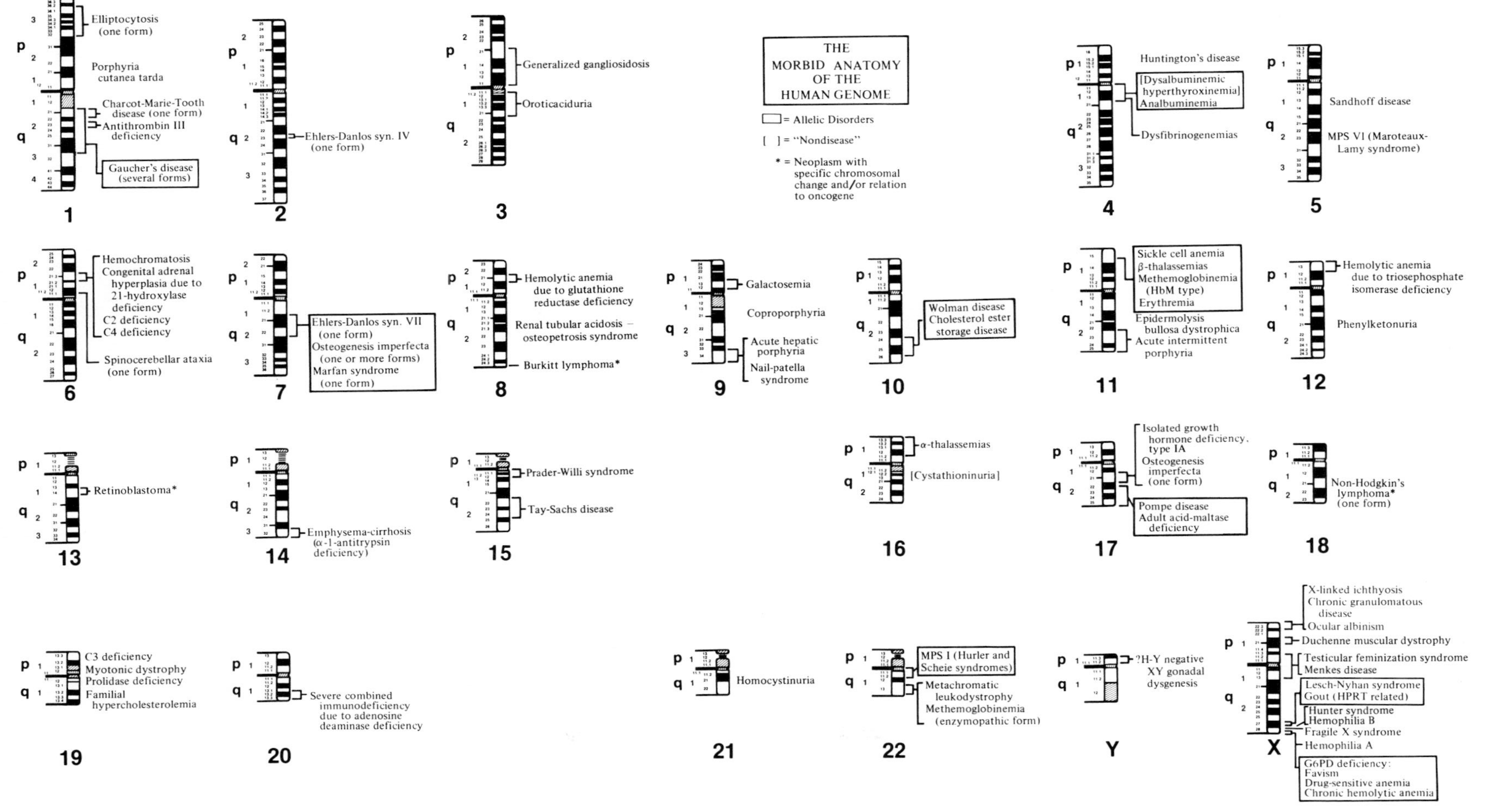

Figure 11–9. Map of the morbid anatomy of the human genome as of November 1984. Courtesy of V. A. McKusick.

the restriction enzyme *Hin*dIII. Each site has two "alleles"; that is, *Hin*dIII will or will not cut the fragment at that site. In summary:

	Allele 1	Allele 2
Site 1	− (0.74)	+ (0.26)
Site 2	+ (0.81)	− (0.19)

+ indicates presence of a cleavage site, − indicates its absence. The bracketed numbers are frequencies; by convention, the more frequent allele is termed allele 1, the less frequent allele 2. Four haplotypes result:

	Site 1	Site 2	Frequency
A	−	+	0.60
B	−	−	0.14
C	+	+	0.21
D	+	−	0.05

Each individual has two haplotypes, one on each chromosome 4. Note that the AD combination (− +/+ −) is not distinguishable from BC (− −/+ +) in agarose gels, but family studies may resolve the uncertainty. The chance that any individual is heterozygous at the locus is 1.00 minus the chance of being homozygous for one of A, B, C or D:

$$1.00 - (.60^2 + .14^2 + .21^2 + .05^2) = .57$$

With a predicted probability of heterozygosity of 57 percent, G8 is quite a good genetic marker, though not ideal; 43 percent of HD patients have homozygous G8 haplotypes, and for them the G8 polymorphism is not helpful.

In the Venezuelan family studied, the lod score was 8.53 at $\theta = 0.0$, indicating very close linkage (though the 99 percent confidence limits of the estimate of the distance between the two loci is 10 cM). No recombinants were observed within the Venezuelan family, but recombination has occasionally been recognized in families studied later.

Clinical Implications

If further studies show that HD is not heterogeneous and if the recombination frequency between the HD and G8 loci can be precisely measured and is found to be very small, it will be possible to use the linkage in informative families to identify individuals who have the Huntington disease gene at a presymptomatic stage, and even prenatally. If additional variation is found in G8 or closely linked to it, the probability of identifying heterozygosity, and thus the informativeness of linkage analysis in HD families, will be improved.

The discovery of a linked marker also raises the possibility of applying recombinant DNA technology to clone and characterize both the gene for Huntington disease and its normal allele, and to use their DNA sequences to determine the amino acid sequence of the normal gene product and the nature of the defect that leads to Huntington disease.

Ethical Implications

Huntington disease is a devastating disorder, and any child of a known patient is at 50 percent risk of inheriting it. An at-risk but asymptomatic member of an HD family faces a serious dilemma if forced to decide whether or not to have a linkage test and resolve the uncertainty. On the other hand, the spouse of such a person may feel entitled to have the information before making reproductive decisions. The fact that not all tested families are informative compounds the difficulty of the situation.

At the time of writing the G8/HD linkage is not yet being used clinically, but the problem of clinical application will have to be faced in the near future. Meanwhile DNA from blood samples of diagnosed family members can be stored and used eventually if required for linkage studies. Banks for long-term storage of these samples have already been established in a number of North American centers.

ALPHA THALASSEMIA

In prenatal diagnosis, a fetus with homozygous $--/--$ α thalassemia (hydrops fetalis) can be identified by absence of the α genes in fetal cells. Since each parent is $--/\alpha\alpha$, α genes are present in both heterozygous and homozygous normal fetuses.

BETA THALASSEMIA

Prenatal diagnosis of β thalassemia is complicated because β thalassemia is the result of a group of different mutations with rather similar consequences, not a single condition. Though homozygous β thalassemia has usually been identified in fetuses by studies of β chain synthesis in fetal blood samples, a number of techniques are available that use either RFLPs or radioactive oligonucleotide probes (short synthetic DNA fragments that will hybridize only to exactly matching DNA sequences).

Application of RFLP technology to prenatal testing for β thalassemia requires careful advance assessment of the parents and grandparents with respect to their normal and mutant β globin genes and the linkage phase of these genes with respect to RFLPs. Since the common polymorphisms vary in ethnic groups, the approach is to look for the polymorphisms most likely to be present in a given family. If the distribution of markers in the family appears to be informative, fetal DNA can be analyzed. Obviously this is a complicated procedure, and at present is available in only a few referral centers.

PHENYLKETONURIA

Several different restriction enzyme polymorphisms have already been detected either within or extremely close to the PKU gene. In informative families in which heterozygous parents have already had a PKU child, polymorphism analysis in a fetus would allow prenatal diagnosis (Woo et al., 1983). So far the method is applicable only to families with a previous affected child.

DUCHENNE MUSCULAR DYSTROPHY (DMD)

At the time of writing, analysis of DNA close to the DMD locus is proceeding quite rapidly, and the account given here describes only the early work.

Figure 11–10 shows two RFLPs that span the Duchenne locus on the short arm of the X chromosome (Davies et al., 1983).

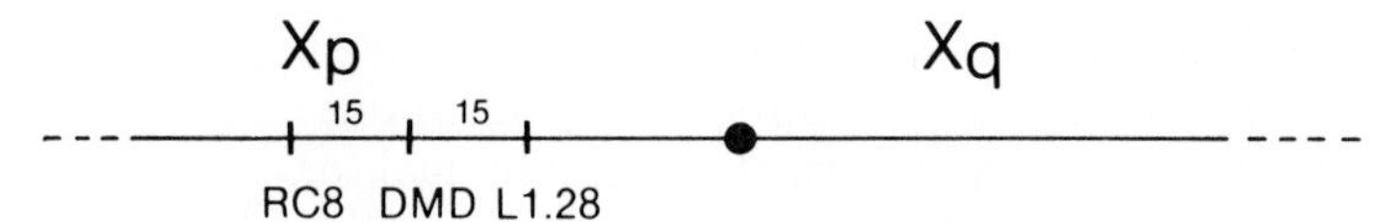

Figure 11–10. The first two restriction fragment length polymorphisms (RFLPs) to be identified within measurable distance of the Duchenne muscular dystrophy locus (DMD), RC8 and L1.28. See text.

RFLP	Frequency of Allele 1	Frequency of Allele 2	Distance from DMD
RC8 ("B")	0.89	0.13	15 cM (distal)
L1.28 ("C")	0.64	0.32	15cM (proximal)

Both these sites are too remote from the DMD locus to be useful for carrier detection or prenatal diagnosis. It is only in the few families in which a mother is heterozygous for both markers and the linkage phase is known that they are of real clinical help (Fig. 11–11). However, they provided a first step in a molecular approach to DMD, a condition in which carrier detection had been unsatisfactory and prenatal diagnosis impossible. Identification of a series of other DNA probes, some of which are closer to the DMD locus than RC8 or L1.28, has made Duchenne carrier detection and prenatal diagnosis feasible for many families (Bakker et al., 1985).

One important outcome of the application of RFLPs to muscular dystrophy families is that it is now clear that the milder Becker form of X-linked muscular dystrophy is also determined by a gene on Xp, which may well be allelic to the Duchenne gene (Kingston et al., 1984).

HEMOPHILIA B

For several years it has been possible to detect hemophilia B (Factor IX deficiency) in utero by Factor IX assay in fetal blood obtained by fetoscopy, though the method is not widely available and not totally reliable technically. More recently, a genomic DNA probe containing part of the gene, as well as a cDNA probe, has become available. Use of the genomic DNA probe has led to the identification of an intragenic polymorphic site at which the frequencies of the two known alleles are estimated to be 0.65 and 0.35 and the proportion of heterozygous females should therefore be $2 \times 0.65 \times 0.35 = 0.46$. Use of this probe allows carrier detection in appropriate families and should also allow prenatal detection in informative families if desired (Giannelli et al., 1984).

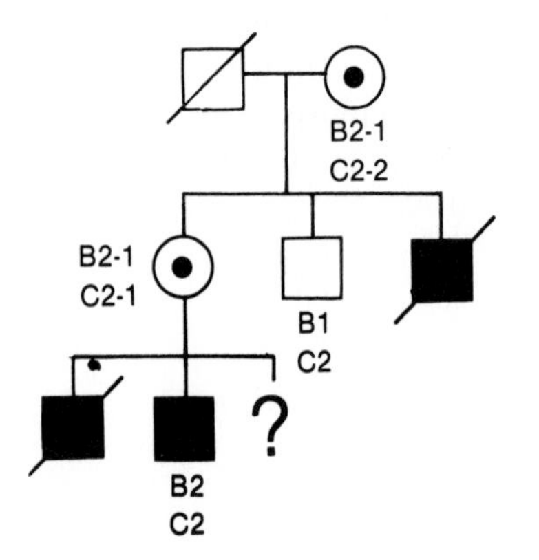

Figure 11–11. Demonstration of the possible use of the linked RFLPs RC8(B) and L1.28(C) in prenatal diagnosis. Here the obligate carrier I–2 has transmitted B2, C2 and the DMD gene to her daughter, II–1, and has given her unaffected son B1, C2 and the normal allele of DMD. II–1 has transmitted B2, C2 and the DMD gene to her son III–2. In her next pregnancy, a male with her B1 and C1 alleles would have close to a 98 percent chance of being unaffected. The chance of a double crossover is only $(0.15)^2$ or 0.0225 (2.25 percent).

GENERAL REFERENCES

Conneally PM, Rivas ML. Linkage analysis in man. Adv Hum Genet 1980; 10:209–266.
McKusick VA. Human gene map 1 December 1984. Clin Genet 1985; 27:207–239.
Race R, Sanger R. Blood groups in man, 6th ed. Oxford: Blackwell, 1975.
Shows TB, Sakaguchi AY, Naylor SL. Mapping the human genome, cloned genes, DNA polymorphisms, and inherited disease. Adv Hum Genet 1982; 12:341–468.
Sparkes RS, Berg K, Evans HJ, Klinger HP, eds. Human gene mapping 7. Seventh international workshop on human gene mapping. Cytogenet Cell Genet 1984; 37(1–4); simultaneous publication in Birth Defects 1984; 20(2).

PROBLEMS

1. a) Question in text re: Figure 11–7
   b) Check the lod score table. For this family, what recombination frequency between the two loci is most likely?

2. Classify each of the following pairs of linked genes as to whether they are in linkage equilibrium or linkage disequilibrium in the population in which these frequencies are found.
   a) HLA system: *A1* 0.17, *B8* 0.11, *A1 B8* 0.09
   b) HLA system: *A3* 0.13, *B7* 0.11, *A3, B7* 0.07
   c) HLA allele *A1* 0.17
   Glyoxalase allele GLO^1 0.013
   d) ABO system: *0* 0.68, *A* 0.26, *B* 0.06
   Gene for nail-patella syndrome 0.0001
   Nail-patella syndrome with gene *A* 0.000026

3. For the examples in question 2 that are in disequilibrium, calculate what the frequency of the combination would be if it were at equilibrium.

4. In a certain population the dominant secretor gene *Se* has a frequency of 0.6 and the Lu^a gene of the Lutheran blood group system has a frequency of 0.04. The two loci are linked on chromosome 19, with a recombination frequency of 0.10.
   a) What is the frequency of $Se\ Lu^a/se\ Lu^b$ people in a population at equilibrium?
   b) What is the frequency of $se\ Lu^a/Se\ Lu^b$ people in the same population?
   c) What would these frequencies be if the recombination frequency were
   1) 0.01 and 2) 0.50?

12 MULTIFACTORIAL INHERITANCE

In Chapter 1 the three major categories of genetic disorders were listed and briefly characterized, and in subsequent chapters single-gene traits and syndromes caused by chromosome abnormalities were treated in some detail. We now turn to the third major type of genetic variation, the type in which inheritance is **multifactorial**. A multifactorial trait is defined as one that is determined by a combination of factors, genetic and possibly also nongenetic, each with only a minor effect. Multifactorial inheritance is less familiar to physicians and medical students (and more difficult to analyze) than the other types of inheritance, but it is believed to account for much of the normal variation in families, as well as for a number of common disorders, including many single congenital malformations. Multifactorial disorders tend to cluster in families, but they do not show any particular genetic pattern in individual families; however, if a series of families is studied, certain common characteristics can be seen. In this chapter the basis of multifactorial inheritance is described, the concept of a **multifactorial-threshold trait** is introduced and the distinguishing features of multifactorial inheritance are given, chiefly as they apply to some common congenital malformations.

The term **polygenic** is often used interchangeably with multifactorial, but is preferably used in a more restricted sense to refer to conditions determined by a large number of genes, each with a small effect, acting additively (Fraser, 1976). In actual experience, it is often hard to judge whether or not environmental factors are concerned in the etiology of a defect, or whether all the genes determining a trait have small and additive effects. More complex causes of genetic variation may be present, for example, a major locus and multiple minor factors simultaneously. However, the family patterns of many normal traits and genetic defects follow the theoretical expectations for multifactorial inheritance.

The idea of multifactorial inheritance arose from genetic experiments with continuously variable, measurable traits. As an example, assume that seed size is determined by three gene pairs and that two inbred lines are homozygous *aabbcc* and *AABBCC* respectively. Further assume that genes *A*, *B* and *C* each add 1 unit to seed size, and are additive, not dominant, in their effects. If the mean size of *aabbcc* seeds is 6 units, that of AABBCC seeds is 12 units. Now if a genetic cross *AABBCC* × *aabbcc* is made, all the F_1 progeny are *AaBbCc* and have an average size of 9 units. If the F_1 are now crossed, the F_2 generation shows a range of seed sizes:

Seed Size	Proportions
6	1
7	10
8	15
9	20
10	15
11	10
12	1

As shown, alleles with small additive effects at just three loci can produce a distribution of seed sizes that approximates a normal curve.

Normal Variation

Many traits have a unimodal distribution in the population. In general this type of distribution is characteristic of multifactorial inheritance, but by no means diagnostic of it; as described earlier (see Fig. 9–7), red cell acid phosphatase activity is determined by combinations of three alleles at a single locus but appears to have a unimodal distribution.

As a rule of thumb, when a trait is distributed normally in the population, measurements more than two standard deviations from the mean in either direction are regarded as "abnormal." The interval between two standard deviations *below* the mean and two standard deviations *above* it ($\bar{x} \pm 2\sigma$) includes more than 94 percent of all observations. Physicians use this idea in many areas; for example, if a child's head circumference is below the third percentile, it is abnormally small; if above the 97th, it is abnormally large. The 3 percent of the population with the lowest intelligence are judged to be mentally retarded; this group includes individuals with retardation of multifactorial origin as well as others in whom the retardation has a specific cause.

Among the many normal traits that are distributed unimodally and have family patterns characteristic of multifactorial inheritance, we will mention three: stature, total fingerprint ridge count (TRC), which is defined in Chapter 17, and intelligence as measured by the intelligence quotient (IQ). Each of these is a metrical trait, that is, a trait that is phenotypically described in terms of measurements (cm of height, number of digital ridges or IQ points).

A population distribution of stature is shown in Figure 12–1. Both genetic and environmental factors are involved in determining height. The distribution may show a slight hump at the lower end of the range, accounted for by people who are exceptionally short, often because of some gene with a major effect on stature (achondroplasia, for example). A few people at the upper end of the range owe their stature to a genetic disorder (for example, the Marfan syndrome or an XXY or XYY chromosomal constitution), but the conditions causing increased stature seem to be too few and too rare to be noticeable in the overall population distribution.

To explore how stature is inherited in families, the most useful measure is the **correlation coefficient**, described in any standard textbook of statistics. If mating is random, in the sense that mates are chosen without regard to genotype, and if environmental differences are not involved, the correlation between parent

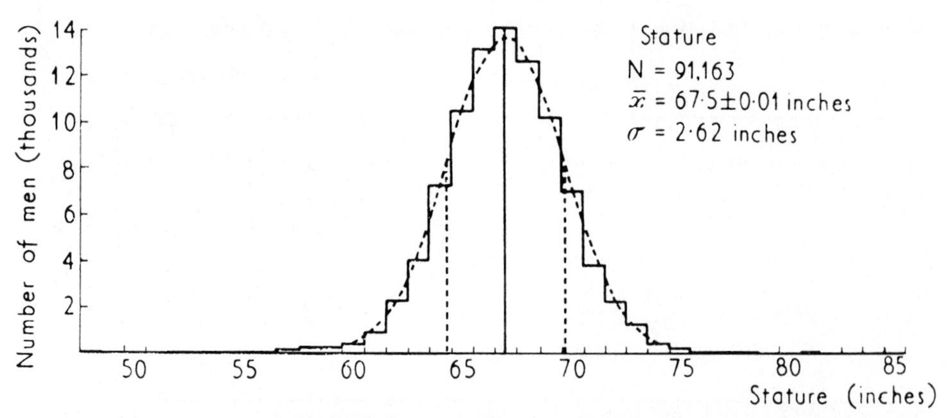

Figure 12–1. Distribution of stature in a sample of 91,163 young English males in 1939. Dotted line is a normal curve with the same mean *(x̄)* and standard deviation *(σ̄)*. From Harrison GA, Weiner JS, Tanner JM, Barnicot MA, Reynolds V. Human biology, 2nd ed. Oxford: Oxford University Press, 1977.

and child, or between sibs, is 0.50. In more general terms, the correlation between relatives is proportional to their genes in common (Fisher, 1918). **Genes in common** are those genes inherited from a common ancestral source.

At this point more terminology must be introduced. Relatives may be classified as **first-degree, second-degree** and so on in terms of how many steps apart they are in the family tree (Fig. 12–2). Relatives of the same degree have the same proportion of genes in common. For analysis of multifactorial inheritance, all relatives within the same degree may be considered together. Table 12–1 lists the closest degrees of relationship, the chief examples of relatives of each degree and the proportion of genes each has in common with a given proband.

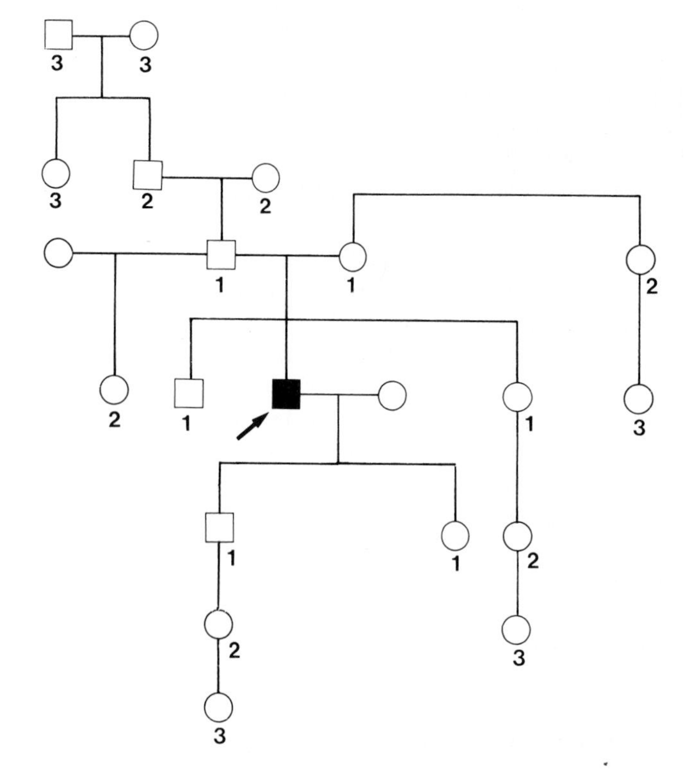

Figure 12–2. Degrees of relationship to a proband (arrow). First-degree relatives are parents, sibs and offspring. Second-degree relatives are half-sibs, uncles and aunts, nephews and nieces, grandparents and grandchildren. Third-degree relatives include first cousins.

Table 12–1. DEGREES OF RELATIONSHIP AND GENES IN COMMON

Relationship to Proband	Proportion of Genes in Common with Proband
MZ twin	1
First-degree relative (parent, DZ twin or other sib, child)	1/2
Second-degree relative (grandparent, uncle or aunt, half-sib, nephew or niece, grandchild)	1/4
Third-degree relative (great-grandparent, great-grandchild, first cousin, etc.)	1/8

Note that the principle of genes in common also applies to single-gene traits; that is, the probability that two relatives share the same gene, inherited from the same source, is the same as the proportion of all their genes that they have inherited from the same source.

Table 12–2 lists the correlation coefficients for the three traits mentioned above—stature, TRC and IQ. For all three, the correlation coefficient for any degree of relationship is close to the theoretical value. Correlations between relatives are not always as close as this for metrical traits, because the correlation coefficient can be affected by numerous factors (for example, by presence of dominant rather than additive genes, environmental variation and nonrandom mating).

As Table 12–2 shows, the correlation coefficients are particularly close to the theoretical values for fingerprint total ridge count (TRC). Mating is likely to be random with respect to TRC (unless TRC is associated with some other characteristic, such as ethnic origin, for which mating is nonrandom). TRC is a particularly good example of a metrical trait determined by multifactorial inheritance.

REGRESSION TO THE MEAN

Regression toward the mean is a common observation in family studies of metrical traits. Consider, as an example, the correlation of height in mother and daughter. A daughter receives half her genes from her mother. Theoretically, for a series of mothers of given height, the range of stature in the fathers is the same as the population distribution. Thus the average height of the daughters should be midway between the mother's height and the population mean.

In everyday terms, the law of regression to the mean accounts for the observation that parents who are especially extreme in terms of their location in the normal curve are likely to have children who, on average, are distinctly less

Table 12–2. CORRELATION BETWEEN RELATIVES FOR CERTAIN HUMAN TRAITS

Trait		Correlation Coefficient*	Genes in Common†
Stature	Sib-sib	0.56	0.50
	Parent-child	0.51	0.50
Fingerprint total ridge count	MZ twins	0.95	1.00
	DZ twins	0.49	0.50
	Sib-sib	0.50	0.50
	Parent-child	0.48	0.50
	Parent-parent	0	0
Intelligence (IQ)	Sib-sib	0.58	0.50
	Parent-child	0.48	0.50

*Data from Bodmer and Cavalli-Sforza, 1976, and Holt, 1968.
†Proportion of genes in common equals theoretical correlation coefficient.

unique than the parents. This can be seen, for example, at both ends of the range for stature or for intelligence. Especially brilliant parents tend to have children less outstanding than themselves though still above average in intelligence, whereas dull normal parents find that their children usually rank higher than themselves, though still below average. Similarly, extremely tall or short parents have children whose mean stature is closer to the population average.

The effect of major deleterious genes or chromosomal defects on intelligence is quite different. Whereas parents of moderately retarded children tend to have below-average intelligence, parents of severely retarded children have a normal range of intelligence. Moderate retardation is usually multifactorial, whereas severe retardation usually has a single major cause.

Heritability

To try to separate the relative roles of genes and environment for multifactorial traits, the idea of heritability has been developed. Heritability is defined as the proportion of the total phenotypic variance of a trait that is caused by additive genetic variance. Students of statistics will recall that variance is a measure of dispersion about the mean, that is, a measure of how much an individual value is likely to vary from the typical value, the mean.

The phenotype of a multifactorial trait (or the liability of an individual to a threshold trait) is taken to be the sum of three values:

- g contribution of additive genetic factors, with variance G
- b contribution of environment within the family (sometimes called "cultural inheritance") with variance B
- e contribution of random environmental factors, with variance E

Heritability (h^2) is then

$$\frac{G}{G + B + E} = \frac{G}{V}$$

Thus in general terms heritability is a measure of whether the role of genetic factors in determining a given phenotype (or liability to a phenotype) is large or small.

One method of estimating heritability is by comparison of the ratio of the concordance rate of a condition in monozygotic (MZ) and dizygotic (DZ) twin pairs with the incidence of the trait in the general population. Concordant twin pairs are those in which both twins have the trait; discordant twin pairs are those in which one twin is affected, the other is not. Some typical data for concordance rates for a number of different conditions in twins are given in Chapter 16. The concept of heritability and ways of measuring it are discussed at length by Vogel and Motulsky and by Emery (see General References to this chapter).

Threshold Traits

Many common congenital malformations result from a failure in a developmental process, so that subsequent developmental stages cannot be reached. The rate of development may be visualized as a continuous distribution determined by many factors, some genetic and others environmental. If a process fails to reach a certain stage at the appropriate time, a malformation may result; in that

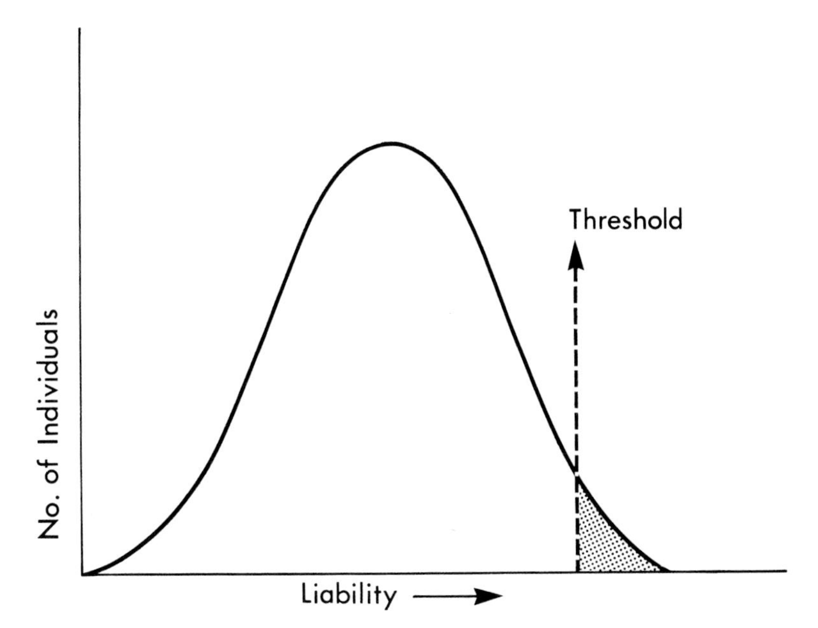

Figure 12–3. The multifactorial/threshold model. Liability to a trait is normally distributed, with a threshold dividing the population into normal and abnormal classes.

case the continuously variable distribution becomes divided into a normal and an abnormal class, separated by a threshold (Fig. 12–3).

The idea of a threshold effect was originally developed by Wright (1934) in connection with a study of the inheritance of polydactyly in guinea pigs. Usually the guinea pig has three toes, but occasionally a four-toed one is born. Wright found that single-gene inheritance could not explain the transmission of polydactyly and decided that it must result from a threshold effect in a trait determined by many genes. The group as a whole has continuously distributed liability to the trait, but the threshold separates it into two classes, one with polydactyly and one without. Later Grüneberg (1952), working with another type of congenital malformation in another animal, the mouse, coined the term **quasicontinuous variation** for this kind of inheritance.

A number of human congenital malformations have been shown to fit the family patterns expected of the multifactorial-threshold model (Table 12–3). The simplest method of analysis is by collection of information on the frequency of the malformation in the general population and in different categories of relatives. From this information, the **empiric risk** of the condition in subsequent family members can be estimated. The empiric risk is based solely on past experience, and does not require knowledge of the nature of the genetic and environmental factors in the pathogenesis of the malformation.

Table 12–3. SOME COMMON CONGENITAL MALFORMATIONS WITH MULTIFACTORIAL INHERITANCE

Cleft lip ± cleft palate	Neural tube defects
Cleft palate	Anencephaly
Club foot (talipes equinovarus)	Spina bifida
Congenital dislocation of the hip	Pyloric stenosis
Congenital heart defects	
Ventricular septal defect	
Patent ductus arteriosus	
Atrial septal defect	
Tetralogy of Fallot	
Others	

Note: many of these disorders are heterogeneous, that is, usually but not invariably multifactorial in etiology. Data from Carter, 1976; Nora, 1968.

CLEFT LIP AND CLEFT PALATE

Cleft lip with or without cleft palate (CL ± CP) is etiologically distinct from cleft palate (CP) alone (Fogh-Andersen 1942, Fraser 1980). (There is one exceptional condition in which either CL or CP alone, or CLP, may occur: the autosomal dominant disorder known as lip pit syndrome, in which mucous cysts of the lower lip accompany clefting.) Both categories are heterogeneous and include numerous single-gene syndromes in which CL ± CP or CP alone is a feature; cases associated with chromosomal syndromes (especially trisomy 13); cases resulting from teratogenic agents (the rubella virus, thalidomide, anticonvulsants); and forms that appear in nonfamilial syndromes. When all these have been removed, the remaining cases, the great majority, are in the multifactorial group.

Cleft Lip With or Without Cleft Palate (CL ± CP).

This disorder originates as a failure of fusion of the frontal prominence with the maxillary process, at about the 35th day of uterine development. About 60 to 80 percent of those affected are males. There is considerable variation in frequency in different ethnic groups: 1 per 1000 in whites, 1.7 per 1000 in the Japanese and 0.4 per 1000 in American blacks. When neither parent is affected, the recurrence risk in subsequent sibs of an affected child is about 4 percent, but after two affected children it is considerably higher, about 9 percent. The risk varies with the severity of the defect; for example, the recurrence risk in the data of Melnick and colleagues (1980) is as follows:

Cleft lip and palate, bilateral	8.0 percent
Cleft lip without cleft palate, bilateral	6.7 percent
Cleft lip and palate, unilateral	4.9 percent
Cleft lip without cleft palate, unilateral	4.0 percent

Cleft Palate (CP)

Failure of fusion of the secondary palate has a population incidence of 1 in 1500 and is more common in females than in males. There is little ethnic variation in incidence. The recurrence risk in sibs is only about 2 percent, apparently regardless of family history, though more data are needed. Microforms of CP include bifid uvula and congenital palatopharyngeal incompetence, a condition in which the individual speaks as though the palate is cleft, even though no cleft is apparent.

Cleft palate has been carefully studied in the mouse. It is clearly multifactorial, and the threshold is easy to define. If the palate is to fuse normally, the palatine shelves must move from a vertical position on either side of the tongue to a horizontal position above it. If the shelves move into position too late to meet and fuse, a cleft results. Many factors contribute to the underlying continuous variation of liability: the time the shelves move to the horizontal position, the width of the head at that time, the size and position of the tongue, the size of the shelves themselves and so forth.

PYLORIC STENOSIS

The multifactorial-threshold model was originally developed to explain the unusual familial incidence of pyloric stenosis (Table 12–4). In this disorder there are severe feeding problems, with projectile vomiting in infancy resulting from narrowing of the pyloric opening caused by hypertrophy of the pyloric sphincter.

Table 12–4. THE INCIDENCE OF PYLORIC STENOSIS IN OFFSPRING OF PATIENTS AS COMPARED WITH THE INCIDENCE IN THE GENERAL POPULATION

Relatives	Risk	Multiple of Population Risk
Sons of male patients	1 in 18	× 11
Daughters of male patients	1 in 42	× 24
Sons of female patients	1 in 5	× 40
Daughters of female patients	1 in 14	× 70

Data from Carter, 1969.

The Ramstedt surgical procedure is performed to relieve the obstruction. Pyloric stenosis is five times as common in boys as in girls (males, 5 per 1000; females, 1 per 1000). Carter (1976) showed that the familial incidence can be explained if pyloric stenosis is a threshold trait (Fig. 12–4).

If the underlying liability to a trait is continuous but there is a lower threshold in males than in females, then affected females are more extreme in range (farther to the right) than are affected males, as shown by the relative area of each curve beyond the threshold. The more extreme the parent, the higher above the population mean is the mean of the offspring; in other words, offspring of affected females have a higher mean liability to pyloric stenosis than do offspring of affected males. However, the threshold is lower for males than for females, so whether the condition has been inherited from the father or from the mother, more sons than daughters are affected. The greatest risk (about 20 percent) is for sons of female patients. The same principles and risk figures apply

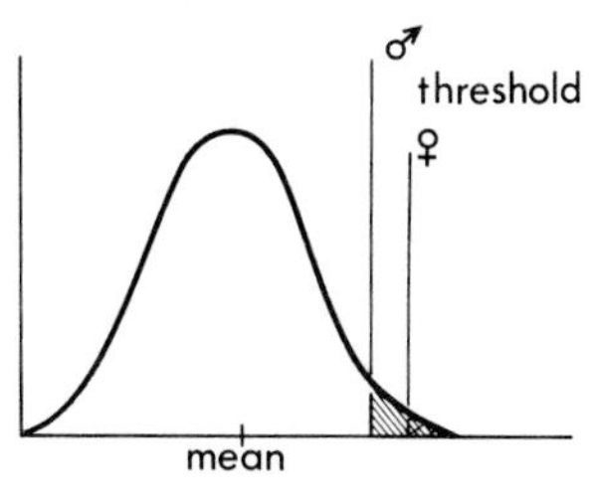

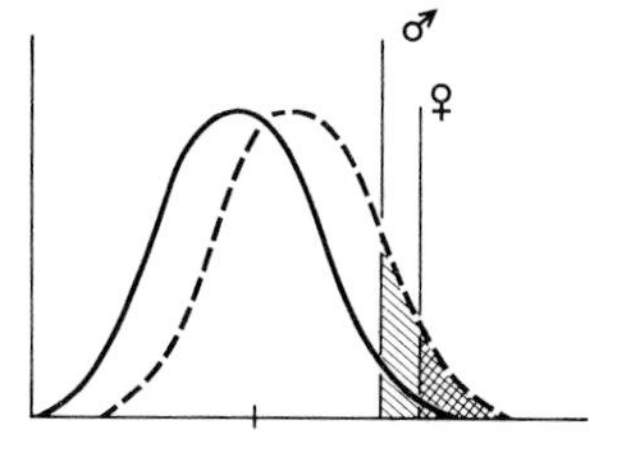

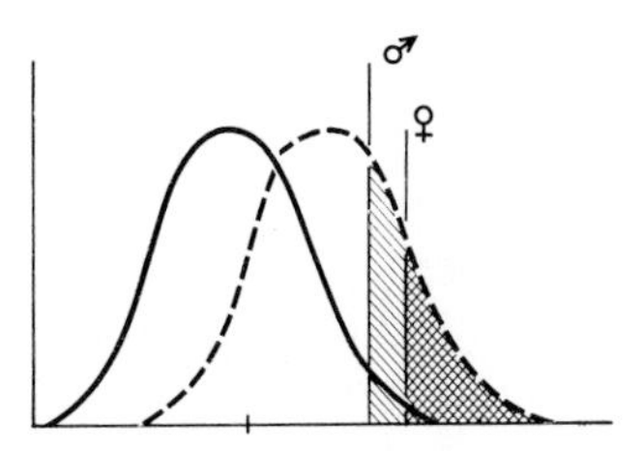

Figure 12–4. An interpretation of the sex ratio and familial distribution of pyloric stenosis in terms of multifactorial inheritance, with a more extreme threshold in females than in males. See text for discussion. Based on Carter CO. The genetics of common malformations. In Fishbein M, ed. Congenital malformations: papers and discussions presented at the second international conference on congenital malformations. New York: International Medical Congress, 1964.

to sibs as to offspring, since both offspring and sib have half their genes in common with the proband.

NEURAL TUBE DEFECTS

Anencephaly and spina bifida are neural tube defects (NTDs) that occur together frequently in families and are considered to have a common origin (Fig. 12–5). A small proportion (less than 10 percent) of NTDs have identifiable specific causes: amniotic bands (fibrous connections between amnion and fetus resulting from early rupture of the amnion); single-gene defects such as Meckel syndrome, an autosomal recessive malformation syndrome; chromosomal syndromes, especially trisomy 13; and aminopterin, a known teratogen. In some instances NTDs are associated with other major malformations, especially defects of the heart, kidneys or genitalia, but more often an NTD occurs as an isolated malformation. The great majority of NTDs are presumed to be multifactorial. As a group they are a leading cause of stillbirth, death in early infancy and handicap in surviving children.

In anencephaly the forebrain, midbrain, overlying meninges, vault of the skull and skin are all absent, and the survival after birth is a few hours at most. About two-thirds of the affected infants are female. Spina bifida is failure of fusion of the arches of the vertebrae, typically in the lumbar region. There are various degrees of severity, ranging from spina bifida occulta, in which the defect is in the bony arch only, to spina bifida aperta, often associated with meningocele (protrusion of the meninges) or meningomyelocele (protrusion of neural elements as well as meninges). The incidence is a little higher in girls than in boys.

Many infants born with spina bifida are so severely abnormal that they do not survive, and others have minimal defects, but there is an intermediate group in which, even if surgical intervention is successful, the child is left with partial or complete paresis of the lower limbs and fecal and urinary incontinence. Thus there is much controversy over the ethical dilemma of whether to provide active treatment or withhold treatment if the child could not survive without surgery.

The incidence of neural tube defects is extremely variable, ranging from close to 1 percent in Ireland to approximately 2 per thousand or less in North America and even lower in other populations. There are several unusual features of the population distribution, such as a lower incidence in children of professional and managerial families than in children of semi-skilled and unskilled workers, seasonal variation, and variation in frequency over time (with a marked decrease in recent years). Though the variation in incidence is unexplained, it is believed that nutritional differences may be contributing factors.

After the birth of one child with a neural tube defect, whether anencephaly or spina bifida, the risk for a subsequent child is approximately equally divided between the two conditions.

Prenatal Diagnosis of Neural Tube Defects

Neural tube malformations rank second only to chromosomal disorders as conditions for which prenatal diagnosis is now possible. In a pregnancy at risk because of a family history of a neural tube defect (in a previous child, one of the parents or perhaps a more remote relative), a high level of α-fetoprotein (AFP) in the amniotic fluid at the 16th week of pregnancy will identify virtually any anencephalic fetus and about 90 percent of those with spina bifida, the undetected cases being the milder ones. Caution must be taken to ensure correct timing, as the means and ranges of AFP levels decline with time and an error in

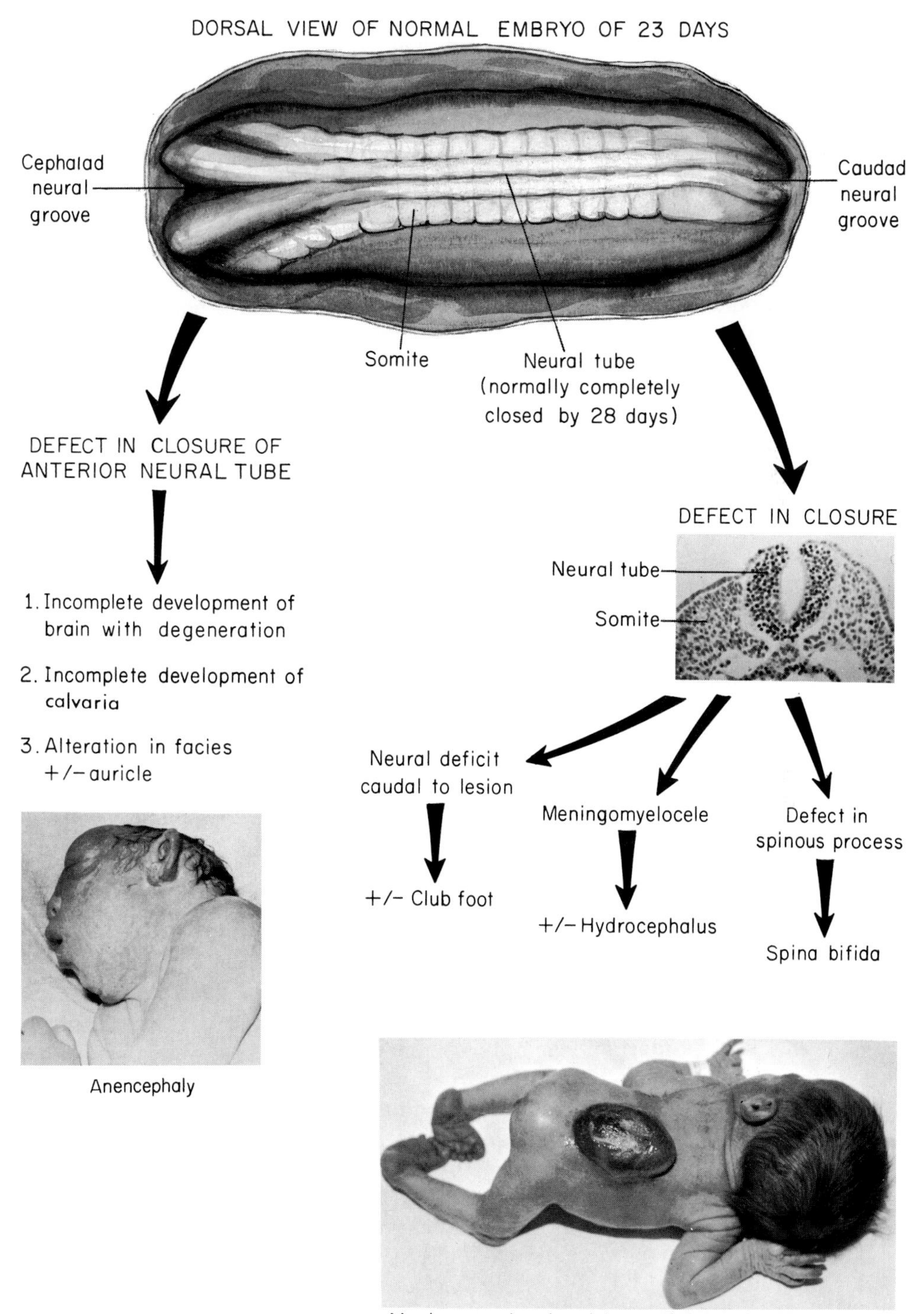

Figure 12–5. The origin of the neural tube defects anencephaly and spina bifida. From Smith DW. Recognizable patterns of human malformation, 3rd ed. Philadelphia: W.B. Saunders Company, 1982.

timing could lead to misinterpretation. If a woman has already had a child with a neural tube defect, there is a 4 percent risk that the next child will be affected. Half this risk (2 percent) is for anencephaly, which will be detected. Half is for spina bifida, and the chance of detecting it is 90 percent of 2 percent, or 1.8 percent. Thus there is only a 0.2 percent chance of missing the diagnosis.

The function of AFP is unknown; it may play a part in inhibiting the maternal immune response, which would otherwise reject the fetus.

In addition to amniotic fluid AFP assay, women at risk usually also receive one or more ultrasound scans. With highly sophisticated equipment and an experienced ultrasonographer, not only anencephaly but also spina bifida can usually be identified at the 16th to 18th week of pregnancy. Since ultrasonography is a noninvasive procedure, it may eventually become more widely used than amniotic fluid AFP assay.

Population Screening for Alpha-Fetoprotein in Maternal Serum

Probably 90 percent of children with neural tube defects are born into families without any previous NTD history. It is therefore gratifying that a method of screening pregnant women by means of radioimmunoassay of maternal serum AFP is currently under development. Maternal serum AFP measurements are considerably less sensitive and specific than amniotic fluid AFP measurements; however, a high level can indicate a problem and the need for amniocentesis or ultrasonography to confirm the diagnosis. Like many other population screening projects, in its early stages maternal serum AFP screening has been beset with problems of organization, timing and interpretation. If these can be resolved and if mothers are willing to participate, the method may prove to be well worthwhile.

Vitamin Supplementation in the Prevention of Neural Tube Defects

Even though prospects for prenatal diagnosis of neural tube defects are excellent, primary prevention would be preferable. The possibility of prevention by periconceptional vitamin supplementation, first raised in 1980 by Smithells and colleagues, has aroused great interest.

Women who had previously given birth to an NTD infant were given a vitamin preparation containing folic acid, ascorbic acid and riboflavin among its ingredients, and were asked to take the preparation for at least 28 days before conception as well as afterward. (Neural tube closure is normally completed by 28 days of uterine development, that is, about 42 days after the beginning of the last pre-pregnancy menstrual period.) Amniocentesis was offered, and in most cases accepted. The results showed that the risk of an NTD child was significantly lowered, from 5 percent in the control group of mothers who did not receive vitamin supplementation to 0.6 percent in the experimental group.

The original study was subject to criticism on a number of counts, and a more extensive study is now under way (Wald, 1984). One unanswered question is whether vitamin supplementation, even though it had an effect in the British study, would be effective in North America, where the incidence is lower and there are dietary and other differences. The overall safety of the vitamin regimen has also been questioned. At present it is too early to claim that vitamin supplementation is effective, but the results of further research are awaited with interest.

Criteria for Multifactorial Inheritance

Several characteristics of multifactorial inheritance have already been mentioned, and others are included in the following summary:

1. The correlation between relatives is proportional to the genes in common. Exceptions may occur if mating is nonrandom, if environmental variation is present or if the genes responsible for the trait do not have small, additive effects.

2. On the average, if mating is random, if environmental differences are unimportant and if the genes concerned are additive, the mean for the offspring is midway between the parental value and the population mean.

3. If a threshold trait is more frequent in one sex than the other, the recurrence risk is higher for relatives of patients of the less susceptible sex.

4. If the population frequency is p, the risk for first-degree relatives is $\sqrt{p}$ (Edwards, 1960). This is not true for single-gene traits. Figure 12–6 is a graph of the proportional change in risk to sibs with changing population incidence; it shows that the range for multifactorial traits is quite different from that for single-gene traits. The lower the population risk, the greater the relative increase in risk for sibs. For example, neural tube defects (anencephaly and spina bifida) have a population frequency of about 1 in 130 and a recurrence risk of 1 in 20 in South Wales, a 7-fold increase, but in London the population frequency is about 1 in 350 and the recurrence risk about 1 in 23, a 15-fold difference (Carter, 1976).

Nora (1968) compared the incidence of several types of congenital heart defects in the general population and in first-degree relatives of patients and found close agreement with this rule (Table 12–5). However, only rather large differences in population frequency make an appreciable difference to the risk in relatives.

5. The recurrence risk is higher when more than one family member is affected. For single-gene defects, in contrast, the risk to subsequent sibs is the same regardless of the number of affected children in the sibship.

6. The more severe a malformation, the higher the recurrence risk. Data for cleft lip and palate given above bear out this rule also. The threshold model predicts that the more extreme an individual is in the range of liability, the greater the severity of the malformation and the greater the risk that other family members will fall beyond the threshold.

7. The risk to relatives is sharply lower for second-degree than for first-degree relatives, and declines at a lower rate for third-degree and more remote relatives. Typical estimates for several conditions are given in Table 12–6. In contrast, for autosomal dominants the risk declines by half with each increasingly remote degree of relationship to the proband.

8. In twin studies, if the concordance rate in MZ pairs is more than twice as high as the rate in DZ pairs, the trait cannot be a simple dominant; if it is more than four times as high the trait cannot be a simple recessive either. The possibility of multifactorial inheritance must then be considered.

9. Although parental consanguinity is usually taken as a strong indicator of autosomal recessive inheritance, it can also signify that multiple factors with additive effects are involved. If recessive inheritance can be ruled out (by showing that the parent-offspring correlation is similar to the sib correlation), multifactorial inheritance is the only other possibility. A further point is that for multifactorial inheritance, the risk to subsequent sibs is higher when the parents are consanguineous than when they are unrelated; for autosomal recessive inheritance the risk is the same whether or not the parents are consanguineous.

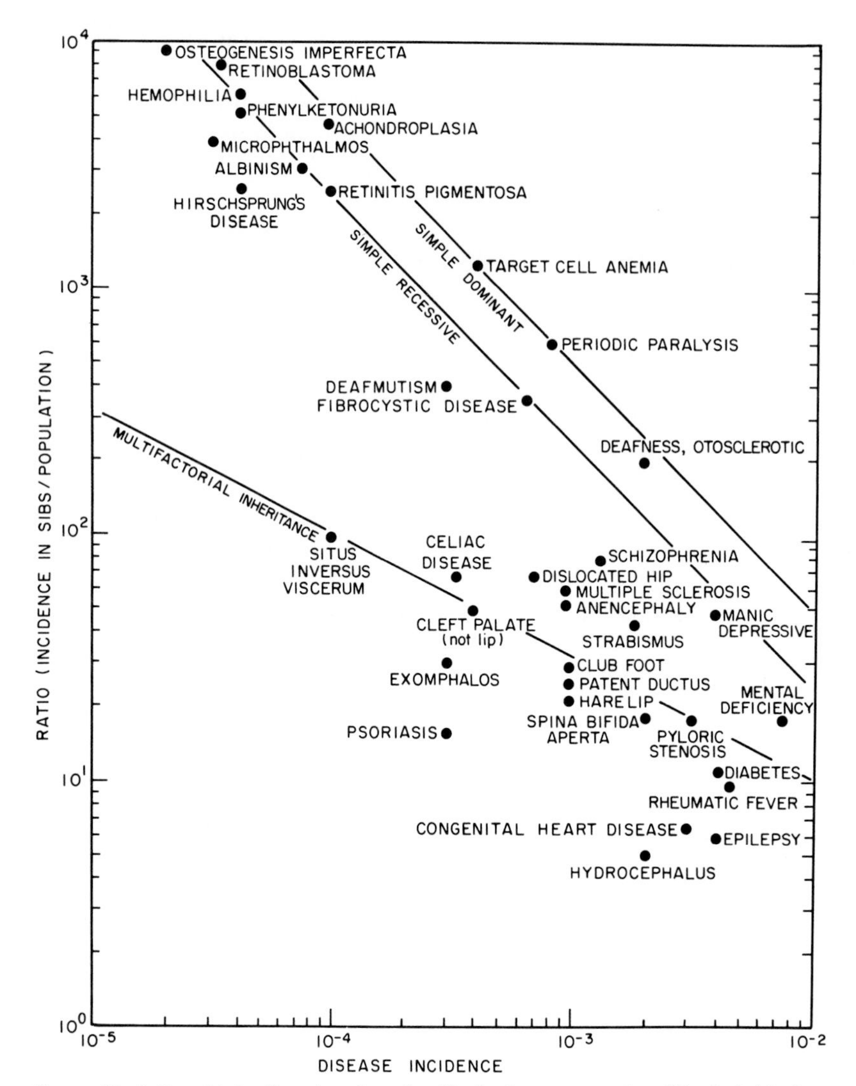

Figure 12–6. The risk to sibs of probands with single-gene and multifactorial traits. On a log scale, the population incidence of each disease is plotted against the sib incidence/population incidence ratio. Single-gene and multifactorial traits fall within separate clusters, because the risk to sibs is much lower for most multifactorial traits. From Newcombe HB. Epidemiological studies: discussion. In Fishbein M, ed. Congenital malformations: papers and discussions presented at the second international conference on congenital malformations. New York: International Medical Congress, 1964.

Table 12–5. FREQUENCY OF SIX COMMON CONGENITAL HEART DEFECTS IN SIBS OF PROBANDS

Anomaly	Frequency in Sibs (%)	Expected Frequency (%)*
Ventricular septal defect	4.3	4.2
Patent ductus arteriosus	3.2	2.9
Tetralogy of Fallot	2.2	2.6
Atrial septal defect	3.2	2.6
Pulmonary stenosis	2.9	2.6
Aortic stenosis	2.6	2.1

Data from Nora, 1968.
*$\sqrt{p}$, where p = population frequency of the specific defect.

Table 12–6. EMPIRIC RISKS FOR SOME CONGENITAL MALFORMATIONS (STATED AS MULTIPLES OF POPULATION RISK)

	Cleft Lip ± Cleft Palate	Talipes Equino-varus	Congenital Dislocation of the Hip (Males Only)	Congenital Pyloric Stenosis (Females Only)
General population	0.001	0.001	0.002	0.005
First-degree relatives	×40	×25	×25	×10
Second-degree relatives	×7	×5	×3	×5
Third-degree relatives	×3	×2	×2	×1.5

Data from Carter, 1969.

OTHER FACTORS AFFECTING RECURRENCE RISK

In addition to the factors mentioned above, several others can change the recurrence risk. If an affected child already has a large number of normal sibs, the chance of recurrence is less than if there are no sibs. If the malformation has previously appeared in both paternal and maternal relatives, the recurrence risk is appreciably higher. If an individual who has had an affected child has a subsequent child by a different partner, the risk for the later-born child is equivalent to the risk for any second-degree relative of a patient, that is, quite low. Table 12–7 gives some recurrence risk estimates for three common congenital malformations, taking into account a number of different family histories.

Common Diseases of Adult Life

A number of common disorders seem to fit the pattern of multifactorial inheritance reasonably well in that they show a definite tendency to run in

Table 12–7. EXAMPLES OF RECURRENCE RISKS (PERCENT) FOR CLEFT LIP ± CLEFT PALATE AND NEURAL TUBE MALFORMATIONS

	Cleft Lip ± Cleft Palate	Anencephaly and Spina Bifida
No sib affected		
Neither parent affected	0.1	0.3
One parent affected	3	4.5
Both parents affected	34	30
One sib affected		
Neither parent affected	3	4
One parent affected	11	12
Both parents affected	40	38
Two sibs affected		
Neither parent affected	0	10
One parent affected	19	20
Both parents affected	45	43
One sib and one second-degree relative affected		
Neither parent affected	6	7
One parent affected	16	18
Both parents affected	43	42
One sib and one third-degree relative affected		
Neither parent affected	4	5.5
One parent affected	14	16
Both parents affected	44	42

Data from Bonaiti-Pellié and Smith, 1974.

Table 12–8. EXAMPLES OF COMMON FAMILIAL DISORDERS OF ADULT LIFE

Cancer, some forms	Hypertension
Coronary heart disease	Manic-depressive psychosis
Diabetes mellitus	Peptic ulcer
Gout	Schizophrenia

families but are not Mendelian (see Table 12–8 for examples). Probably most such disorders are heterogeneous. It is important to attempt to resolve the underlying heterogeneity and to distinguish individual entities, for which genetic risk estimates can then be made with improved accuracy; otherwise the risk estimates can be wildly inaccurate for specific patients if they are based on population averages. To a limited extent this has been accomplished for diabetes mellitus and the hyperlipoproteinemias (see below), but attempts to elucidate the genetic basis of schizophrenia, manic-depressive disorders and cancer (other than a few Mendelian types) have not so far met with much success.

The role of HLA haplotypes in helping to determine susceptibility to a variety of medical problems, chiefly of adult life, was outlined earlier. Clearly HLA types or genes closely linked to the HLA locus can play a part in susceptibility to certain diseases. The associations of blood groups with diseases, also described earlier, point to genetic factors in susceptibility to those diseases as well. The HLA associations are thought to imply a genetic defect in the immune response to environmental agents, but most blood-group associations are difficult to explain.

Probably the familial disorders of adult life that have received the most attention so far are the hyperlipoproteinemias associated with coronary artery disease. A single-gene defect, familial hypercholesterolemia, accounts for about 10 percent of cases of coronary artery disease with onset before the age of 50 years, but most patients with hypercholesterolemia have a multifactorial form to which hypertension, obesity, exercise, smoking, stress and various medical conditions can contribute. Several other phenotypes among the hyperlipoproteinemias, some single-gene and others multifactorial, have been defined. It is likely that further analysis of the relative roles of genetic and nongenetic factors in the hyperlipoproteinemias will help physicians to distinguish cases in which appropriate adjustment of life style can be helpful in prevention from those in which a pronounced genetic component makes environmental manipulation relatively ineffectual (Nora et al., 1980).

The analysis of disorders of complex inheritance has necessitated the use of sophisticated mathematical models with powerful computer technology to match, as well as extensive data that allow heterogeneity to be uncovered and recurrence risks to be determined. The growing need for such studies has led to the definition of a new field, **genetic epidemiology**, at the interface of human genetics and epidemiology. Genetic epidemiologists are "slaves to the concept of multifactorial causation" (Neel, 1984); their methodology allows approaches to the role of genetics in many disorders previously impervious to genetic analysis.

Conclusion

The multifactorial-threshold model has not been accepted without criticism. In brief summary, the criticisms are based on the difficulty of testing some of the underlying assumptions and on the fact that family data for some conditions will fit other models, especially if heterogeneity or reduced penetrance is invoked. However, multifactorial inheritance does explain many of the formerly puzzling features of the family distributions of a number of disorders, and has stimulated research into the nature of the underlying factors that contribute to their development.

GENERAL REFERENCES

Carter CO. Genetics of common disorders. Br Med Bull 1969; 25:52–57.
Cavalli-Sforza LL, Bodmer WF. The genetics of human populations. San Francisco: Freeman, 1971.
Emery AEH. Methodology in medical genetics. Edinburgh: Churchill Livingstone, 1976.
Fraser FC. Evolution of a palatable multifactorial threshold model. Am J Hum Genet 1980; 32:796–813.
Fraser FC. The genetics of common familial disorders—major genes or multifactorial? Canad J Genet Cytol 1981; 23:1–8.
Morton NE. Outline of genetic epidemiology. New York: Karger, 1982.
Vogel F, Motulsky AG. Human genetics: problems and approaches. New York: Springer-Verlag, 1979.

PROBLEMS

1. For a certain malformation, the recurrence risk in sibs of affected individuals is the same as the recurrence risk in offspring, 10 percent. The risk in nieces and nephews is 5 percent, and in first cousins 2.5 percent.
   a) Is this more likely to be an autosomal dominant trait with reduced penetrance or a multifactorial trait?
   b) What other information might support your conclusion?

2. In a certain clinic, prenatal diagnosis by measurement of α-fetoprotein in amniotic fluid is available to any woman whose fetus has a risk of 1 percent or higher of a neural tube defect (anencephaly or spina bifida). Which of the following women would qualify? Arrange your answer in order of risk.
   a) A woman who has had a child with anencephaly and no other children.
   b) A woman who has had a child with anencephaly and four normal children.
   c) A woman who has had two anencephalic offspring.
   d) A maternal aunt of an anencephalic.
   e) A wife of a paternal uncle of an anencephalic who herself has a sib with spina bifida.

3. A baby girl has pyloric stenosis. What is the recurrence risk for
   a) her brother?
   b) her sister?

4. A series of children with a particular type of congenital malformation includes both male and female affected members. In all cases, the parents are normal. How would you determine that the malformation is more likely to be autosomal recessive than multifactorial?

5. A large sex difference in affected individuals is often a clue to X-linked inheritance. How would you establish that pyloric stenosis is multifactorial rather than X-linked?

13 GENETIC ASPECTS OF DEVELOPMENT

It is a matter of concern that, although prenatal diagnosis, which was first introduced about 1967, is now widely used for prevention of the birth of genetically defective children, the genetics of human development is still very poorly understood. Of course, there are numerous practical difficulties that hamper progress in this area. Early human embryos and fetuses are simply not available in quantity or in good condition for laboratory studies; there are serious ethical problems about the use of even those few that are available. Moreover, the approach of the experimental geneticist, who can breed specific mutants and obtain them for study at a sequence of stages, is entirely proscribed in human subjects. Consequently, much of our knowledge of development, normal and abnormal, has had to come from studies in other animals.

The human embryo develops from the zygote by a long and complex process involving cell **proliferation**, **differentiation** of cells into a variety of specialized types and **morphogenesis**, the development of shape. Development requires differential gene action; although we can describe at least some of the different patterns of gene activity that occur, we really know next to nothing about the mechanism by which, at programmed times, particular genes are switched on in specific tissues while different genes are activated in other tissues even though the genome is the same in virtually all cells of the organism. The process of development must not only allow the embryo to undergo a sequence of changes to achieve its eventual form, but also permit it to continue to function while the changes are going on—it is "business as usual during alterations". In this chapter we will describe some recent progress in the understanding of gene activity during development, and we will return to the problem of congenital malformations, which has already been considered from the standpoint of multifactorial inheritance in Chapter 12.

The Development of Hemoglobin

Globin switching, as it is called, is a classic example of the complex regulation of gene expression during development. The genes in each of the two globin gene clusters are expressed sequentially, and there is equimolar production of the α-like and β-like globin chains.

The synthesis begins in red cell precursors early in fetal development with the production of the transitory embryonic hemoglobins containing the α-like ζ and β-like ϵ globins, and the formation of fetal hemoglobin ($\alpha_2\gamma_2$), the γ chain being unique to fetal hemoglobin. During late fetal life and in the first few months after birth, the production of γ chains declines until fetal hemoglobin comes to represent only about 2 percent of the total hemoglobin, a level that is

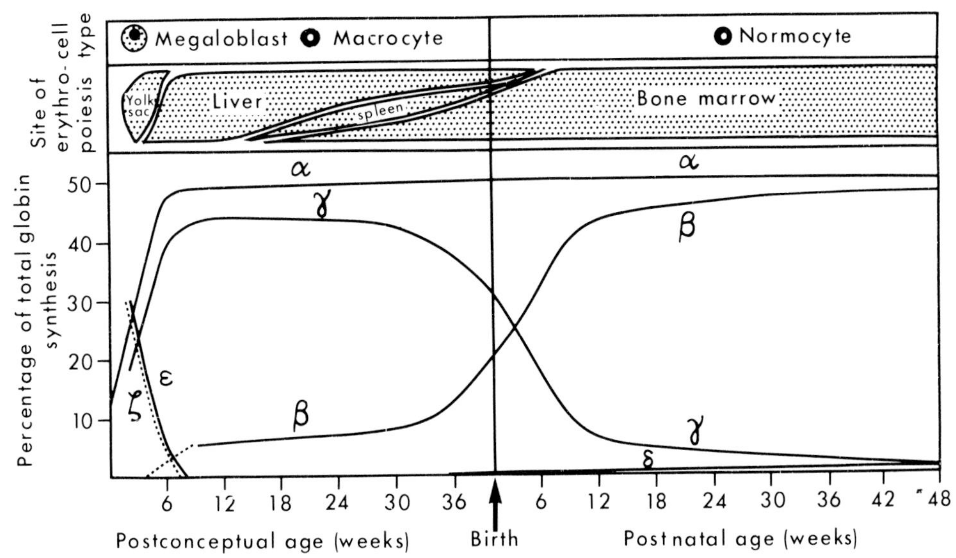

Figure 13–1. Development of erythropoiesis in the human fetus and infant. Types of cells responsible for hemoglobin synthesis, organ(s) involved and types of globin chain synthesized at successive stages are shown. Redrawn from Wood WG. Haemoglobin synthesis during human fetal development. Br Med Bull 1976; 32:282–287. Reproduced by permission of The Medical Department, The British Council.

maintained thereafter. Meanwhile, the synthesis of β chains has begun, and it increases steadily until almost all the hemoglobin present is of the adult type, $\alpha_2\beta_2$. Concurrently, in late fetal life and after birth, δ chain synthesis has begun, though the hemoglobin of which it forms a part (Hb A2) is never more than a small proportion of adult hemoglobin. These changes are shown in Figure 13–1. The mechanism of regulation of chain production is still unclear, though molecular models are currently being proposed and tested (Lavett, 1984).

Hereditary Anemias in the Mouse

Much can be learned about a normal developmental pathway by examination of the effect of a mutation that disrupts that pathway at some point. One of the best examples of this principle is provided by the analysis of a number of mutations that affect the development of hemopoiesis in the mouse. Elizabeth Russell, who has performed extensive studies of hereditary anemias of the mouse, has reviewed the effects of different anemia-producing mutations on prenatal and postnatal hemopoiesis (Russell 1979, 1984). Though a complete account of the action of the various mutations on blood formation is outside our scope, perhaps some concept of the complexity of the developmental pathways concerned can be provided.

Two kinds of anemic, sterile, black-eyed white mice are of particular interest. Alleles of the *W (dominant spotting)* series exert their action in very early embryonic stages, and lead to lifelong macrocytic anemia in mice of homozygous or compound genotypes. These genes also have independent, pleiotropic effects on germ cells and pigment cells. Alleles of the *Sl (Steel)* series have almost identical phenotypic effects, but the result is due to some gene-determined factor in the cellular environment, and is not intrinsic to the hemopoietic stem cells themselves as in *W* mice. An array of other mutations act at various later stages of blood formation and may independently affect such aspects as iron transport, heme synthesis, globin synthesis or formation of the red cell membrane. Some have pleiotropic effects differing from the effects of *W* and *Sl*. The effect on viability varies from prenatal lethality for some genotypes to transitory anemia in fetal and early postnatal life in others. The study of genetic birth defects in humans cannot possibly compare in depth with the detailed analysis of hereditary anemias possible in the mouse, in which the investigator can set up experimental matings, perform fetal research and profit from short gestation time and relatively inexpensive upkeep. The message for medical genetics from the study of murine anemias is that long, diverse and complex pathways may exist between the original gene action and the final phenotype.

Chimeras

In mythology, a chimera is an animal with the head of a lion, the body of a goat and the tail of a dragon. In genetics, the term is used more prosaically to refer to an individual with at least two different cell lines of genetically different origin (thus differing from a mosaic, in which the different cell lines originate from the same zygote).

Experimental geneticists use chimeras for the analysis of cell lineage and cell interaction in mammalian development. Not only can chimeric mice be made by fusion of two or even three early embryos, but they can also be made from mouse teratoma cells combined with mouse blastocysts, as described below.

Liveborn interspecific chimeras between goats and sheep have now been produced (Fehilly et al. 1984, Meinecke-Tillmann and Meinecke 1984). As would be expected, gestation of the chimeras in the female sheep or goat recipient leads to immunological problems related to maternal-fetal incompatibility, but experimental means of circumventing the mother's immune response have been devised.

Differentiation of Transplanted Teratocarcinoma Cells

A **teratocarcinoma** is an embryonic neoplasm arising spontaneously in ovarian or testicular tissue and containing a variety of differentiated somatic tissues. Cell lines established from mouse teratocarcinoma stem cells are used experimentally for analysis of differentiation. Because these cells are totipotent (able to differentiate into all types of tissue), can be genetically manipulated in vitro, and, like other cell cultures, can be grown in large quantities for analysis, they have been valuable tools for studies of differentiation.

One important experiment with a teratocarcinoma cell line is shown in Figure 13–2 (Stewart and Mintz, 1981). The experiment showed that cells of this culture line were capable not only of normal somatic differentiation but also of germ line differentiation, even after numerous generations in culture and freezing and thawing. The tumor cells actually became grandparents.

The technique can be briefly described. Cells derived from a malignant teratocarcinoma of a particular inbred mouse strain were cultured, and a euploid XX cell line was established. The mice of this strain have the agouti coat color. After propagation for nine or more passages in culture, the cells were frozen in liquid nitrogen for storage. After several additional passages in culture, the cells were injected into blastocysts of inbred mice of a strain with black coat color and with differences from the teratocarcinoma cell line in a number of markers that could be analyzed by electrophoresis. (The blastocyst stage is a multicellular

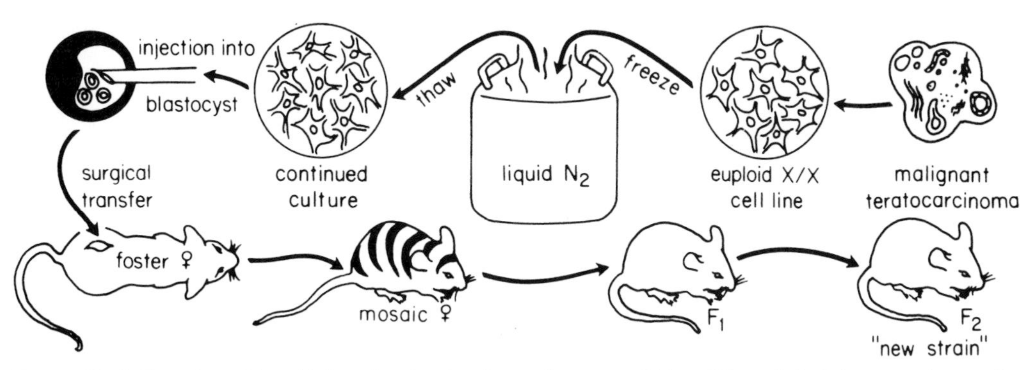

Figure 13–2. Experiment in which cells of a teratocarcinoma cell line differentiate into normal somatic and germ line cells. See text for details. From Stewart TA, Mintz B. Successive generations of mice produced from an established culture line of euploid teratocarcinoma cells. Proc Nat Acad Sci USA 1981; 78:6314–6318, by permission.

stage that follows the morula stage of the mammalian embryo. By this stage, the embryo has entered the uterus and has developed a fluid-filled cavity.) Blastocysts were surgically transferred to the oviducts of pseudopregnant albino foster mothers and allowed to develop to term. A few of the offspring had mosaic agouti and black coats, indicating that they were chimeras with tumor-derived (agouti) cells as well as normal cells from the original embryo (black). One such female, mated to a black male, produced some F_1 progeny with agouti coats. When agouti F_1 mice were mated, some of the F_2 offspring were homozygous for genes of the tumor "grandparent."

Incorporation of Foreign Genes into Fertilized Ova

A different type of experiment involves microinjection of multiple copies of specific cloned genes into fertilized eggs, followed by search for their presence and expression in the mice at embryonic stages or after birth. The cloned genes, obtained in plasmids by recombinant DNA technology, are usually injected into the male pronucleus of fertilized mouse eggs. The fertilized eggs are then surgically implanted into the oviduct of a foster mother to continue development. Later the DNA is analyzed to see whether the foreign DNA has been incorporated; the expression of the microinjected gene is looked for; and eventually the animals are bred to learn if the gene has entered the germ line. Both human and viral cloned genes introduced into mouse egg pronuclei by microinjection have been found in the DNA at later fetal or postnatal stages. Typically, the foreign DNA is integrated into a chromosome of the host early in development and thus is eventually transmitted through the germ line. Mice so treated are said to be **transgenic**.

Although not all foreign genes introduced by microinjection are expressed, there are some examples of expression, one of which is particularly striking. Palmiter and colleagues (1982) constructed a DNA fusion fragment containing the promoter (regulatory sequences) of the mouse metallothioneine I (MT) gene (the structural gene for a metal-binding protein) attached to the coding sequences of the rat growth hormone (GH) gene. Growth hormone is synthesized in the anterior pituitary gland and performs its major function in the liver, where it stimulates the production of somatomedins, which in turn stimulate the growth of mesodermal tissues. Some forms of dwarfism are due to growth hormone deficiency. When the fusion gene was used in a microinjection experiment, the

Figure 13–3. Comparative size of a normal mouse and a giant mouse of the same age. The larger mouse was derived from a fertilized egg that had been microinjected with growth hormone genes from the rat or human fused to the promoter of the mouse metallothioneine I gene. Based on Palmiter RD, Norstedt G, Gelinas RE, Hammer RE, Brinster RL. Metallothioneine–human GH fusion genes stimulate growth of mice. Science 1983; 222:809–814.

transgenic mice that developed grew rapidly, reached giant size (Fig. 13–3) and had greatly elevated GH levels. The methodology has potential for increasing growth rate and ultimate size in animal husbandry breeding, and may well have broad implications for medical research and therapy if its safety can be established.

Cloning

A clone is a population of cells all derived from a single ancestral somatic cell. All cells of a clone have the same genetic constitution as the original cell (unless somatic mutation has occurred). The term **"cloning"** is now also used to describe the procedure of taking a segment of DNA and multiplying it in bacteria.

The idea of cloning human beings holds fascination for many people and has been the subject of fiction in recent years, but it seems unlikely that cloning of humans will soon move from the realm of fiction to the realm of science, though cloning of frogs has been achieved. In experimental work with frog eggs, if the pronuclei of a fertilized ovum are removed and a new nucleus taken from a somatic cell (such as the nucleus of an intestinal cell) is slipped in, development continues. This is true only if the nucleus has come from a cell of a young embryo and consequently is still totipotent. Nuclei from a single early embryo can be used to replace the pronuclei of a number of fertilized frog eggs, thus producing a clone of genetically identical frogs.

In Vitro Fertilization and Embryo Transfer

The year 1978 saw the birth of the first "test-tube baby," the result of fertilization of a human ovum in vitro followed by replacement in the mother's uterus, successful implantation and continued normal development (Edwards 1981; Edwards and Steptoe 1983). The technique soon came into general use, especially for women who are sterile because of tubal occlusion.

STEPS INVOLVED

1. Induction of follicle growth and ovulation. Oocytes must be harvested during the final stages of maturation, just before natural ovulation. They are then

in an advanced stage of meiosis. If ovulation is primed hormonally by means of exogenous gonadotropins or clomiphene, several follicles usually ripen in a single cycle. Alternatively, the approach of natural ovulation may be monitored closely and a single ovum obtained.

2. Aspiration of oocytes. Laparoscopy is performed and the preovulatory oocytes are collected by aspiration.

3. Fertilization and cleavage in vitro. Sperm and oocytes are combined in culture medium. With normal sperm, the rate of successful fertilization is high. Cleavage follows: the cleaving embryos are maintained in vitro to the 2- to 8-cell stage, then introduced transcervically into the uterus.

4. Establishment of pregnancy. Implantation is the least successful step in the process. Only 20 to 30 percent of embryos that are replaced become implanted. The incidence of implantation declines with maternal age, especially after age 40.

The probability of a successful pregnancy is enhanced by transfer of two or even three embryos into the uterus; thus, multiple pregnancies with their associated risks are not uncommon among women who have had embryo transplants. The incidence of spontaneous abortion is high, over 25 percent as compared with about 15 percent in other pregnancies.

PREGNANCY FOLLOWING CRYOPRESERVATION

The procedures used to induce ovulation frequently result in the production of more oocytes than can be used immediately. A method of preservation of 4- to 8-cell embryos frozen in liquid nitrogen (cryopreservation) with successful establishment of pregnancy after thawing and transfer has been reported (Trounson and Mohr, 1983).

CHROMOSOME ABNORMALITIES IN HUMAN EMBRYOS AFTER FERTILIZATION IN VITRO

One possible reason for the high failure rate of implantation is chromosome abnormality. In one of the few reports of chromosome studies following fertilization in vitro, Angell and colleagues (1981) described the chromosomes of several 8-cell embryos; they found a high incidence of abnormality, in particular haploidy and aneuploidy of nondisjunctional origins. Chromosome abnormalities of the types seen in these embryos would be lethal early in development, and may contribute to the high failure rate of implantation. At present there are no comparative data on the chromosomes of very early embryos following fertilization in vivo.

Congenital Malformations

A congenital malformation is a malformation present at birth; the term congenital neither connotes nor excludes genetic etiology. Congenital malformations are often referred to as **birth defects**, though, broadly speaking, birth defects also include biochemical abnormalities manifest at or near the time of birth whether or not they are associated with dysmorphism. In this section we will consider only traits that are malformations in the anatomical sense, determined by some disorder of morphogenesis.

Morphogenesis is an elaborate process during which many complex interactions must be performed in orderly sequence. We have little knowledge of the genes and gene products that bring about morphogenesis, though at times we

Table 13–1. CAUSES OF CONGENITAL MALFORMATIONS

Cause	Frequency (%)
Single-gene defects	7.5
Chromosomal aberrations	6.0
Multifactorial traits	20.0
Maternal factors	
Infection (toxoplasmosis, rubella, cytomegalovirus, herpes simplex)	2.0
Diabetes	1.4
Other illnesses	0.1
Anticonvulsant drugs	1.3
Other	1.7
Unknown	60.0
	100.0

Data from Kalter and Warkany, 1983.

can see the consequence of the absence of normal gene products in mutations that interfere with normal development. For example, we know at least 10 mutant genes that affect the normal differentiation of the digits; if these are not allelic, there must be at least 10 gene loci responsible for the normal development of digit number and shape.

THE EFFECT OF GENES ON DEVELOPMENT

As we have seen in earlier chapters, dysmorphism can arise by a variety of mechanisms. In single-gene disorders, such as achondroplasia, a single major error in the genetic blueprint underlies the abnormality or abnormalities. Genes acting during development often have pleiotropic effects which may be independent or may represent a sequence of effects from a single primary defect. In chromosome disorders malformations are virtually always present and usually severe; they may be visualized as the result of the developing organism's efforts to cope with a partially duplicated or incomplete blueprint. In multifactorial inheritance there may be no major error or actually false instructions, but a number of minor faults that combine to produce a defective end-product; in much the same way, a number of otherwise trivial defects in material and workmanship can result in a car that is a so-called lemon. In multifactorial inheritance one or more of the multiple factors may be environmental, but there are situations in which a fault of the environment alone is enough to cause a malformation, or more often a malformation syndrome, even where the genotype is normal. We must therefore define a fourth kind of causative agent, an environmental teratogen.

Table 13–1 gives an estimate of the relative importance of single-gene defects, chromosomal aberrations, multifactorial traits and factors of the maternal environment as causes of congenital malformations.

TERATOGENS

A teratogen is any agent that can produce a malformation or raise the incidence of a malformation in the population. Most known teratogens are infectious agents, radiation or drugs. Some examples are given in Table 13–2.

Probably the best-known teratogen is thalidomide, which gained deserved notoriety when it was found to cause major limb malformations (phocomelia) and other malformations in children whose mothers had been given the drug thalidomide during early pregnancy. During the short period in the early 1960's when this drug was used in pregnancy as a tranquilizer and antiemetic, all the

Table 13–2. DRUGS WITH ESTABLISHED TERATOGENIC EFFECTS

Established Teratogens		Possible Teratogen
Alcohol	Streptomycin	Isotretinoin
Anticonvulsant drugs	Tetracyclines	
Chemotherapeutic agents	Thalidomide	
Folic acid antagonists	Thiourea compounds	
Inorganic iodides	Trimethadione	
Lithium	Warfarin	
Sex steroids		

Data chiefly from Golbus, 1980.

mothers who received the drug at a "sensitive stage" of early pregnancy, when differentiation of the embryo was proceeding rapidly, produced babies with some degree of phocomelia. Thus it appears that susceptibility to teratogenesis from thalidomide is a uniform trait of the human species that unfortunately is not shared by the experimental animals in which the drug was tested.

Thalidomide remains the only example of a teratogen for which the introduction of a drug led to a dramatic rise in the incidence of a specific type of malformation, and withdrawal of the drug was immediately followed by virtual disappearance of the deformity. None of the many other teratogens that have been examined, or suspected, in "epidemics" of malformations has been proved to have a definite cause-and-result connection with a given type of abnormality, or an effect on all fetuses exposed at a particular stage. However, failure to demonstrate such a connection is not proof that no connection exists. There is a strong suspicion that isotretinoin, a vitamin A analogue used in the treatment of acne, may affect all exposed fetuses.

In the analysis of teratogenicity, four factors must be considered:

1. Time of exposure to the teratogen. This point has been mentioned above in connection with thalidomide. Another example is provided by rubella embryopathy; after maternal rubella in pregnancy, if the infection occurs in the first four weeks of development half of the infants have major malformations, in the third month only one-sixth and thereafter close to zero. Teratogens exert their effect during the stage of active differentiation of an organ or tissue. Because a woman may not even realize she is pregnant at this early stage, it is wise for her to avoid any possibly teratogenic drugs until her pregnancy status has been determined. The general rule is that a woman who might be pregnant should consider that she is until proved otherwise.

2. Dosage. There is little information on dosage in human studies, but work with experimental animals has shown the expected dose-response relationship. The dose affecting the fetus is of course partly determined by the mother's ability to metabolize the chemical, so the maternal genotype can be a factor in the response.

3. Genotype of the embryo. Thalidomide is an example of a teratogen to which all humans are susceptible, but there are many proven examples in experimental animals, and a number of suspected ones in humans, of genetic differences in response to a teratogen.

4. Genotype of the mother. As mentioned above, the mother's ability to detoxify certain drugs is a possible factor in determining the teratogenicity of any drug to her fetus. At present, anticonvulsant drugs used in the control of epileptic seizures and maternal alcohol ingestion are both under active investigation as causes of congenital malformations. Both have been indicted in connection with the birth of infants with a combination of dysmorphic features, prenatal and postnatal growth retardation and impaired mental development, but only a minority of users of anticonvulsants or of alcohol during pregnancy

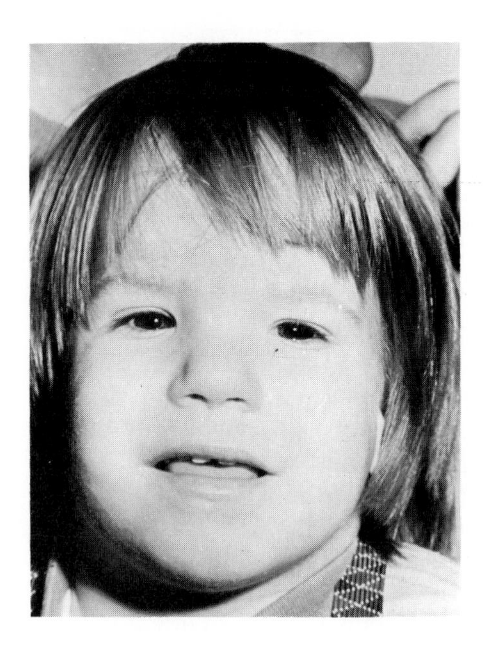

Figure 13–4. Facial appearance of a child aged two-and-one-half years, born to an alcoholic mother. Note particularly the short palpebral fissures and mid-face hypoplasia. From Smith DW. Recognizable patterns of human malformation, 3rd ed. Philadelphia: W. B. Saunders, 1982.

have children with such problems. So far almost nothing is known about the possible effect of timing of administration, dosage or maternal or fetal genotype on the expression of these malformation syndromes. A child with the features of fetal alcohol syndrome is shown in Figure 13–4.

The incidence of congenital malformations is somewhat increased in the offspring of diabetic mothers, but probably it is not the diabetes itself but inadequate management of it that causes these occasional malformations. This is a field requiring further study.

DYSMORPHOLOGY

Although the diagnosis of malformation syndromes is made on the basis of an overall pattern of anomalies, the pattern of any syndrome varies to some extent from patient to patient. Dysmorphology is the branch of clinical genetics concerned with the diagnosis and interpretation of patterns of structural defects. Dysmorphologists distinguish three types of structural defects:

1. **Malformation**. A morphologic defect resulting from an intrinsically abnormal developmental process.

2. **Disruption**. A morphologic defect resulting from breakdown of, or interference with, an originally normal developmental process. Examples include abnormalities resulting from amniotic bands, and limb reduction defects caused by vascular anomalies.

3. **Deformation**. Abnormality in form or position of a body part caused by a nondisruptive mechanical force. Deformations may be due to mechanical constraint in the uterus or to defects (neurological, myopathic or connective tissue defects, for example) in the fetus itself.

Sometimes either a structural defect or a mechanical factor can lead to multiple secondary defects. The total picture is then termed a **sequence**, and is distinguished from a **syndrome**, in which the multiple anomalies are thought to be independent rather than sequential but are derived from a single cause (a gene defect, chromosome abnormality or teratogen). Making the distinction between sequence and syndrome may not always be possible, given our limited knowledge of early human morphogenesis. The term **association** is reserved for

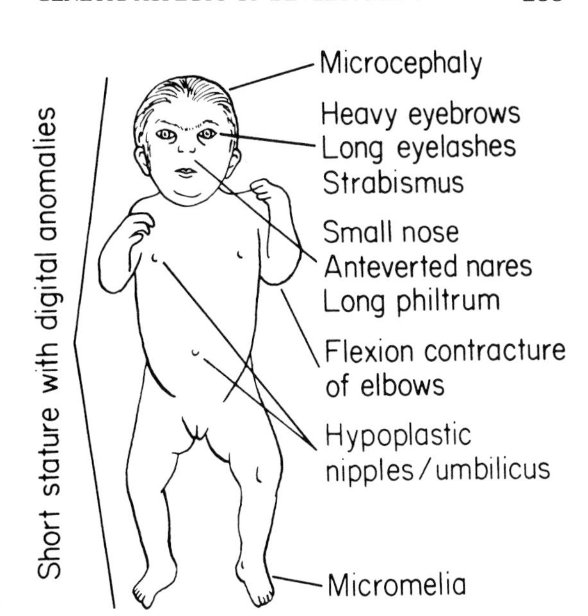

Figure 13–5. The characteristic anomalies seen in Cornelia de Lange syndrome, a malformation and retardation syndrome of unknown cause. From Goodman RM, Gorlin RJ. The malformed infant and child. An illustrated guide. New York: Oxford University Press, 1983.

the nonrandom occurrence in a patient or group of patients of severe morphological defects not identified as a sequence or syndrome; it seems likely that most associations will eventually be reclassified as either sequence or syndrome when their etiology becomes clear.

Malformation syndromes may be of known cause (genetic, chromosomal or teratogenic), or unknown origin. There are numerous examples of syndromes that occur frequently enough to be recognizable, but for which the cause is completely unknown at present (Fig. 13–5).

GENERAL REFERENCES

Golbus MS. Teratology for the obstetrician: current status. Obstet Gynecol 1980; 55:269–276.
Goodman RM, Gorlin RJ. The malformed infant and child. An illustrated guide. New York: Oxford University Press, 1983.
Kalter H, Warkany J. Congenital malformations. Etiologic factors and their role in prevention. New Engl J Med 1983; 308:424–431 and 491–497.
Shepard TH. A catalog of teratogenic agents, 4th ed. Baltimore: Johns Hopkins, 1983.
Smith DW. Recognizable patterns of human deformation. Philadelphia: W. B. Saunders Co., 1981.
Smith DW. Recognizable patterns of human malformation, 3rd ed. Philadelphia: W. B. Saunders Co., 1982.

PROBLEMS

You are given a number of definitions. What term (used in this chapter) fits each definition?

1. Individual with two or more different cell lines originating from a single zygote.
2. Tumor of embryonic origin, usually occurring in testicular or ovarian tissue, containing a variety of differentiated tissues.
3. Morphological defect due to intrinsic developmental abnormality.

4. Condition in which there are several abnormalities of apparently independent origin.

5. Individual with two or more different cell lines originating from separate zygotes.

6. Attachment of the blastocyst to the endometrium of the uterus.

7. Term used for multiple effects of a single gene.

8. Agent that can produce congenital malformation.

9. Condition in which there are several abnormalities arising as secondary consequences of a single underlying problem

10. An abnormality due to a mechanical force that does not actually disrupt tissues.

14

ELEMENTS OF MATHEMATICAL GENETICS

Though not all problems in medical genetics require a mathematical approach, some cannot be handled in any other way. A medical geneticist need not be an expert mathematician, but should at least be aware of potential pitfalls and of the importance of giving attention to the biomathematical aspects of research problems, from the planning stage through the analysis of the observations.

Mathematical genetics, aided by computerization, has become a highly specialized and very productive field that can elucidate many complex problems relevant to medical genetics. The simple examples given in this chapter, however, do not require a mathematical background other than elementary algebra. Students with a special interest in mathematics will find a much more complete and sophisticated treatment of the subject in several excellent textbooks, some of which are listed in the General Reference to this chapter.

Several important mathematical-genetic problems are discussed in other chapters (see Chapters 11 and 15). This chapter is concerned with problems that may arise in the analysis of family data to demonstrate the inheritance pattern and recurrence risk of specific conditions, and with Bayesian methods of estimating risks.

For any genetic disorder, one of the important considerations is the **recurrence risk,** the likelihood that the condition will recur in some other family member—a sib, a child of the patient or some more distant relative. Initially, the recurrence risk is found by collecting a series of families with the same condition, counting the number of affected and unaffected members of each category of relative and examining the pedigree pattern in the families. Of course, recurrence risks for many genetic disorders have already been established and can be found in the medical literature.

Collection of a suitable series of families gives rise to a number of problems, such as ensuring that all the families really do have the same disorder, classifying family members accurately as affected and unaffected and so forth. Among these problems, the one we are most concerned with here is **bias of ascertainment,** the tendency for families not to be randomly ascertained.

Bias of ascertainment, though defined here in genetic terms, applies to many other clinical situations as well. A child has both Down syndrome and cleft lip; does this mean that cleft lip is more common in Down syndrome than in the general population? A man whose wife has had several early miscarriages is karyotyped and found to be XY/XXY; does this mean that if an XY/XXY male is fertile, his offspring are likely to be inviable? In these examples, the proband who has brought the family to the attention of geneticists must be omitted from calculation of the frequency of cleft lip in Down syndrome or of reproductive failure in mosaic Klinefelter syndrome. The problem that unusual cases are much more likely than routine ones to come to attention pervades not only genetics but many other areas of medical research.

Small family size, long generation time and the impossibility of making test matings complicate medical genetic analysis. If human families were large enough for Mendelian ratios to be approximated (say, with 12 or more sibs), many of the problems resulting from ascertainment bias would disappear. Other complications would vanish if geneticists lived longer than the subjects of their research and could extend their observations over several generations. (Here pediatricians have an advantage over other clinicians, since they normally see two generations of a family at a time.) Since test matings cannot be arranged, it is necessary to accumulate the results of nature's own experiments; this is sometimes a laborious process, and can be complicated by the problem of genetic heterogeneity in the accumulated data.

Probability

Genetic transmission from parent to child is much like tossing coins. In both, there are two possibilities: the coin turns up heads or tails; the parent transmits one or the other of each pair of alleles. In either case, the outcome is determined by chance.

Statisticians define probability, the mathematical expression of chance, as the ratio of the number of occurrences of a specified event to the total number of all possible events. The specified event may be that a tossed coin turns up heads. As there are two possible events, heads and tails, then for a single toss:

$$\frac{\text{Possibility of heads}}{\text{Total number of possibilities}} = \frac{1}{2}$$

Similarly, the probability that a child whose father is of blood group AB will receive the *A* allele rather than the *B* allele is:

$$\frac{\text{Possibility of } A}{\text{Total number of possibilities}} = \frac{1}{2}$$

PROBABILITY OF COMBINATION OF TWO INDEPENDENT EVENTS

A child receives his genotype from two parents; the probability of receiving certain combinations of alleles from the two parents can be simulated by tossing two coins at once. The probability of heads is 1/2 for each coin, and the probability of heads turning up on both coins simultaneously is $1/2 \times 1/2 = 1/4$. Similarly, the chance that a child whose parents are both of blood group AB will receive the *A* allele from each parent is $1/2 \times 1/2 = 1/4$.

There are many genetic applications of this simple but important principle. Perhaps one of the most useful to the physician has to do with sex ratios in human families. The probability of a boy as the firstborn child is 1/2. (Actually it is about 0.51, but 1/2 is close to the precise figure and simplifies the arithmetic.) The next child is an "independent event" since gametes, like coins, have no memory, so the probability that the second-born child will be another boy is again 1/2. The probability that the first two children in a family will both be boys is $1/2 \times 1/2 = 1/4$. This procedure can be continued to obtain the probability of sons in any number of successive births; for example, the probability of a sequence of six sons is $(1/2)^6$ or 1/64.

THE BINOMIAL DISTRIBUTION

The seventeenth century mathematician Jakob Bernoulli was a member of a family noted for its many brilliant mathematicians, but his genetic significance as an example of hereditary genius is outweighed by his contribution to genetic theory: the binomial distribution (Bernoulli distribution). Though the binomial formula was originally discovered by Sir Isaac Newton in 1676, it was Bernoulli who provided its first rigorous proof.

Suppose the problem to be solved is not the probability of a boy at each of two births, but the distribution of boys and girls in two successive births. This is an extension of the problem of the probability of coincidence of random events.

If the probability of a boy is 1/2 and the probability of a girl is also 1/2, then the distribution of families with no boys, one boy and two boys among all the two-child families is given by the expansion of $(1/2 + 1/2)^2$. In general terms, when there are two alternative events, one having a probability of p and the other having a probability of $q = 1 - p$, in a series of n trials the frequencies of the different possible combinations of p and q are given by the expansion of the binomial $(p + q)^n$.

Thus, for two-child families:

$$(p + q)^2 = p^2 + 2pq + q^2$$

$$p = q = 1/2$$

$$p^2 = \text{families of 2 boys} = 1/4$$

$$2pq = \text{families of 1 boy and 1 girl} = 1/2$$

$$q^2 = \text{families of 2 girls} = 1/4$$

For three-child families:

$$(p + q)^3 = p^3 + 3p^2q + 3pq^2 + p^3$$

$$p^3 = \text{families of 3 boys} = (1/2)^3 = 1/8$$

$$3p^2q = \text{families of 2 boys, 1 girl} = 3(1/2)^2(1/2) = 3/8$$

$$3pq^2 = \text{families of 1 boy, 2 girls} = 3(1/2)(1/2)^2 = 3/8$$

$$q^3 = \text{families of 3 girls} = (1/2)^3 = 1/8$$

The difference between a simple multiplication of the separate probabilities and the use of the binomial expansion is that the binomial expansion includes all the possible combinations of the two events. For example, in three-child families there are eight possible sequences:

BOY BOY BOY
BOY BOY GIRL
BOY GIRL BOY
BOY GIRL GIRL
GIRL BOY BOY
GIRL BOY GIRL
GIRL GIRL BOY
GIRL GIRL GIRL

The probability of each sequence is 1/8, but among families with two sons and a daughter, the daughter might be born first, second or last; the three possible sequences account for the total 3/8 probability of having two sons and a daughter

Table 14–1. DISTRIBUTION OF BOYS AND GIRLS IN FAMILIES

Number of Children in Family	$(p + q)^n$	Distribution
1	$(1/2 + 1/2)^1$	1/2(1♂) + 1/2(1♀)
2	$(1/2 + 1/2)^2$	1/4(2♂) + 1/2(1♂:1♀) + 1/4(2♀)
3	$(1/2 + 1/2)^3$	1/8(3♂) + 3/8(2♂:1♀) + 3/8(1♂:2♀) + 1/8(3♀)
4	$(1/2 + 1/2)^4$	1/16(4♂) + 4/16(3♂:1♀) + 6/16(2♂:2♀) + 4/16(1♂:3♀) + 1/16(4♀)
5	$(1/2 + 1/2)^5$	1/32(5♂) + 5/32(4♂:1♀) + 10/32(3♂:2♀) + 10/32(2♂:3♀) + 5/32(1♂:4♀) + 1/32(5♀)

among three children. The expected distribution of males and females in families of one to five children is summarized in Table 14–1.

Pascal's triangle is a convenient way to arrive at the coefficients of the binomial expansion. Each new coefficient in the triangle is made up by adding the two coefficients nearest it in the line above; for example, on the fourth line, the 3 of $3p^2q$ was obtained by adding the 1 of p^2 and the 2 of $2pq$.

n	Expansion
0	1
1	$1p + 1q$
2	$1p^2 + 2pq + 1q^2$
3	$1p^3 + 3p^2q + 3pq^2 + 1q^3$
4	$1p^4 + 4p^3q + 6p^2q^2 + 4pq^3 + 1q^4$
5	$1p^5 + 5p^4q + 10p^3q^2 + 10p^2q^3 + 5pq^4 + 1q^5$

Note that the number of terms in the expansion is always $n + 1$. For families of size n, there can be 0, 1, 2...n sons; hence there are $n + 1$ classes of family. If there are four children in a family, there can be five different boy:girl distributions: 4:0, 3:1, 2:2, 1:3, 0:4.

Mathematicians may prefer to use the general term of the binomial expansion:

$$\frac{n!}{m!\,(n-m)!}\,p^m q^{n-m}$$

n = total number in the series
$n!$ = ("n factorial") is $n(n-1)\,(n-2)...1$
p = probability of a specified event
$q = 1 - p$ = probability of the alternative event
m = number of times p occurs (in other words, the exponent of p)

For example, the probability of having three sons and two daughters in a five-child family is:

$$\frac{5 \times 4 \times 3 \times 2 \times 1}{(3 \times 2 \times 1)\,(2 \times 1)}\left(\frac{1}{2}\right)^3\left(\frac{1}{2}\right)^2 = \frac{10}{32}$$

In the examples used so far, the values of p and q have been equal ($p = q = 1/2$), but the binomial distribution can be used for other values of p and q. The same method is applied, for example, to give the distribution of a recessive trait in the progeny of two heterozygous parents, but now the probability of the specified event (that a child will be affected) is 1/4, and the probability that a child will be unaffected is 3/4. We will return to this topic in the following section when discussing tests of genetic ratios. Gene frequencies in populations also depend upon simple binomial distributions (see Hardy-Weinberg law, Chapter 15).

THE SEX RATIO

The sex ratio is defined as the ratio of the number of male births to the number of female births. Since sex is normally determined by whether the sperm contributes an X or a Y chromosome to the zygote, and since X-bearing and Y-bearing sperm are theoretically formed in equal numbers at meiosis, the expectation is that the primary sex ratio (ratio at fertilization) should be 1.00. However, in all parts of the world there is an excess of male babies; currently in North America the secondary sex ratio (ratio at birth) is about 1.05 (105 boys per 100 girls). The sex ratio in spontaneous abortion is about 1.3, significantly higher than values typically reported for newborns and indicating a higher risk of abortion in male than in female fetuses (Hassold et al., 1983). After birth, males have a higher mortality rate than females. By about age 25 the sex ratio reaches 1.00; thereafter the excess of females becomes more and more pronounced. The reason for the higher death rate in males is not known, but it may be due in large part to the fact that males have only one X chromosome, so that harmful X-linked recessive genes are exposed to selection in males but not in females.

What effect does parental preference for one sex or the other have on the sex distribution in families and in the population? As an extreme example, consider the situation in which parents either have no more children after they have had one daughter, or continue to have children until they have a daughter. In this case, all one-child families consist of one girl; all two-child families have a boy and a girl, in that order; all three-child families have two boys and a girl, in that order, and so on. The distribution of the sexes within these families is therefore quite different from the binomial proportions set out in Table 14–1. But since at each birth equal numbers of boy and girl babies are born, the overall sex ratio in the population remains unchanged.

Of course, the sex ratio could be drastically altered if parents were to exercise preference for one sex before birth, terminating pregnancies in which the fetus was of the unwanted sex. In general, since males are usually preferred, this would raise the sex ratio. The long-term effects on the population could be dramatic and perhaps unexpected. Since population control may be the major problem facing our species, perhaps there is something to be said for lowering the relative number of female births. The result, in the next generation, would be a corresponding decrease in the total number of births, because population growth depends not on the total population, but on the number of fertile females it contains.

Population data on sex distribution in sibships of different sizes fit the binomial expectation reasonably well. There are, however, a few possible or real exceptions, such as excesses of unisexual sibships and a preponderance of males in some pedigrees and of females in others. In experimental animals there are some unusual genetic mechanisms that upset the normal sex ratio, for example, a "sex ratio" gene in the fruit fly that leads to almost 100 percent female offspring. However, on the whole it is accurate to tell prospective parents that there is an even chance that the baby will be of the desired sex.

Tests of Genetic Ratios

An essential step in the analysis of a genetic disorder is the demonstration of its pattern of inheritance, since this may suggest its mechanism, indicate its recurrence risk and perhaps give clues leading to its prevention. Sometimes inspection of a collection of pedigrees will make the pattern obvious, or tests for heterozygote detection may be applied; even then mathematical tests may be required to alert the investigator to discrepancies caused by such factors as abnormal segregation ratios, different prenatal viability in the two sexes, unsuspected heterogeneity and so forth.

BIAS OF ASCERTAINMENT

In pedigree analysis, the proband through whom the pedigree is ascertained must be omitted from calculations of the proportion of family members affected. As a demonstration of the effect of the bias introduced by including the proband, consider the sex ratio of the families of a class of medical students. This class, composed of 197 men and 53 women, reported the number of males and females in their sibships to be 379 males and 226 females. The sex ratio of 1.7 deviated significantly from the expected 1.0. But when the data were broken down separately for male and female students it was found that:

The 197 men had 132 brothers and 127 sisters.

The 53 women had 50 brothers and 46 sisters.

Thus the sex ratio in the sibs of the probands was 182 males:173 females or 1.06, which is close to the expected ratio. The discrepancy in the first results was caused by the inclusion of the probands.

The same kind of bias can arise in genetic studies if the proband is included. If, as usually happens, a series of families is selected because there is at least one affected member in each family, the expected ratio of normal to abnormal members is not observed unless an appropriate correction is made.

TESTS FOR AUTOSOMAL RECESSIVE INHERITANCE

Proof of autosomal recessive inheritance from pedigree data has special difficulties not encountered with other pedigree patterns. An autosomal recessive trait appears in one fourth of the offspring of two carrier parents. This is a simple and well-known rule, but to prove the ratio by means of family data is not always easy. Since carrier parents are usually identifiable only because they have produced an affected child, collections of families with recessive traits usually consist of parent-child sets in which both parents are phenotypically normal and at least one child is affected. Thus, carrier parents are not ascertained if all their children are normal. Consequently, the proportion of affected children in the families that are actually observed is well above the expected one-fourth.

The number of families missed because they include no affected child decreases with family size. Seventy-five percent of 1-child families are not ascertained, but only 3 percent of the 12-child families are missed. For families of typical sizes, the correction is large enough to be important.

Types of Ascertainment

The terminology followed here for the different types of ascertainment used in genetic studies is that of Morton (1959). The reader is cautioned, however, that some textbooks use different terminology or definitions.

COMPLETE AND INCOMPLETE ASCERTAINMENT

If every pair of parents heterozygous for the condition under study, or a truly random sample, could be identified and all their offspring could be included, ascertainment would be **complete.** This may be possible for autosomal dominants, in which families of sibs are ascertained through an affected parent, but complete ascertainment is rarely possible for autosomal recessives. In actual

practice ascertainment is almost always **incomplete;** that is, only those sibships that contain at least one affected child are identified.

TYPES OF INCOMPLETE ASCERTAINMENT

Given incomplete ascertainment, the next point to be determined is how the sibships for study have been selected from the population. Selection is **truncate** if every affected child in the population is included in the survey as a proband (or if the group studied is a true random sample of the total group of affected children). If the population is fully screened, any one affected child is equally likely to be identified as a proband regardless of the number of affected children in his sibship. Selection is **single** if there is no chance that any one sibship will be ascertained more than once; thus, there is only one proband per sibship and the chance that a sibship will be ascertained is proportional to the number of affected members it contains.

In actual practice, more often than either of these two extremes there is **multiple** ascertainment. Some sibships are ascertained more than once, through more than one of their affected members, and multiplex families (with more than one affected child) have a higher chance than simplex families of being ascertained. As an example, assume that the probability of ascertaining a particular child is only 80 percent. Then 20 percent of all sibships with one affected child will be lost to the study, but only 4 percent (0.20^2) of sibships with two affected members and less than 1 percent (0.20^3) of sibships with three affected members are missed.

To illustrate possible situations in which truncate, single and multiple selection might apply, consider how a study of the genetics of cystic fibrosis (CF) might be designed.

Truncate Ascertainment. Every child with CF in a given area might be independently identified, regardless of the number of affected sibs, age, clinical status or other factors that might affect the child's likelihood of being ascertained. Total ascertainment could involve an exhaustive search through clinics, hospital records, school health records, private physicians' practices and so forth. If the search is successful in finding every case of CF in the community, ascertainment is truncate, every affected child is a proband, and in the collected family data for sibships of a given size, those with one, two, three or more affected members are distributed binomially, except for the class with no affected members. (Thus the distribution is truncate, which means "shortened by cutting off a part." The cut-off part is the group of families with heterozygous parents but no affected children.)

Single Ascertainment. Suppose our study of the genetics of CF is limited to a survey of all the children entering Grade 1 in a given year. It is very unlikely that any sibship with affected members would be encountered more than once. However, the more affected children a family contains, the more likely that one of them will be entering Grade 1. In this situation, single selection methods are appropriate.

Multiple Ascertainment. If the case load of a particular CF clinic at a pediatric hospital is the source of the probands, a good many affected children in the community might be missed because they do not attend the clinic. Some of these might be treated by private physicians, others might be in a hospital for chronically ill children, and still others might be managed at home without regular medical attention. Among the children attending the clinic, there would be some with sibs who also were attending; in other words, some sibships would contain more than one proband, and the probability that a particular family would be included would depend partly, but not entirely, upon the number of cases of CF in the sibship. This is the multiple ascertainment situation. If it is clear which affected members are probands and which are not, the families are counted once for each proband.

Corrections for Bias of Ascertainment

THE PROBAND METHOD OF WEINBERG

In 1912 Weinberg proposed a simple method of correction for ascertainment bias by discarding the proband from the calculations and determining the incidence of the disorder in the proband's sibs. The proband is used only as the individual through whom the sibship has been ascertained. This is the kind of correction applied to the data on sex ratio in sibships given earlier. It is especially suitable to the single ascertainment situation but can be used for multiple ascertainment if each proband (not each affected person, but each person through whom a given sibship has been ascertained) is considered separately. The proband method is often useful for a preliminary analysis and may be sufficient for all practical purposes.

THE METHOD OF DISCARDING THE SINGLES

Li and Mantel (1968) devised a convenient test for recessive inheritance, which is appropriate to use when ascertainment is truncate. It is only necessary to count the total number of individuals in all the sibships (T), the number of affected individuals in all the sibships (R) and the number of "singles" (that is, the number of patients who are the only affected members of their sibships [J]). Then calculate using the following formula:

$$\frac{R - J}{T - J}$$

The value can be compared with the expected 0.25 with the help of tables given by Li and Mantel for calculation of variance and standard error.

THE APERT OR A PRIORI METHOD

If it is suspected that a specific disease is inherited as an autosomal recessive, one method of testing is to assume that recessive inheritance is present; in other words, to set up an a priori expectation of recessive inheritance, then to test the data for agreement with this hypothesis. Since the probability that a family with two heterozygous parents will be missed varies with family size, the calculation must be made separately for each size of sibship. The method can be used for either truncate or single ascertainment, but the expected proportion of affected children is different in the two cases, as will be shown.

Truncate Ascertainment. In truncate ascertainment, the assumption is that every case in the population is identified, regardless of the number of affected in the family; in other words, every case is a proband. As an example, let us begin by considering 16 two-child families, each with both parents heterozygous. One-quarter of the children are expected to be affected. How are these eight children distributed in their respective sibships?

One-fourth of the families have an affected child at the first birth, and one-fourth at the second birth:

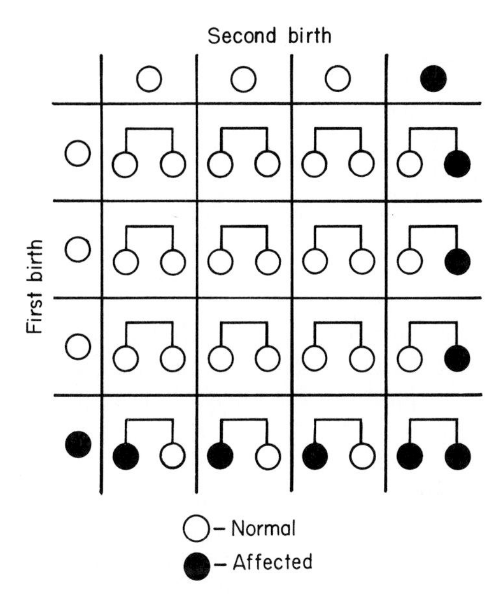

The chance distribution produces 1 family with 2 affected children, 6 with 1 affected child each and 9 with no affected children (which are not ascertained because they are not distinguishable from the rest of the population). The incidence of affected children in the observed families is 8/14 or 0.57.

Note the mathematical relationship: if the probability of a normal child is 3/4 and the probability of an affected one 1/4, then the distribution of the trait in two-child families is given, as in the general situation, by the expansion of $(p + q)^n$. Here $p = 3/4$, $q = 1/4$ and $n = 2$.

Now consider 64 sibships of three children each, with heterozygous parents. The proportion of families with 1, 2 and 3 affected children can again be calculated from the expansion $(p + q)^n$, but now $n = 3$.

The population distribution of such families is:

$$(3/4)^3 + 3(3/4)^2(1/4) + 3(3/4)(1/4)^2 + (1/4)^3 = 27/64 + 27/64 + 9/64 + 1/64$$

27/64 have no affected child and are not ascertained.
27/64 have 1 affected child.
9/64 have 2 affected children.
1/64 have 3 affected children.

Since 27 of the 64 families are not ascertained, then among the 37 that can be studied the following proportions are found:

27/37 (0.730) have 1 affected child.
9/37 (0.243) have 2 affected children.
1/37 (0.027) have 3 affected children.

The expected incidence of affected children in the ascertained sibships can now be calculated:

1/3 of 0.730 = 0.243
2/3 of 0.243 = 0.162
3/3 of 0.027 = 0.027
Expected proportion of affected children = 0.432

Table 14–2. EXPECTED PROPORTIONS AND NUMBERS OF AFFECTED MEMBERS IN SIBSHIPS OF SIZES 1–6, UNDER TRUNCATE AND SINGLE ASCERTAINMENT

	Truncate Ascertainment		Single Ascertainment	
Sibship Size	*Expected Proportion*	*Expected Number*	*Expected Proportion*	*Expected Number*
1	1.0000	1.0000	1.0000	1.0000
2	0.5714	1.1428	0.6250	1.2500
3	0.4324	1.2972	0.5000	1.5000
4	0.3657	1.4628	0.4375	1.7500
5	0.3278	1.6390	0.4000	2.0000
6	0.3041	1.8246	0.3750	2.2500

This method of correction can be extended to sibships of any size (Table 14–2). When the expectations in a given sample have been calculated, the expected and observed numbers of affected sibs can be compared by the χ^2 method, described later.

Single Ascertainment. Whereas under truncate ascertainment the assumption is that every affected child is a proband, under single ascertainment the assumption is that each family is ascertained only once and has only one proband, though there may be affected members who are not probands. The probability that any one sibship will be included is then directly proportional to the number of affected persons it contains.

Consider again the distribution of a recessive trait in three-child sibships. Since the probability that any sibship is ascertained is now proportional to the number of affected sibs, each term of the expansion is multiplied by the number of affected members, and the expression becomes:

$$o(p)^3 + 1(3p^2q) + 2(3pq^2) + 3(q)^3$$
$$= 3p^2q + 6pq^2 + 3q^3$$

If we now divide by the common term $3q$, the distribution becomes

$$p^2 + 2pq + q^2$$

which is $(p + q)^2$. In other words, the distribution in the ascertained families is binomial, with n = the sibship size minus 1. This is equivalent to omitting the proband from the calculations and counting only the sibs.

Table 14–2 summarizes the expected proportions of affected members under truncate and single ascertainment for sibships of up to six members. Unfortunately, collections of data strictly suitable for analysis by either method are hard to acquire, since some intermediate situation (multiple ascertainment) is usually a better representation of reality. Other methods of dealing with more complicated situations have been devised, requiring advanced mathematical knowledge and computer technology. Differences in ascertainment are discussed here chiefly to alert the reader to the importance of paying attention to how family data are collected.

The mathematical methods of demonstrating patterns of inheritance described above are becoming less important for autosomal recessive conditions that can be detected in carriers. However, they still have a place in medical genetics for analysis of the many disorders that are not identifiable in carriers and to point up curiosities that indicate complexities. In many studies of cystic fibrosis, for example, a segregation ratio of slightly over 0.25 has been found. This has led to an ongoing, as yet unresolved, discussion about whether two loci or more, rather than a single locus, could be involved, and about whether there might be some heterozygote advantage in being a carrier.

The χ^2 Test of Significance

Statistical tests of significance are used to determine whether a set of data conforms to a certain hypothesis. Textbooks of statistics and of genetics describe a variety of such tests and the circumstances under which they can appropriately be applied. One of the most useful is the χ^2 (Chi square) test, which is applicable to many problems in genetics because it does not require that the data analyzed be more or less normally distributed, but only that the numbers in the different categories be known.

The calculation of χ^2 is quite simple. If O is the observed number in each category and E is the number expected in that category on the basis of the hypothesis being tested, then χ^2 is the square of the difference between O and E, divided by E, summed over all the categories; in other terms:

$$\chi^2 = \Sigma \left[\frac{(O - E)^2}{E}\right]$$

The probability associated with a given value of χ^2 can be obtained from χ^2 tables originally prepared by Fisher and reprinted in many textbooks (see General References). Before using these tables, it is necessary to know the number of degrees of freedom available; the example given subsequently should show how this can be determined. A value of χ^2 associated with a probability of less than 0.05 is considered to be significant, that is, to indicate a significant disagreement between the observations and the hypothesis being tested. Table 14–3 shows examples of significant χ^2 values at the 0.05 and 0.01 levels, for different degrees of freedom.

The hypothesis being tested may be a null hypothesis, in the sense that the observed values do not differ from the expected, or that two variables being compared are not in any way associated. If the observed and expected values are significantly different, the null hypothesis is disproved.

Example 1. A series of patients with congenital pyloric stenosis includes 25 boys and 5 girls. Does this distribution differ from the normal sex ratio?

	Number observed	Number expected
Boys	25	15
Girls	5	15
Total	30	30

$$\chi^2 = \frac{(25 - 15)^2}{15} + \frac{(5 - 15)^2}{15} = 6.7 + 6.7 = 13.3$$

Table 14–3. EXAMPLES OF SIGNIFICANT χ^2 VALUES

Degrees of Freedom	Significance Level	
0	0.05	0.01
1	3.84	6.64
2	5.99	9.21
10	18.31	23.21

The expected values are calculated on the basis of a 1:1 sex ratio. There is one degree of freedom; since the total must remain fixed at 30, only one of the two values can be freely assigned. Consulting a table of χ^2, we see that for one degree of freedom, this value of χ^2 would occur by chance with a probability of < 0.01, that is, less than once in 100 times. The acceptable level of significance is $p = 0.05$. Thus, it can be concluded that in this series the observed excess of boys is statistically significant.

***Example* 2.** In a series of twins, at least one member of each pair having congenital dislocation of the hip, the defect was concordant in 34 of 86 MZ pairs and in 5 of 79 like-sexed DZ pairs. Does this evidence suggest that genetic factors are involved in the etiology of this disorder? Since congenital dislocation of the hip is about six times as frequent in females as in males, it is appropriate to compare like-sexed pairs rather than all DZ pairs.

	Concordant	Discordant	Total
MZ pairs	34	52	86
Like-sexed DZ pairs	5	74	79
Total	39	126	165

In this example, the expected proportion in each category can be calculated on the basis of the null hypothesis, which assumes that the degree of concordance is the same for MZ and DZ pairs. In the whole series, the proportion of concordant twin pairs is $39/165 = 0.24$. The expected numbers then are:

$$\text{MZ concordant } 0.24 \times 86 = 20.64$$

$$\text{MZ discordant } 0.76 \times 86 = 65.36$$

$$\text{DZ concordant } 0.24 \times 79 = 18.96$$

$$\text{DZ discordant } 0.76 \times 79 = 60.04$$

$$\chi^2 = \frac{(34 - 20.64)^2}{20.64} + \frac{(52 - 65.36)^2}{65.36} + \frac{(5 - 18.96)^2}{18.96} + \frac{(74 - 60.04)^2}{60.04}$$

$$= 24.91$$

Since the marginal numbers in the table above must remain fixed, only one of the numbers of twin pairs can be filled in at random; thus there is one degree of freedom. Again, the χ^2 table shows the calculated value of χ^2 to be beyond the level of significance ($p < 0.001$). The conclusion is that the excess of concordant MZ pairs is statistically significant. The genetic interpretation is that congenital dislocation of the hip is in part genetically determined.

Bayesian Methods in Risk Estimation

A number of different genetic problems can be approached by the use of Bayesian analysis. Bayes' theorem, first published in 1763, gives a method of assessment of the relative probability of each of two alternatives. It is used in many problems in clinical decision-making as well as in the kinds of applications shown here.

X-LINKED PEDIGREES

Murphy and Chase (1975—see General References for this chapter) have discussed the application of Bayesian analysis to estimation of carrier risk in

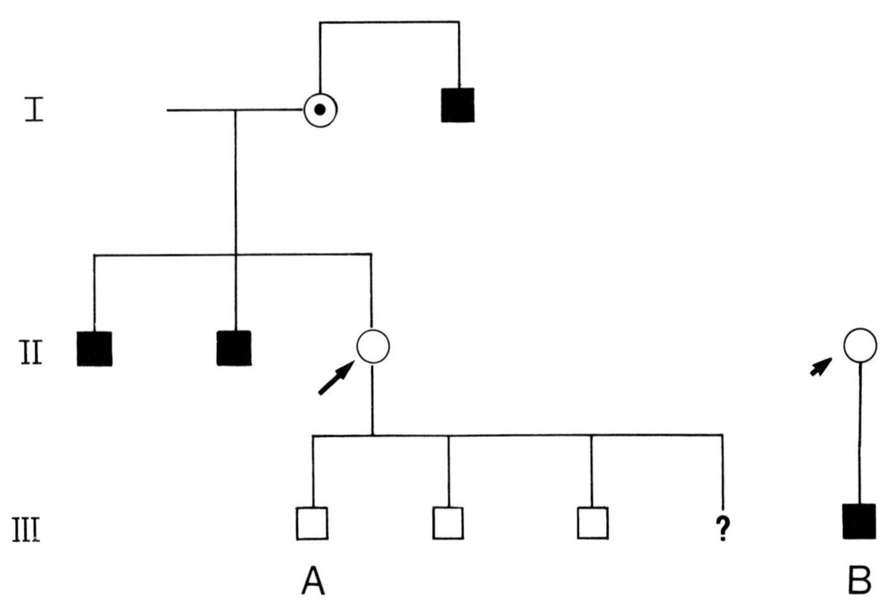

Figure 14–1. Two pedigrees of Duchenne muscular dystrophy. This figure is used to illustrate the application of Bayesian methodology to determine the probability that the mother in each pedigree (arrow) is a carrier.

pedigrees of X-linked lethal diseases such as Duchenne muscular dystrophy (DMD). To illustrate, we will consider the pedigrees shown in Figure 14–1.

In pedigree A of this figure, II–3 is the daughter of an obligate carrier of the Duchenne gene. Thus the prior probability that she is a carrier is 1/2 and the prior probability that she is not a carrier is also 1/2.

II-3 has three normal sons. If she is a carrier, the conditional probability that all three would be normal is $1/2 \times 1/2 \times 1/2 = 1/8$; if she is not a carrier, the conditional probability that all three sons would be normal is 1 (or very close to 1, since she might have a new mutant son).

We now consider the joint probability, which is the product of the prior and conditional probabilities; the joint probability that she is a carrier is $1/2 \times 1/8 = 1/16$, and the joint probability that she is not a carrier is $1/2 \times 1 = 1/2$. The other probabilities (that she is a carrier with 1, 2 or 3 affected sons) have not happened and so can be omitted from consideration.

The posterior probability that she is a carrier is therefore:

$$\frac{1/16}{1/16 + 1/2} = \frac{1}{9}$$

and the posterior probability that she is not a carrier is 8/9. Tabulating the above:

	II–3 a Carrier	II–3 Not a Carrier
Prior probability	1/2	1/2
Conditional probability	1/8	1
Joint probability	1/16	1/2
Posterior probability	1/9	8/9

Now the posterior probability that II–3 is a carrier can be applied to genetic counseling. The risk that her next child will be an affected male is $1/9 \times 1/4 = 1/36$. This is appreciably below the prior probability of 1/8 estimated when the genetic evidence provided by her children is not taken into consideration. Figure

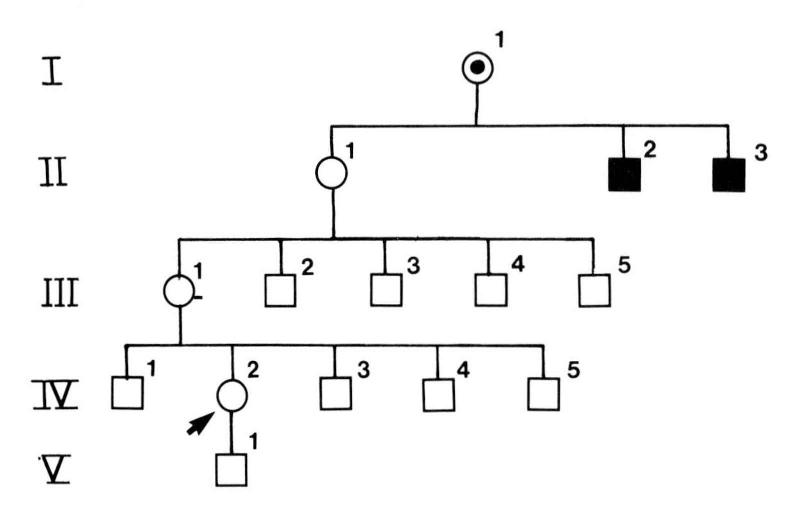

Figure 14–2. A pedigree of an X-linked recessive lethal (Duchenne muscular dystrophy) in which the prior probability that IV-2 is a carrier is 1/8, but Bayesian analysis gives a probability of 1/2115, close to the population risk. For solution, see Answers to Problems. From Murphy EA: Probabilities in genetic counseling. Birth Defects 1973; 9:19–33. Reprinted in Murphy and Chase, 1975.

14–2 is an additional example, in which information from several generations of normal males is used and the posterior probability that IV–2 is a carrier is 1/2115, close to the probability that any woman in the general population is a carrier (about 1/2500).

If a carrier test is used, the results can be included to give an even lower risk. Returning to pedigree A of Figure 14–1, assume that the mother is given a carrier test, serum creatine kinase (CK) determination, and is found to have a normal CK value, whereas two-thirds of carriers have abnormally high CK values but one-third have normal CK. The posterior probability for II–3 can now be revised:

	II–3 a Carrier	II–3 Not a Carrier
Prior probability	1/2	1/2
Conditional probability:		
of 3 normal sons	1/8	1
of normal CK	1/3	1
Joint probability	$1/2 \times 1/8 \times 1/3 = 1/48$	1/2
Posterior probability	1/25	24/25

Risk that the next child will be an affected son $= 1/4 \times 1/25 = 1/100$.

If the patient is an isolated case who may or may not be a new mutant, as in pedigree B of Figure 14–1, the prior probability that the mother is a carrier is quite different. For any woman in the general population, the prior probability that she is a DMD carrier is 4μ (μ = mutation rate). This probability is calculated as follows: the chance that she received a new mutation from one of her parents is μ for each parent, and the chance that she received the gene from a carrier mother is $1/2 \times 4\mu = 2\mu$, thus $\mu + \mu + 2\mu = 4\mu$ in all. Tabulating the calculations:

	I–1 a Carrier	I–1 Not a Carrier
Prior probability	4μ	$1 - 4\mu \cong 1$
Conditional probability		
of one affected son	1/2	μ
Joint probability	$4\mu \times 1/2 = 2\mu$	μ
Posterior probability	$\dfrac{2\mu}{2\mu + \mu} = 2/3$	1/3

Thus the probability that a mother of an isolated case of DMD is a carrier, when there is no family history of either normal or affected males, is 2/3, and the corresponding probability that the patient is a new mutant is 1/3. Additional family information or carrier test results can be incorporated into the risk calculation as shown in the previous example.

Haldane (1935) showed that the proportion of all patients with an X-linked recessive disease who are new mutants (m) is given by the rule:

$$m = \frac{(1 - f)\,\mu}{2\mu + \nu} \qquad \begin{aligned} f &= \text{fitness} \\ \mu &= \text{mutation rate in females} \\ \nu &= \text{mutation rate in males} \end{aligned}$$

In DMD, the fitness is zero (that is, virtually no patients reproduce), and the mutation rate of the gene appears to be the same in ova and sperm, that is, $\mu = \nu$; thus 1/3 of all cases should be new mutants. Note that this calculation gives the same result as the one reached by Bayesian analysis.

INCOMPLETE PENETRANCE

To estimate the recurrence risk for disorders with incomplete penetrance, the probability that an apparently normal person actually carries the gene concerned must be considered.

Figure 14–3 shows a pedigree of split-hand deformity, an autosomal dominant abnormality with incomplete penetrance. An estimate of penetrance can be made from the pedigree if it is large enough, or by reviewing published pedigrees; we will use 40 percent in our example.

The pedigree shows several people in whom the defect is not penetrant: I–1 or I–2, and II–3. The other unaffected family members may or may not have the gene.

If III–4 is the consultand, her risk of a child with split-hand deformity can be calculated as follows:

	Heterozygous	Normal
Probability of III–4's genotype	1/2	1/2
Conditional probability of her normal phenotype	6/10	1
Joint probability	3/10	5/10
Posterior probability	3/8	
Probability of III–4's child's genotype	3/16	13/16
Conditional probability child's phenotype will be abnormal	4/10	~0
Joint probability	12/160 = 0.075	~0

Thus, the chance that III–4 will have a child with the split-hand deformity is 7.5 percent.

INTERPRETATION OF TEST RESULTS

A common problem in risk estimation is that of judging test results when false-positive and false-negative results can occur. As an example, consider the

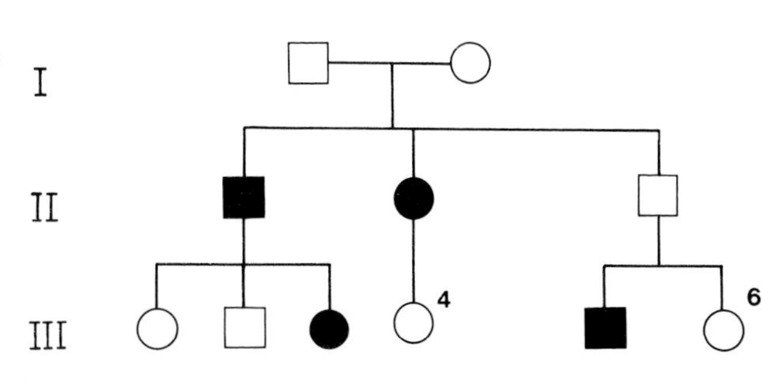

Figure 14–3. Pedigree of split-hand deformity, with lack of penetrance. For discussion, see text.

following data obtained by screening 100,000 pregnant women for the presence of a fetal neural tube defect (NTD) by means of a single serum α-fetoprotein assay, in a population in which the incidence of NTD is 2 per 1000 births. In general, studies have shown that 80 percent of women carrying fetuses with open NTD have elevated serum α-fetoprotein, when "elevated" is defined as a level higher than that in 95 percent of women with unaffected fetuses.

	Maternal Serum α-Fetoprotein (MSAFP)		
	Normal	*Elevated*	*Total*
Fetus normal	94,810	4990	99,800
Fetus with NTD	40	160	200
Total	94,850	5150	100,000

What is the probability that a woman with no family history of NTD and an abnormally high serum AFP is actually carrying a fetus with an open neural tube defect?

	Fetus Normal	Fetus with NTD
Prior probability (population risk)	0.998	0.002
Conditional probability of elevated AFP	$\frac{4990}{99{,}800} = 0.05$	$\frac{160}{200} = 0.80$
Joint probability	~ 0.05	0.0016
Combined probability of elevated AFP with NTD	$\frac{0.0016}{0.05 + 0.0016} = 0.031$ (3.1%)	

Given the population incidence and the findings shown here, the probability that a fetus of a mother with one abnormally high serum AFP measurement is actually affected (that is, the predictive value of an abnormal test) is only 3.1 percent. At the end of the chapter is a problem in which the prior probability is 0.05 because a first-degree relative is affected, and the population incidence is different from 2 per 1000; in this case the predictive value of an abnormal test is considerably higher.

The specificity of a test is defined as the proportion of all normal individuals who have normal test results—in the example above, 94810/99800 = 0.95. The sensitivity of a test is the proportion of all affected individuals who have an abnormal test result—160/200 or 0.80. The incidence must be determined, and the specificity and sensitivity of the test must be established for each laboratory by experience. The predictive value of an abnormal result is much higher when the prior probability of an abnormal result is also high.

AUTOSOMAL RECESSIVE PEDIGREES

A couple in which one or both parents are members of an autosomal recessive pedigree may be concerned about the risk of affected children. As in the X-linked pedigrees mentioned above, the probability of a subsequent affected child is reduced if a child or children already born to the couple are unaffected.

In the pedigree shown in Figure 14–4, II–2 and II–3 each have a relative with a particular recessive disease (for example, cystic fibrosis).

Probability II–2 is a carrier = 2/3
Probability II–3 is a carrier = 2/3
Probability both are carriers = 2/3 × 2/3 = 4/9

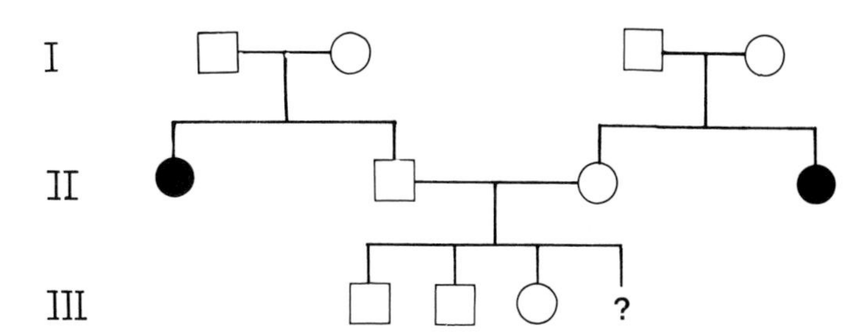

Figure 14–4. Pedigree of cystic fibrosis. For discussion, see text.

Now, considering their children:

	Both Carriers	Not Both Carriers
Prior risk for parents	4/9	5/9
Conditional probability (3 normal children)	$(3/4)^3$	1
Joint probability	3/16 = 0.19	5/9 = 0.56
Relative probability	$\frac{0.19}{0.75} = 0.25$	$\frac{0.56}{0.75} = 0.75$

The probability that the next child will be affected is now $1/4 \times 1/4 = 1/16$, appreciably below the original risk of $1/4 \times 4/9 = 1/9$.

LATE ONSET AGE

The use of Bayesian probability in Huntington disease is illustrated in Chapter 4. As an alternative way of reaching the same conclusion, the probability that an individual of any age is heterozygous can be derived by reading from the onset age figure (see Figure 4–25) the cumulative risk of being affected by that age. The chance that the individual is heterozygous can be expressed as:

$$\frac{(1 - \text{chance of being affected at present age})}{1 + (1 - \text{chance of being affected at present age})}$$

DETERMINATION OF TWIN ZYGOSITY

Yet another application of Bayesian analysis in genetics is the method used for twin zygosity determination, described in Chapter 16.

GENERAL REFERENCES

Cavalli-Sforza LL, Bodmer WF. The genetics of human populations. San Francisco: Freeman, 1971.
Emery AEH. Methodology in medical genetics. An introduction to statistical methods. Edinburgh: Churchill Livingstone, 1976.
Fisher RA. Statistical methods for research workers, 14th ed. New York: Hafner Press, 1973.
Goldberg MF, Oakley GP. Interpreting elevated amniotic fluid alpha-fetoprotein levels in clinical practice: use of the predictive value positive concept. Am J Obstet Gynecol 1979; 133:126–132.
Morton NE. Genetic tests under incomplete ascertainment. Am J Hum Genet 1959; 11:1–16.
Morton NE. Outline of genetic epidemiology. Basel: Karger, 1982.
Murphy EA, Chase GA. Principles of genetic counseling. Chicago: Year Book Medical Publishers, 1975.
Stempel LE. Eenie, meenie, minie, mo . . . What do the data *really* show? Am J Obstet Gynecol 1982; 144:745–752.
Vogel F, Motulsky AG. Human genetics: Problems and approaches. Berlin: Springer-Verlag, 1979.

PROBLEMS

1. If a coin is flipped six times, what is the chance of getting:
   a) six heads?
   b) three heads and three tails?

2. In a four-child family, state the probability that:
   a) The fourth child will be a boy.
   b) All four children will be boys.
   c) At least one child will be a girl.

3. Two parents carry the same recessive gene for congenital deafness. What is the probability that:
   a) Their first child will be deaf?
   b) Their five children will be normal?
   c) Two of their five children will be deaf?

4. A man with Huntington disease has three children. What is the chance that:
   a) None of the three will develop the disease?
   b) The first child will be affected and the next two normal?
   c) One child will be affected and two normal?

5. In a series of two-child families in each of which there is at least one child with cystic fibrosis, what proportion of all the children in the sibships will be affected:
   a) Under truncate selection?
   b) Under single selection?

6. See legend of Figure 14–2.

7. A woman has one child with a neural tube defect and is in the 16th week of her second pregnancy. In her population, the incidence of neural tube defects is about 1 in 400 births. The risk to first degree relatives of patients is the square root of the population frequency, or about 5 percent. (This is typical of multifactorial traits.) She is given a serum α-fetoprotein assay. In the laboratory performing the test, the specificity of the test is 0.95 and the sensitivity is 0.80. She is found to have elevated serum AFP (defined here as a level above that in 95 percent of women with unaffected fetuses). What is the probability that her fetus is affected?

8. A man and woman each have a history of autosomal recessive limb-girdle muscular dystrophy in a first cousin.
   a) What is the chance that their first child will have this disease?
   b) Their first two children are unaffected. How does this affect your estimate of the risk to their child?

9. A man aged 50 has a father with Huntington disease. His sons aged 30 and 28 and his grandson aged 5 are unaffected. What is the probability that he himself will never develop Huntington disease?

10. In a particular family with 8 sibs, a trait that might be either autosomal recessive or X-linked is segregating.
    a) Three of the four males and none of the four females is affected. What is the probability of this distribution if inheritance is X-linked? autosomal recessive?
    b) If one of the four males and two of the four females were affected, how would this change the probabilities?

15

POPULATION GENETICS

Population genetics is the study of the distribution of genes in populations and of how the frequencies of genes and genotypes are maintained or changed. Because of its concern with environment as well as heredity, population genetics extends into epidemiology; it has given rise to the field of genetic epidemiology, which deals with the etiology, distribution and control of disease in kindreds and with heritable causes of disease in populations.

Gene Frequencies in Populations: The Hardy-Weinberg Law

Under certain ideal circumstances, genotypes are distributed in proportion to the gene frequencies in the population and remain constant from generation to generation. If gene frequencies and the corresponding genotype frequencies remained fixed, evolution could not proceed; hence population genetics encompasses factors concerned in human evolution.

The cornerstone of population genetics is the **Hardy-Weinberg law,** which is named for its discoverers, the English mathematician G. H. Hardy and the German physician W. Weinberg, who independently defined it in 1908. It is reported that Hardy, when presented with the problem of why a dominant trait does not automatically increase until it replaces the recessive, worked it out immediately and thought it too trivial to publish; as it happens, he is remembered more for this piece of work than for his fundamental contributions to mathematical theory.

FREQUENCY OF AUTOSOMAL GENES

For any gene locus, in a population with random mating, the genotype frequencies are determined by the relative frequencies of the alleles at the locus. When we speak of the population frequency of a gene, we have in mind a **gene pool,** in which are collected all the alleles at that particular locus for the whole population. For an example using only two different alleles, we can think of the gene pool as a beanbag containing beans of two colors, white and black. (Population genetics is sometimes called "beanbag genetics.") The chance that in two draws a person will draw any one of the three possible combinations of beans (two white, two black or one of each color) depends on the frequency of each color in the bag. If p is the frequency of white beans and $q = 1 - p$ is the frequency of black beans, then the relative proportions of the three combinations are as follows:

$$p^2 \text{ (2 white)} + 2pq \text{ (1 white, 1 black)} + q^2 \text{ (2 black)}$$

Table 15–1. RELATIONSHIP OF GENE FREQUENCY TO GENOTYPE FREQUENCY

Gene Frequencies		*Genotype Frequencies*		
p (*T*)	q (*t*)	p^2 (*TT*)	$2pq$(*Tt*)	q^2(*tt*)
0.5	0.5	0.25	0.50	0.25
0.6	0.4	0.36	0.48	0.16
0.7	0.3	0.49	0.42	0.09
0.8	0.2	0.64	0.32	0.04
0.9	0.1	0.81	0.18	0.01
0.99	0.01	0.98	0.02	0.0001

Returning to the example of the taster/nontaster traits used in Chapter 4, now let the white beans represent the allele *T* (taster) and the black beans the allele *t* (nontaster). The relative frequencies of the genotypes for some arbitrarily selected values of p and q are shown in Table 15–1.

If the different genotypes in a population are present in these proportions, the population is said to be in Hardy-Weinberg equilibrium. Note that this is a very simple application of the binomial expansion $(p + q)^n$, with $p = q = 1/2$ and $n = 2$.

Hardy-Weinberg equilibrium can be disturbed by various factors, including nonrandom mating, mutation, selection and migration (see later in this chapter).

An important consequence of the Hardy-Weinberg law is that the proportions of the genotypes do not change from generation to generation. The next generation following random mating of a population in which the genotypes *TT, Tt, tt* are present in the proportions $p^2{:}2pq{:}q^2$, so that the mating types are in the proportions of the expansion $(p^2 + 2pq + q^2)^2$, is shown in Table 15–2.

The chief clinical application of Hardy-Weinberg equilibrium is for the calculation of gene and heterozygote frequency, if the frequency of a trait is known. If the frequency of an autosomal trait $(q^2) = 0.0001$, the frequency of q is $\sqrt{0.0001} = 0.01$, the frequency of p is $1 - 0.01 = 0.99$, and the heterozygote frequency $(2pq)$ is 0.02.

Because for rare disorders p (the frequency of the normal allele) is usually close to 1, the heterozygote frequency is usually close to $2q$ (twice the gene frequency).

If the three genotypes are phenotypically distinguishable, the gene frequencies can be determined simply by counting, as in the following example:

In a certain random-mating group, 16 percent of the members are of blood group M, 48 percent are MN and 36 percent are N. What are the frequencies of genes *M* and *N*?

0.16 of the members are M, with two *M* genes each
0.48 have 1 *M* and 1 *N* gene
0.36 have two *N* genes each

Table 15–2. FREQUENCIES OF MATING TYPES AND OFFSPRING

Mating Types	Frequency	Offspring *TT*	*Tt*	*tt*
TT × *TT*	p^4	p^4		
TT × *Tt*	$4p^3q$	$2p^3q$	$2p^3q$	
TT × *tt*	$2p^2q^2$		$2p^2q^2$	
Tt × *Tt*	$4p^2q^2$	p^2q^2	$2p^2q^2$	p^2q^2
Tt × *tt*	$4pq^3$		$2pq^3$	$2pq^3$
tt × *tt*	q^4			q^4

Sum of *TT* offspring $= p^4 + 2p^3q + p^2q^2 = p^2(p^2 + 2pq + q^2)$
Sum of *Tt* offspring $= 2p^3q + 4p^2q^2 + 2pq^3 = 2pq(p^2 + 2pq + q^2)$
Sum of *tt* offspring $= p^2q^2 + 2pq^3 + q^4 = q^2(p^2 + 2pq + q^2)$

The common factor $p2 + 2pq + q^2$ can be dropped, and the proportions of the three genotypes are then seen to be $p^2{:}2pq{:}q^2$ as in the parental generation.

Thus in 100 members with 200 genes, there are $(16 \times 2) + 48 = 80$ M genes and $48 + (2 \times 36) = 120$ N genes. In other words, the frequency of M is 0.40 and the frequency of N is 0.60.

FREQUENCY OF X-LINKED GENES

In Chapter 4 the Xg blood group system was described as an example of X-linked inheritance. Here the Xg system will be used to show how the frequencies of X-linked genes relate to genotype frequencies.

Often the first clue that a trait is X-linked is the difference in its frequency in males and females. For the Xg blood system, 67 percent of males are Xg(a+) and 33 percent are Xg(a−), whereas for females 89 percent are Xg(a+) and only 11 percent are Xg(a−).

Since males are hemizygous for X-linked genes, there are only two male genotypes.

Genotype	Phenotype	Frequency
Xg^aY	Xg(a+)	$p = 0.67$
XgY	Xg(a−)	$q = 0.33$

The frequencies of the two alleles are given directly by the frequencies of the two phenotypes.

Females have two X's, therefore two Xg alleles. Three genotypes are possible:

Genotype	Phenotype	Frequency
$Xg^a\,Xg^a$	Xg(a+)	p^2 } $= 0.89$
Xg^aXg	Xg(a+)	$2pq$
$Xg^{a-}Xg$	Xg(a−)	$q^2 = 0.11$

From the male data, we see that $q = 0.33$. The square root of the frequency of the Xg(a−) phenotype in females gives the same result ($\sqrt{0.11} = 0.33$). Substituting the male values of p and q, we find that Xg(a+) females have an expected frequency of 0.89, which is very close to actual observation.

Note that for deleterious X-linked recessives the theoretical ratio of affected males to affected females is q/q^2, or $1/q$. This relationship demonstrates that males are far more frequently affected.

For X-linked dominants, the ratio of affected males to affected females is

$$\frac{q}{2pq + q^2}$$

For rare traits (that is, when q is very small) this ratio is very close to 1/2, which is to be expected since males have only one X (one draw from the gene pool) and females have two. For more common traits the ratio is greater than 1/2.

Systems of Mating

Hardy-Weinberg equilibrium is maintained only if there is **random mating**—that is, if any one genotype at a locus has a purely random probability of combining with any other genotype at that locus—so that the frequencies of the different kinds of matings are determined only by the relative frequency of the genotypes in the population. The random mating requirement is probably rarely fulfilled in practice. Within any population there are many subgroups differing genetically (for example, members of different ethnic groups each with characteristic gene frequencies) and nongenetically (for example, with respect to religion). Members of such subpopulations are more likely to mate within their own subgroup than outside it. If mating is nonrandom, it is said to be **assortative**; it is a common observation that within any population mating tends to be positively assortative with respect to ethnic background, intelligence, stature, economic status and many other traits, some of which are determined genetically or partly genetically. Assortative mating may be **positive**, as when there is a tendency for persons who resemble one another to marry more frequently than chance alone would indicate, or **negative**, in the sense that "opposites attract."

Consanguinity

Consanguineous mating is a special form of assortative mating. It can disturb Hardy-Weinberg equilibrium by increasing the proportion of homozygotes at the expense of heterozygotes. Of course, inbreeding alone does not affect the proportions of alleles in the next generation, but only their assortment into genotypes. However, inbreeding may expose recessive genes to selection and loss, thus permanently altering the gene frequencies in the population.

In North America today, the incidence of consanguineous matings is lower than in older countries, at least partly because much of the continent is so recently settled that there are fewer cousins among whom to find a mate. The low birth rate also reduces the number of cousins available. In many countries the improvements in communication of the last century have had the effect of breaking down isolates (enclaves within which the members intermarried much more frequently than they married outsiders). Nevertheless, there are many subgroups within which consanguineous mating is still relatively common.

Of course, given true random mating, a certain small proportion of marriages would be consanguineous. For that matter, all marriages are at least distantly consanguineous. Estimating about three generations per century, at the time of the Declaration of Independence (1776) the young adults of today each had about 64 (2^6) ancestors, and these may well have been 64 different people; but at the time of the Magna Carta (1215) the number was about 2^{24}, or 17 million, and it is very unlikely that there were 17 million different people involved. Probably many people are homozygous for rare genes that have come down in different lines from one heterozygous ancestor. Few people can trace their ancestry in all its ramifications beyond the first few generations. French Canadians who can do so and whose progenitors have been in Canada since the 17th century sometimes find themselves to be inbred through 10 or more different lines of descent.

THE MEASUREMENT OF CONSANGUINITY

The measurement of consanguinity is relevant to medical genetics because consanguineous marriages have an above-average risk of producing offspring homozygous for some deleterious recessive gene. The risk is proportional to the closeness of the relationship of the parents concerned. For practical purposes, only first-cousin and second-cousin matings, or incestuous matings, occur often enough and carry a sufficiently increased risk to be of practical importance. Some types of consanguineous matings are shown in Figure 15–1, and the corresponding coefficients of inbreeding in Table 15–3.

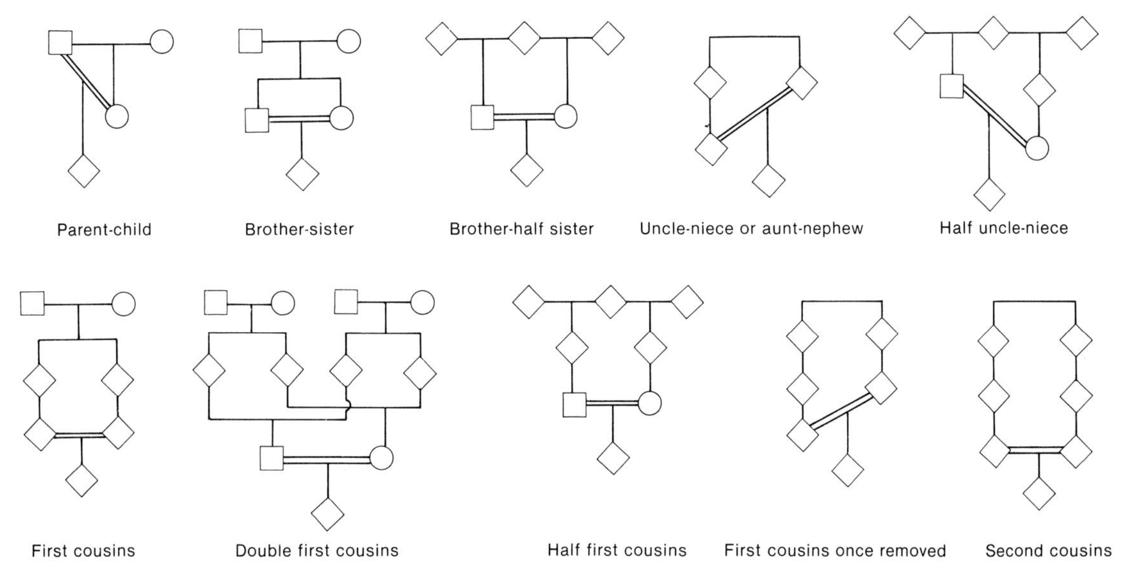

Figure 15–1. Types of consanguineous mating. Incestuous matings are those in which marriage of the partners is illegal.

The **coefficient of inbreeding** F is the probability that an individual has received both alleles of a pair from an identical ancestral source, or the proportion of loci at which he is homozygous. In Figure 15–2, IV–1 is the offspring of a first-cousin mating. As described earlier (Chapter 4), for any gene the father possesses, the chance that the mother also possesses it is one-eighth. For any gene the father gives his child, the chance that the mother has the same gene and will transmit it is $1/8 \times 1/2 = 1/16$ (0.06). This is the coefficient of inbreeding for the child of first cousins. It signifies that the child has a 1/16 chance of being homozygous at any one locus, or that he is homozygous at 1/16 of his loci. An alternative way of reaching the same conclusion: each of the four alleles at the A locus in generation I has a 1/64 chance of being homozygous in IV–1, thus IV–1's probability of being homozygous for one of the four is $4 \times 1/64 = 1/16$.

Some examples of the coefficient of inbreeding for certain human populations are listed in Table 15–4. In some inbred communities, the average coefficient of inbreeding is as high as 0.03, indicating that on the average the members of the groups are as closely related as half-first cousins.

Table 15–3. CONSANGUINEOUS MATINGS

Type	Degree of Relationship	Genes in Common	F (Coefficient of Inbreeding of Child)
MZ twins	—	1	–
Parent-child	1st degree	1/2	1/4
Brother-sister (including DZ twins)	1st degree	1/2	1/4
Brother-half sister	2nd degree	1/4	1/8
Uncle-niece or aunt-nephew	2nd degree	1/4	1/8
Half uncle-niece	3rd degree	1/8	1/16
First cousins	3rd degree	1/8	1/16
Double first cousins	2nd degree	1/4	1/8
Half first cousins	4th degree	1/16	1/32
First cousins once removed	4th degree	1/16	1/32
Second cousins	5th degree	1/32	1/64

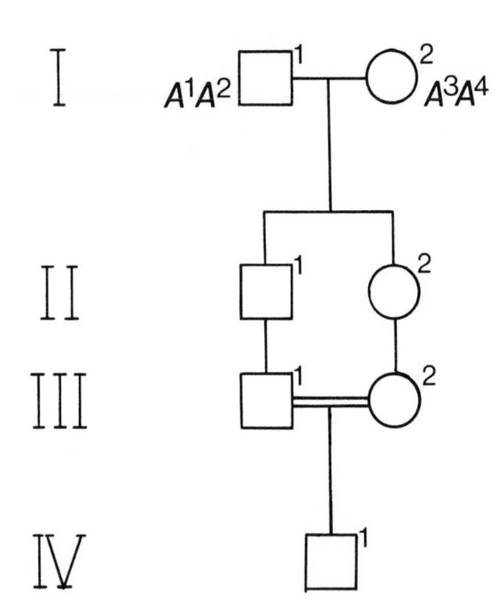

Figure 15–2. A cousin marriage, used in the text to demonstrate how F, the coefficient of inbreeding of the child IV-1, is calculated.

Mutation

New hereditary variations arise by mutation; the new gene (or the individual who carries it) is called a mutant. Mutations are the only source of the material of evolution, upon which natural selection acts to preserve the fit and to eliminate the less fit.

In a sense, any change in the genetic material is a mutation, but here we will follow the usage of Crow (1961) and define a mutation as a change that cannot be shown to depend upon a detectable chromosomal rearrangement or some sort of recombination mechanism. Though small chromosomal rearrangements or deletions are very difficult or impossible to distinguish from actual mutations in human material, this discussion will be restricted as much as possible to point mutations, that is, mutations involving changes in base sequence in individual codons. As described earlier (Chapter 3), mutations can occur in regions outside as well as within the coding sequences of a gene. Moreover, DNA analysis shows that small deletions may be a relatively common cause of abnormal genes.

A mutation usually leads to loss or change of the function of a gene. Since a random change is unlikely to lead to an improvement, the majority of mutations are deleterious, though some are advantageous and become established. The frequency of a gene in the population represents a balance between the mutation rate of the gene and the effect of selection. If either the mutation rate or the effectiveness of selection is altered, the gene frequency changes.

Table 15–4. EXAMPLES OF COEFFICIENT OF INBREEDING (F) FOR SOME HUMAN POPULATIONS

Population	F
Canada:	
Roman Catholic	0.00004–0.0007
United States:	
Roman Catholic	0–0.0008
Hutterites	0.02
Dunkers (Pennsylvania)	0.03
Latin America	0–0.003
Southern Europe	0.001–0.002
Japan	0.005
India (Andhra Pradesh)	0.02
Samaritans	0.04

Many mutations are lethal, leading to the death of the persons who receive them. A mutation that leads to failure of an adult to reproduce is equally lethal in the genetic sense to one that causes an early abortion of an embryo. Others may be sublethal. Most of those actually observed in man are less severely detrimental, since a mutation cannot be identified at all unless it is compatible with life, at least to the time of birth.

The number of generations a mutation persists before failing to be passed on (undergoing genetic death) is closely related to the selection rate against it. For example, a mutation that causes a reduction in reproductive fitness of 20 percent survives, on the average, for 1/0.20 = 5 generations.

The load of genetic damage in man is not accurately known, but it is estimated that at least 6 percent of all persons born have some tangible genetic defect that will be apparent later in life if not at birth. Perhaps a quarter of all genetic defects are caused by single-gene mutations.

Though it is very difficult to know how many of these mutations are new and how many have been inherited from previous generations, it is estimated that on the average each individual carries about three to five **lethal equivalents**. A lethal equivalent is defined as one gene which, if homozygous, would be lethal, or two genes which, if homozygous, would be lethal in half the homozygotes, and so on.

Even a slight **heterozygote advantage** can lead to the preservation and increase in frequency of a gene that is severely detrimental when homozygous. The classic example is the resistance to malaria afforded by the sickle cell trait, described below. Hb C, thalassemia, G6PD deficiency and the *Fy* allele of the Duffy blood group system are also thought to provide protection against malaria. It has been suspected that a similar but unknown advantage might account for the high Caucasian frequency of the cystic fibrosis gene, but there is no reasonable suggestion as to what the advantage might be. Human genes evolved under very different environmental conditions from those that exist today; infectious diseases have periodically decimated populations over the entire known world, even in recent generations. It seems unlikely that gene frequencies and the factors affecting them can be traced, even in the relatively recent history of man.

In experimental organisms, some genes that are detrimental in homozygotes are transmitted through the sperm to far more than the usual 50 percent of the offspring of heterozygotes. This property is known as **meiotic drive**. A detrimental gene that confers an advantage on the gamete has a greatly enhanced chance of survival and increase.

MUTATION RATE

The spontaneous mutation rate varies for different loci, the limits of the observed range being about 1 in 10,000 and 1 in 1,000,000 per locus per gamete per generation, with an average of about 1 in 100,000. Some loci, such as the blood group loci, are virtually never observed to mutate. At the opposite end of the range, neurofibromatosis and Duchenne muscular dystrophy have mutation rates among the highest measured in man, in the range of 1 in 10,000. The mutation rate of fragile X syndrome may be close to 1 in 1,000 (Sherman et al., 1984). Measurements of mutation rates in man are usually not highly accurate, but a value much greater than 1 in 100,000 may suggest that mutations at several different loci with similar phenotypic effects are being measured as though all occurred at a single locus, or that some other characteristic (reduced penetrance, for example) is interfering with the accuracy of the measurement.

For rare dominant mutations, there is a direct method of measurement. The number of cases, n, of a disorder that have been born to normal parents in a defined area over a defined time span, and the total number of births in the same area during the same period, N, are determined. Then, since a new dominant trait requires a mutation in only one of two gametes, the mutation rate is n/2N. In practice, allowance must be made for unascertained cases, unclassified parents, reduced penetrance and, especially, genetic heterogeneity. If the same phenotype can be produced by mutation at more than one locus or by environmental factors, accurate measurement of the mutation rate at a given locus may be impossible.

The mutation rate of recessive genes is much more difficult to determine, because it is usually impossible to know whether the gene in question is a new mutation or has been inherited from a heterozygous parent. If there is a test for heterozygotes, the answer will be clear, but many recessives have no demonstrable heterozygous expression. In any case, the probability is always in favor of inheritance rather than mutation, since the population frequency of recessive alleles is usually of the order of 0.003 to 0.01, far above the mutation frequency. If heterozygote advantage is present, it may lead to a gross overestimate of the mutation rate.

One would expect that the older a parent is at the time of conception, the more likely that he or she has accumulated mutations that the child might inherit. For fathers of children with some fresh dominant mutations, such as achondroplasia, this has been repeatedly demonstrated (Karp, 1980; Stoll et al., 1982). As shown in Figure 15–3, fathers of children with neurofibromatosis are on the average a few years older than other fathers in the population (Riccardi et al., 1984). The relationship may not hold for all dominant mutations and is not present, or at least has not been recognized, in older mothers.

For X-linked genes, the mutation rate is difficult to assess directly unless the trait determined by the mutation is a genetic lethal, in which case the typical frequency of the trait in males is three times the mutation rate (see Chapter 14).

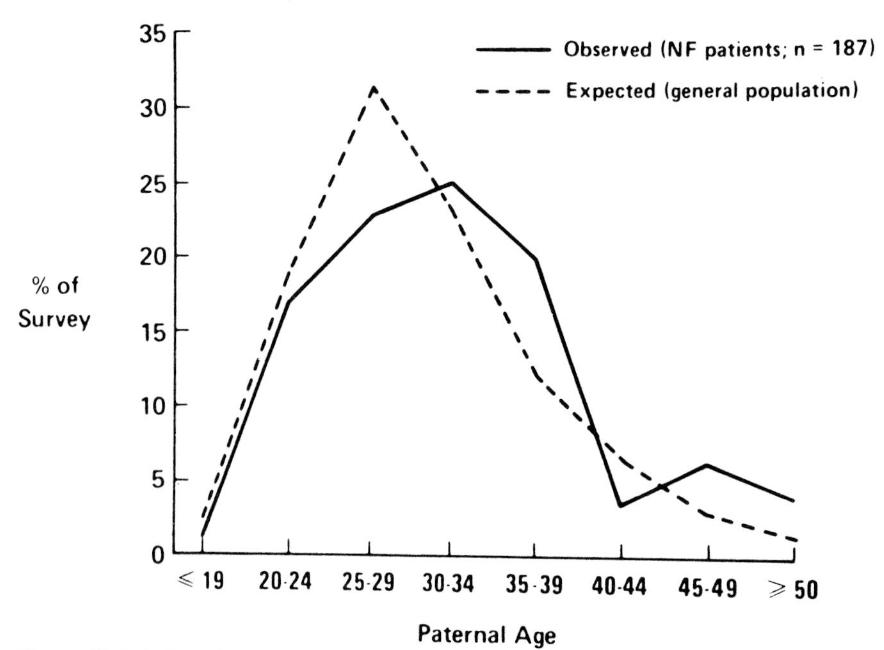

Figure 15–3. Paternal age in neurofibromatosis as compared with the general population. From Riccardi VM, Dobson CE, Chakraborty R, Bontke C. The pathophysiology of neurofibromatosis: IX. Paternal age as a factor in the origin of new mutations. Am J Med Genet 1984; 18:169–176, by permission.

ENVIRONMENTAL MUTAGENS

The mutation rate can be increased by a number of environmental agents, especially radiation and many chemicals. It has been very difficult to study the magnitude of such effects directly in man; for the most part we have had to rely on experimental animals, especially fruit flies and mice. Methods are now being developed to assay for mutagenic effects in other systems. One method is the Ames assay, using microorganisms that are injected into mutagenized test animals, exposed there to the mutagen, then recovered for measurement of their mutation rate (Ames et al., 1973). Another method measures the incidence of abnormally shaped sperm in mice exposed to mutagens (Wyrobek and Bruce, 1978). Both assays have been used to test a variety of substances for mutagenicity.

Radiation

Ionizing radiation is by far the most potent mutagen known. The effect of X-rays in increasing the rate of mutation was first observed by Muller (1927) and earned him the 1946 Nobel prize. Man is exposed to a certain background level of radiation from cosmic rays and other natural sources. This exposure can be measured in terms of the roentgen (r), the unit of exposure dose; the rad, the unit of absorbed dose; or the rem (roentgen-equivalent man), defined as the quantity of any ionizing radiation that has the same biological effect as 1 rad of X-rays. It is, of course, only the dose absorbed by the gonad that is genetically significant. Natural radiation provides perhaps 95 millirads (mrads) of genetically significant exposure per year, while man-made radiation (chiefly medical and dental X-rays) adds another 40 or so mrads per year, bringing the total to about 135. Over an estimated 30-year period, to the end of the reproductive years, the total exposure is perhaps 3r.

A common measure of the effect of radiation in inducing mutations is the **doubling dose**, the dose that would be required to double the spontaneous mutation rate. In the mouse, this is in the 30–40 rem range for acute exposure and the 100 rem range for low-dose, low dose-rate exposure.

The gonadal exposure from diagnostic and therapeutic X-rays varies greatly with the quality of the equipment used and the shielding provided; therefore, we will not attempt to provide an estimate of the gonadal dose for different radiological procedures. In mice, the relation between exposure and the number of mutations produced is linear, except possibly at very low intensity (Russell et al., 1959). Thus it seems wise to reduce gonadal exposure to the lowest possible level in persons who have not completed their reproductive years.

Evidence for the production of mutations by ionizing radiation in man has been sought from a number of sources, but especially from studies of the offspring of survivors of the atomic bombings of Hiroshima and Nagasaki. Ever since the atomic bombings, the potential genetic effects have been closely monitored. In a reappraisal based on data from children born to survivors of the bombings (Schull et al., 1981), four indicators were used: incidence of untoward pregnancy outcome, childhood deaths, sex chromosome aneuploidy, and mutation resulting in an electrophoretic variant. In no case was there a statistically significant effect of exposure, but in all cases the observed effect was in the direction expected if the parents' exposure to atomic radiation from the bombings had caused genetic damage. The average doubling dose (that is, the amount of radiation that would double the spontaneous mutation rate in the preceding generation), was estimated to be about 156 rems. This estimate of the doubling dose is four times higher than that estimated in the mouse from experimental studies. An even higher doubling dose would be expected in man if the exposure were chronic and at low levels, rather than acute and at high levels as at Hiroshima and Nagasaki. Obviously, this study has important implications for setting permissible limits to human radiation exposure.

Chemical Mutagenesis

Many chemicals are known to be carcinogenic (cancer-inducing); most of these are now known to be mutagenic as well. The first example, found by Charlotte Auerbach in 1941, was mustard gas, an alkylating agent that produces single-base substitutions in DNA. To be mutagenic, a substance must permeate the nucleus and react with DNA, influencing either its stability or its replication. Thus the types of chemical mutagens known include, in addition to alkylating agents, substances related to the dye acridine orange, and base analogues, all of which are capable of interfering with DNA replication, though by different mechanisms.

Caffeine is a base analogue and so may be a mutagen, but so far it has not been shown to have a mutagenic effect in mammals.

DELAYED MUTATION

Occasionally pedigrees are seen in which a condition that was previously unknown in a family appears in several branches of the family at much the same time and thereafter shows typical Mendelian transmission. Some of these pedigrees may be explained by incomplete penetrance or simply by the coincidence of two or more independent mutations, but another possible explanation is delayed mutation. This interpretation was suggested in 1956 by Auerbach, who noted that in experimental animals a delayed effect is characteristic of many chemical mutagens. Auerbach observed that when fruit flies (*Drosophila*) were treated with mustard gas, many of the mutations that were induced occurred as labile premutations. A premutation might revert to the wild type, remain as a premutation, or undergo delayed mutation to a new mutant form.

$$\text{Wild type gene} \rightleftarrows \text{premutation} \rightleftarrows \text{mutant gene}$$

Thus an individual carrying a premutation could be a **gonadal mosaic**, with three possible cell lines: a true-breeding wild type line, a true-breeding mutant line and a line carrying the premutation. Auerbach proposed that a similar explanation might apply to certain human pedigrees in which several new mutants of the same type appear simultaneously in different branches.

Delayed mutation has been invoked to explain unusual pedigrees of achondroplasia, the Beckwith-Wiedemann syndrome of exomphalos, macroglossia and gigantism (Fig. 15–4), and certain other conditions, as well as the pedigree of split-hand deformity that originally led Auerbach to propose this mechanism (Fig. 15–5). The difference between delayed mutation and complete mutation is clinically important with respect to interpretation of pedigree data and recurrence risk estimation (Herrmann and Opitz, 1977). It should be kept in mind as a possible explanation of pedigrees in which there are distantly related cases of a disorder connected by unaffected "carriers."

SEX DIFFERENCE IN MUTATION RATES

For many years there has been a controversy over whether the mutation rate is the same in male and female germ cells. This question can be examined in X-linked mutations and has been tested for several disorders including classic hemophilia, Duchenne muscular dystrophy and Lesch-Nyhan syndrome. For

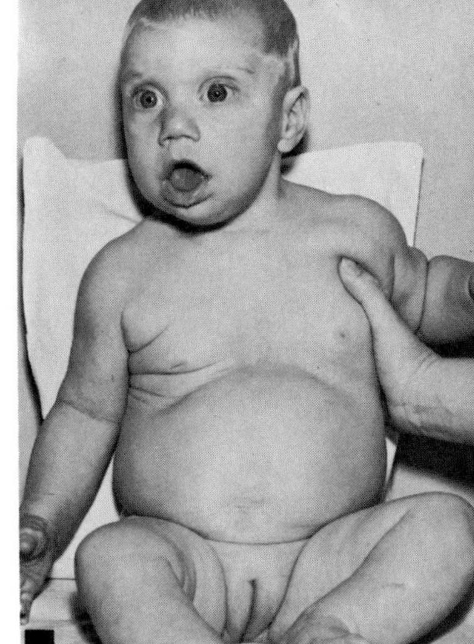

Figure 15–4. The Beckwith-Wiedemann syndrome, characterized by omphalocele, macroglossia and gigantism. Delayed mutation has been suggested as a possible explanation of unusual pedigrees of this syndrome. From Smith DW. Recognizable patterns of human malformation, 3rd ed. Philadelphia: W. B. Saunders Company, 1982.

both hemophilia and Duchenne muscular dystrophy, the mutation rate appears to be the same in male and female germ cells, but in Lesch-Nyhan syndrome there is some evidence for a higher mutation rate in male germ cells. Although the much greater number of DNA replications between primordial germ cell and gamete in males as compared with females certainly would explain an excess of mutations in males, at present no consistent difference has been found; even when it does exist the magnitude of the effect seems smaller than would be expected. The question of a sex difference in mutation rates remains unresolved.

Selection

Darwin postulated **natural selection** as the factor of importance in evolution. In modern terms, survival of the fittest is interpreted as taking place through the

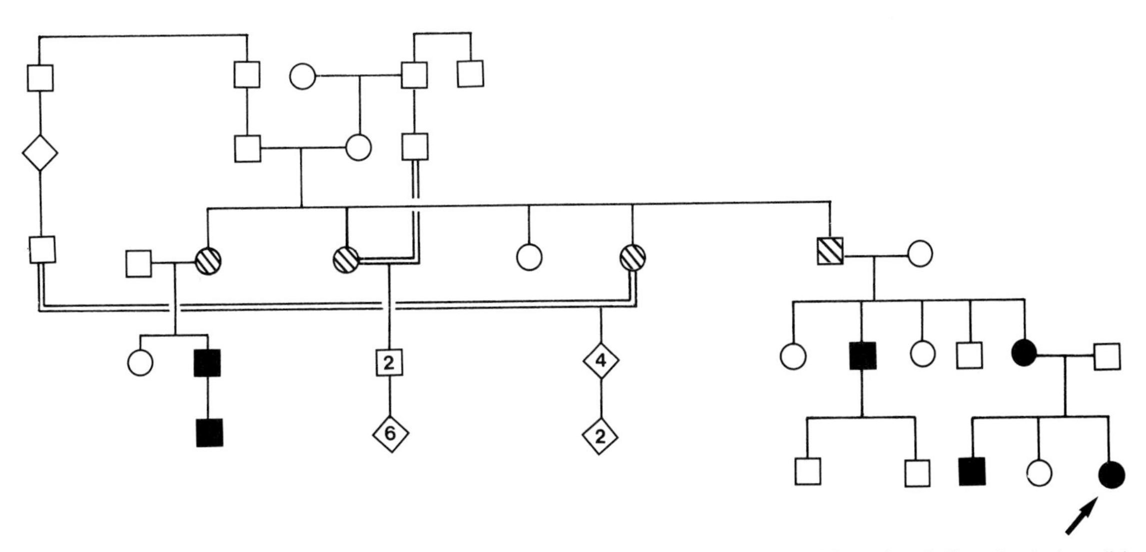

Figure 15–5. Part of a pedigree of split-hand deformity possibly explained by delayed mutation. Symbols: solid black, affected; cross-hatched, X-rays normal. Redrawn from Graham JB, Badgley CE. Split-hand with unusual complications. Am J Hum Genet 1955; 7:44–50.

action of selection upon new genotypes that have arisen by mutation or recombination.

In the biological sense, the term **fitness** has no connotation of superior endowment except in a single respect—the ability to contribute to the gene pool of the next generation. The many factors that affect fitness can operate at any stage of the life cycle, at least until the end of the reproductive period. Perhaps selection can operate even after the reproductive period; there may be some selective value in a life span that lasts no longer than the productive years.

Dominant deleterious genes are openly exposed to selection, in contrast to autosomal recessives, most of which are hidden in heterozygotes. Consequently, the effects of selection are more obvious and can be more readily measured for dominants than for recessives.

SELECTION AGAINST AUTOSOMAL DOMINANT MUTATIONS

A harmful dominant mutation, if fully penetrant, is expressed in any individual who carries it. Whether or not it is transmitted to the succeeding generation will depend on how deleterious it is. If it prevents reproduction and accordingly is not represented at all in the next generation, its relative fitness is zero. If it is just as likely as the normal allele to be represented in the next generation, its fitness is 1. Most deleterious dominant mutations have a fitness value between zero and 1. Many lethal multiple congenital defects may be due to new dominant mutations, but it is usually impossible to be certain since the affected children do not reproduce and molecular analysis of the patient's and parents' DNA is rarely practicable as yet.

If the mutation is deleterious but affected persons are fertile, they may contribute fewer than the average number of offspring to the next generation; that is, their fitness may be reduced. The mutation will be lost through selection at a rate proportional to the loss of fitness of heterozygotes. The coefficient of selection, s, is the measure of the loss of fitness. Fitness is measured as $f = 1 - s$.

Achondroplastic dwarfs have about one-fifth as many children as do normal members of the population. Thus, the fitness is 0.20, and the coefficient of selection is 0.80. In the subsequent generation, the frequency of the achondroplasia alleles passed on from the current generation is 20 percent. The remaining 80 percent required to maintain the gene frequency at equilibrium are added through new mutation. The observed gene frequency in any one generation represents a balance between loss of alleles through selection and gain through recurrent mutation.

Selection against a dominant trait can lower the frequency of the dominant gene precipitately. If no heterozygotes for the dominant gene for Huntington disease reproduced, the incidence of the disease would fall in one generation to a level determined by the mutation rate, because the only Huntington disease genes remaining in the population would be the new mutations.

SELECTION AGAINST AUTOSOMAL RECESSIVE MUTATIONS

Selection against harmful recessive genes is less effective than selection against dominants. Even if there is complete selection against homozygous recessives, it takes 10 generations to reduce the gene frequency from 0.10 to 0.05; the lower the gene frequency, the slower the decline. Removing selection (for example, by successful medical management of children with cystic fibrosis or Tay-Sachs disease, so that they could survive and reproduce at a normal rate) raises the gene frequency just as slowly.

SELECTION AGAINST X-LINKED RECESSIVES

Harmful X-linked recessive genes are exposed to selection in hemizygous males but not in heterozygous females. In X-linked recessives that are genetic lethals in the sense that affected males do not reproduce, or, in other words, have a fitness of zero, only the genes in the carrier females are passed on to the next generation. An important consequence is that one third of all cases of the disorder are new mutants, born to genetically normal mothers who have no risk of having subsequent children with the same disorder. This rule applies for a population at equilibrium when the mutation rate is the same in male and female germ cells. Its application to carrier risk estimation for genetic counseling is discussed in Chapter 14. In less severe disorders such as classic hemophilia, where fitness is reduced but is above zero, the proportion of new mutants is less than one third; at present the treatment of hemophilia is improving so rapidly that the fitness is increasing and the gene frequency can be expected to rise and to stabilize at a new level several generations in the future (unless, of course, prenatal diagnosis and selective abortion become widely used and reduce the frequency of the hemophilia gene).

SELECTION AGAINST HETEROZYGOTES

If selection occurs at the gamete stage, single alleles are selected against. More frequently, selection takes place at the diploid stage; loss of one allele involves loss of its partner also. If the two alleles are different (that is, if the individual concerned is a heterozygote), the relative frequency of the two alleles (or all alleles, if there are more than two possibilities) is altered, and a new balance is achieved at a different level but with a lower frequency of the rarer allele of the pair.

An example of selection against heterozygotes is provided by the Rh blood group system, though the removal of selection by improved medical management of Rh-negative mothers will alter this picture in future generations. The loss of a child due to Rh hemolytic disease removes both a recessive r gene and a dominant allele (usually R^1 or R^2). The loss of these two genes from the gene pool of the next generation has a greater effect on the frequency of r than on the frequency of the non-r alleles. Over many generations, the frequency of r should fall to a level maintained by recurrent mutation. However, in North American whites at present r has a frequency of about 0.40, compared with a frequency of 0.60 for the various non-r genes; in view of the rather severe selection against r in the past it is surprising that its frequency is still so high.

SELECTION FOR HETEROZYGOTES

Sickle cell anemia is a classic example of a situation in which the heterozygote is more fit in a particular environment that is either type of homozygote.

Recall that there are three genotypes with relation to normal adult hemoglobin and sickle cell hemoglobin. In certain parts of Africa the frequency of the Hb S allele (specifically, an allele for an abnormal β globin chain with a specific change in amino acid sequence) is higher than can be expected on the basis of recurrent mutation of an allele that, in homozygotes, has a fitness of zero.

The less favorable allele is maintained at its comparatively high frequency by **heterozygote advantage**. Heterozygotes are resistant to the malaria organism *Plasmodium falciparum*. This is a parasitic protozoan that spends a part of its life cycle in the red cells of vertebrates, to which it is introduced by the bite of the vector, the *Anopheles* mosquito. Thus, in any region where malaria is endemic, normal homozygotes are susceptible to malaria and probably are almost all infected and relatively unfit; sickle cell homozygotes are selected against because of their anemia. However, heterozygotes, whose red cells are not hospitable to the malaria organism and whose hemoglobin is quite adequate for normal environmental conditions, are relatively fit and reproduce at a higher rate than either homozygote.

A selective advantage in one environment may not be advantageous, or may even be disadvantageous, in a different environment. Sickle cell heterozygotes do not have a particular advantage in North America, where malaria is not endemic. There is some evidence that the frequency of the sickle cell gene is now dropping in North American blacks.

When selective forces operate in both directions, toward the preservation of an allele and toward its removal, a **balanced polymorphism** is said to exist. Removal of one of the selective pressures would be expected to allow a rapid change in the frequency of the sickle cell allele. If malaria could be eradicated, large changes might occur in the frequency of the sickle cell gene and several other genes that appear to play a part in protection against malaria. Unfortunately, at present malaria seems far from being controlled.

STABILIZING SELECTION

Population geneticists distinguish three main types of selection with respect to quantitative traits: stabilizing, directional and disruptive. The kinds of changes each of these types of selection bring about in regard to phenotypic distribution in the population are illustrated in Figure 15–6.

Stabilizing selection favors an intermediate optimum phenotype and selects against the **phenodeviants** at either extreme. The more extreme the deviant, the less its genetic contribution to the next generation. **Directional selection** is selection directed toward a new optimum, not at the mean of the population, in response to a new environmental challenge. **Disruptive selection**, rather than favoring any one phenotype, favors two quite different forms and selects against the intermediates; this may be viewed as directional selection of two separate subpopulations, in response to two different sets of environmental conditions. Stabilizing selection tends to maintain the status quo. For many human characteristics, an intermediate value is the favored optimum. An obvious example is birth weight, for which the effect of stabilizing selection in eliminating the more extreme phenodeviants is clear: babies much smaller or larger than the average are less likely to survive the perinatal period than those in the intermediate range.

There are few obvious examples of directional selection within the human

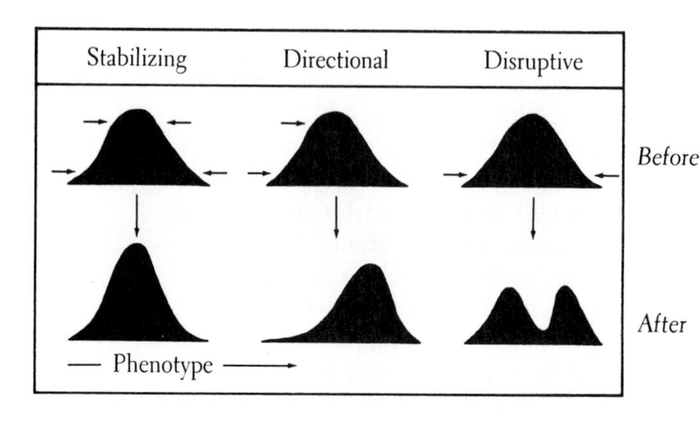

Figure 15–6. The three main types of selection: stabilizing, directional and disruptive. From Wallace B. Topics in population genetics. New York: Norton, 1968.

species. For example, except for the most recent generations, height appears to have changed little during the last 40,000 generations. Directional selection was basic to Darwin's concept of the mechanism of evolution, and is widely used by agricultural geneticists. Directional selection can be a very powerful mechanism; many scientists and writers have speculated about what might happen if the techniques of animal breeders were applied to man, as enthusiasts for eugenics advocated at one time.

MIGRATION

GENETIC DRIFT

Changes in gene frequencies often occur when new settlements are formed by members of an older group. There are two possible explanations. Perhaps migrants are themselves a separate subpopulation, differing genetically from the population as a whole. Of more importance, however, is the fact that the gene frequencies of the migrants will probably not be representative of the population from which they come; the smaller the migrant group, the greater the discrepancy is likely to be. Chance fluctuation in gene frequency in small populations is called genetic drift.

Striking demonstrations of genetic drift are provided by studies of the Old Order Amish of Pennsylvania and the Hutterites of the western United States and Canada. In these groups, social custom has provided an excellent situation in which to observe drift. The members of a group intermarry; few if any genes are added to the population by marriage with outsiders. As a community grows, it is the custom for a few families to leave to set up a new colony elsewhere. Among one Amish group, by chance a founder was heterozygous for a rare form of dwarfism with polydactyly, known as Ellis-van Creveld syndrome. There are numerous cases of the syndrome among members of this group, and none in other Amish communities descended from different founders. Thus, genetic drift can favor the establishment of genes that are not favorable or even neutral, but actually harmful. Another well-established example of genetic drift is that of tyrosinemia in a remote area of Quebec. The socially isolated Amish and the geographically isolated French-Canadian groups illustrate the **founder principle** of Mayr (1963): if among a small number of founders who form a community there is a member with a rare recessive allele, the frequency of the allele is much higher within the community than outside it; the small number of the original group of founders allows for a large effect of drift.

GENE FLOW

In contrast to the random variation in gene frequencies in small populations resulting from genetic drift is the more gradual change in frequency in larger populations resulting from gene flow. A classic example is the steady decline in the frequency of the *B* allele of the ABO blood group system from about 0.30 in Eastern Asia to about 0.06 in Western Europe (Fig. 15–7). The flow of "white" genes into American blacks is another example; by several different measures, such as comparison of the frequency of the strictly African R^0 allele in African and American blacks, it has been shown that a large percentage of the alleles now carried by American blacks are of white origin. Reed (1969), using data for the "Caucasian" allele Fy^a of the Duffy blood group system and other alleles differing in frequency in blacks and whites, found the proportion of "white" genes in American blacks in the United States to be lower in southern areas than

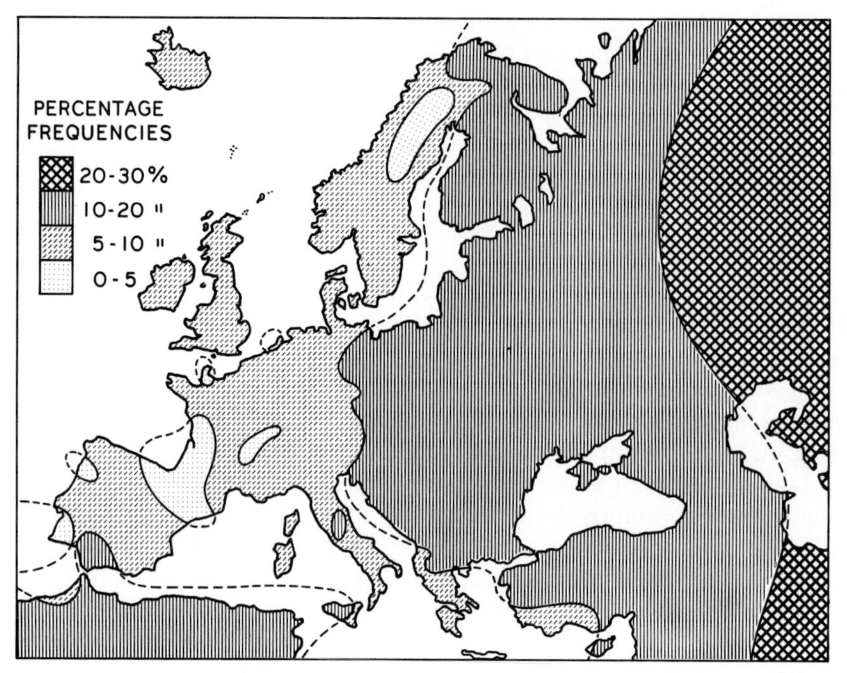

Figure 15–7. Cline of distribution of blood group B across Europe. After Mourant AE. The distribution of the human blood groups. Oxford: Blackwell Scientific Publications, 1954.

elsewhere (4 percent in Charleston, South Carolina, but 22 percent in Oakland, California and 26 percent in Detroit).

Eugenics

The term eugenics, coined by Francis Galton in 1883, refers to the improvement of a population by selecting its best specimens for breeding. This has been a principle of plant and animal breeding since ancient times. During the late nineteenth century, Galton and others began to promote the idea of using selective breeding to improve the human species, thus initiating the so-called eugenics movement, which was widely supported for the next 50 years or more. The movement had its origin partly in a growing concern that differential fertility (a tendency for the less intelligent to have larger families than the more intelligent) would lead to an overall deterioration in intelligence. Coming at a time when science and scientists were achieving great public prestige and when the nature of heredity was just beginning to be understood, it drew the attention and approval of a wide segment of the population. In the United States, the Eugenics Records Office (forerunner of the Cold Spring Harbor Laboratory), headed by Charles B. Davenport, became the leading center for the study of mental defects and the promotion of eugenic ideas. Under the influence of Davenport and his students, during the heyday of the movement, laws preventing the procreation of mentally deficient persons by means of compulsory sterilization were enacted in many parts of the United States and Canada, and remained in force in some states and provinces until after World War II. Eventually, a strong reaction developed against eugenics; this was partly because eugenics was used in Nazi Germany as a false justification for racism and prejudice, but also because of a growing appreciation of the theoretical and practical problems of carrying out eugenics programs.

Perhaps the major theoretical difficulty is that there is little agreement about what hereditary characteristics are desirable. A second problem, already familiar to plant and animal breeders, is that selective breeding reduces genetic diversity and thus limits the capacity of a species to adapt to changing conditions.

In the medical context, the impact of improvement in medical care for genetic disorders might be said to be dysgenic (that is, to have deleterious genetic effects on the population as a whole). If children with cystic fibrosis (CF) were enabled to survive and reproduce at a normal rate, and if parents who had had a CF child also had families of average size, the population incidence of CF would inevitably rise. However, considerations of this kind do not influence attempts to find effective treatment for CF, and do not influence the genetic counseling given to parents of CF children. This illustration could be multiplied many times. For example, the discovery of insulin, a major medical advance, could conceivably lead to a pronounced increase in the incidence of diabetes in the population, because diabetics who previously would not have survived have lived to pass on their diabetes susceptibility genes to their offspring.

A number of activities in genetics other than treatment per se have eugenic implications. As the use of prenatal diagnosis becomes widespread, significant numbers of fetuses with genetic disorders are likely to be selectively aborted, thus reducing the frequency of the disorders concerned. Huntington disease, for example, could be eradicated—or at least could have its incidence reduced to the level maintained by new mutation—within a generation if all potential patients were identified by molecular linkage analysis, and if their affected offspring were then diagnosed prenatally. Artificial insemination has strong eugenic implications, though the great problem of just what traits are desirable and should be favored remains unresolved. Recently, publicity has been given to a sperm bank that has solicited Nobel Prize winners to become donors; so far, the eugenic consequences of this program are unknown. In vitro fertilization and embryo transfer have provided an opportunity for a small number of previously infertile women to become mothers. We do not know to what extent the infertility was genetically determined. If it always had an environmental cause there would be no genetic implications, but if it was sometimes genetically caused, in the long term the incidence of female infertility would rise. Medical advances in treatment have allowed phenylketonuric women to reproduce, but their offspring have been mentally retarded; this example, disconcerting though it has been, teaches us that we still have much to learn about the interaction of genes and environment before we should presume to devise eugenics programs.

GENERAL REFERENCES

Several of the General References listed at the end of Chapter 1 have excellent discussions of population genetics, for example, Bodmer and Cavalli-Sforza, 1976; Cavalli-Sforza and Bodmer, 1971; Harris, 1980; Vogel and Motulsky, 1979.

Kevles DJ. In the name of eugenics. New York: Knopf, 1985.

PROBLEMS

1. In a certain population, three genotypes are present in the following proportions: *AA* 0.81, *Aa* 0.18, *aa* 0.01.
   a) What are the frequencies of *A* and *a*?

b) What will their frequencies be in the next generation?
c) What proportion of all the matings in this population are *Aa* × *Aa*?

2. In screening programs to detect carriers of Tay-Sachs disease among Ashkenazi Jews, the incidence was found to be about 0.035. Calculate
   a) the frequency of matings that could produce an affected child.
   b) the incidence of Tay-Sachs disease (or, if prenatal diagnosis is used, affected fetuses) in the Ashkenazi population.

3. In an isolated population of 800 individuals, all members are of blood group O. In another population, all are of blood group A. If 200 members of the second population are added to the first, what will the frequencies of the two blood groups be after a generation of random mating?

4. Two sisters marry two brothers. The son of one couple marries the daughter of the other couple. What is the coefficient of inbreeding of their children?

5. Monozygotic twin sisters marry dizygotic twin brothers. What is the coefficient of relationship of the children of one couple to the children of the other couple (that is, what proportion of their genes are in common)?

6. a) In a population in which the incidence of brachydactyly is 1 in 10,000, what is the frequency of the gene for brachydactyly (an autosomal dominant)?
   b) In the same population, the frequency of phenylketonuria is 1 in 10,000. What is the gene frequency? The carrier frequency?
   c) The frequency of classic hemophilia in males is 1 in 10,000. What is the gene frequency? The frequency of heterozygous females?

7. Which of the following populations is in Hardy-Weinberg equilibrium?
   a) *AA* 0.70, *Aa* 0.21, *aa* 0.09
   b) Blood groups M 0.33, MN 0.34, N 0.33
   c) 100 percent MN
   d) *AA* 0.32, *Aa* 0.64, *aa* 0.04
   e) *AA* 0.64, *Aa* 0.32, *aa* 0.04

8. For each of the populations in question 7, what are the expected proportions of the three genotypes after one generation of random mating?

9. a) X-linked dominant incontinentia pigmenti is lethal in males before birth, but is not selected against in females. What proportion of patients are likely to be new mutants?
   b) If reproductive compensation occurred (that is, if the mothers tended to replace the aborted fetuses by having another pregnancy), what would happen to the gene frequency? To the proportion of cases that are new mutants?

10. The probe G8 recognizes two polymorphic sites close to the Huntington disease locus. For linkage analysis, heterozygotes are required.
    a) Referring to Chapter 11, determine:
       –the frequency of heterozygotes at site 1
       –the frequency of heterozygotes at site 2
       –the frequency of individuals heterozygous at both sites simultaneously.
    b) What is the frequency of individuals heterozygous for at least one of these variants?
    c) Do the frequencies of the four G8 haplotypes shown in Chapter 11 indicate that the two linked polymorphic sites are in equilibrium?

16

TWINS IN MEDICAL GENETICS

Twins have a special place in human genetics because of their usefulness for comparison of the effects of genes and environment. Diseases caused wholly or partly by genetic factors have a higher concordance rate in monozygotic than in dizygotic twins. Even if a condition does not show a simple genetic pattern, comparison of its incidence in monozygotic and dizygotic twin pairs can reveal that heredity is involved; moreover, if monozygotic twins are not fully concordant for a given condition, nongenetic factors must also play a part in its etiology. The importance of twin studies for comparison of the effects of "nature and nurture" was originally pointed out by Galton in 1875.

Monozygotic and Dizygotic Twins

There are two kinds of twins, **monozygotic** (MZ) and **dizygotic** (DZ), or, in common language, identical and fraternal. Monozygotic twins arise from a single fertilized ovum, the zygote, which divides into two embryos at an early developmental stage, that is, within the first 14 days after fertilization. Because the members of an MZ pair normally have identical genotypes, they are like-sexed and alike (concordant) in their genetic markers. They are less similar in traits readily influenced by environment; for example, they may be quite dissimilar in birth size, presumably because of differences in prenatal nutrition. Phenotypic differences between MZ co-twins may be produced by the same factors that cause differences between the right and left sides of an individual; for example, cleft lip may be bilateral or unilateral in an individual, and may be concordant or discordant in an MZ pair.

Dizygotic twins result when two ova, shed in the same menstrual cycle, are fertilized by two separate sperm. DZ twins are just as similar genetically as ordinary sib pairs, having, on the average, half their genes in common. Phenotypic distinctions between the members of a DZ pair reflect their genotypic dissimilarities as well as differences arising from nongenetic causes.

RELATIVE FREQUENCY OF MZ AND DZ TWINS

There is a simple way of estimating how many of the twin births in a population are MZ and how many are DZ. MZ twins are always like-sexed, while approximately half of the DZ twin pairs are boy-girl sets. Therefore, the total number of DZ pairs is twice the number of unlike-sexed pairs, and the number of MZ pairs can be found by subtracting the number of unlike-sexed pairs from the total number of like-sexed pairs.

$$\frac{\text{All twin pairs} - 2\ (\text{unlike-sexed pairs})}{\text{All twin pairs}} = \text{Frequency of MZ pairs}$$

For precision, a small correction is required because the sex ratio is not exactly 1:1, but the simple method described gives a close approximation. Among white North Americans, approximately 30 per cent of all twins are MZ, 35 percent are like-sexed DZ and 35 percent are unlike-sexed DZ. In some black populations the proportion of DZ twins is considerably higher; in some Asian populations it is lower.

Comparison of the ratio of like-sexed to unlike-sexed pairs in populations with varying frequencies of twin births has shown that the proportion of MZ births relative to all births is much the same everywhere, about 1 in 300 births, but that the proportion of DZ births varies with ethnic group, maternal age and genotype (see below).

Frequency of Multiple Births

Among white North Americans, about one birth in 87 is a twin birth. Thus, two of every 88 white babies are born as twins. Triplets are born about once in $(87)^2$ births, quadruplets about one in $(87)^3$ births and so on, although, because of a much higher mortality, this mathematical relationship (known as Hellin's law) is less close for multiple sets of four or more. Since twins have a higher infant mortality that singletons, the incidence of twins in the general population is somewhat lower than 2 percent.

The frequency of twin births varies with ethnic origin, in the order Africans>Caucasians>Asians, chiefly because of variation in the DZ frequency. One of the highest frequencies (one twin birth in 20 to 30) is reported from Nigeria; one of the lowest (one in 150) from Japan. The frequency of DZ twins rises sharply with maternal age to age 35 and later declines, but the frequency of MZ twins is only slightly affected by the mother's age.

The tendency for twins, especially DZ twins, to run in families is well known. Whether the father plays any part in the occurrence of twinning among his children has been the subject of dispute for many years. The multiple ovulation that accounts for DZ twins is a purely maternal event, related to the mother's FSH (pituitary follicle-stimulating hormone) level. The data of White and Wyshak (1964) from the genealogical records of the Mormon church showed that female DZ twins produced twins at a rate of 17.1 sets per 1000 maternities, whereas the rate in offspring of male DZ twins was only 7.19. Sisters of DZ twins also produced twins at a high rate; brothers did not, but their daughters did. In other words, a disposition to multiple ovulation appears to be an inherited trait expressed only in females. Parisi and colleagues (1983) have published data that support this conclusion, but their data also suggest a slight familial propensity to MZ twinning, in the maternal line only. The recurrence rate for DZ twins in a sibship is about three times the population risk.

Twins are somewhat more common among individuals with chromosomal aneuploidy and their families than in the general population. This is especially true of the Klinefelter and Turner syndromes (Hansmann, 1983). To some extent, the factor of late maternal age, which is related to both an increased risk of trisomy and an increased incidence of DZ twinning, could explain the observations in Klinefelter syndrome, but not in Turner syndrome. The observations suggest a common causative mechanism for twinning and aneuploidy, or selective survival of aneuploid fetuses in twin pregnancies, but the exact mechanism is still unclear.

When women are treated for sterility with drugs such as clomiphene, which stimulates the release of the pituitary gonadotropins FSH and LH (luteinizing hormone), to induce ovulation, several ova may ripen at once. Multiple births (quintuplets, sextuplets and even septuplets) have been reported following clomiphene treatment. These multiple sets are, of course, DZ.

Determination of Twin Zygosity

It is often useful to know the zygosity of a twin pair (or of a higher multiple birth, in which the members may be an MZ set, all from separate zygotes, or a mixture). Accurate classification is a prerequisite if twins are to be used in research; it may be medically useful, for example, if one twin develops diabetes, or needs a transplant. Zygosity may be determined on the basis of the type of placenta and fetal membranes and by looking for genetically determined similarities and differences in a twin pair.

PLACENTA AND FETAL MEMBRANES

A developing fetus is invested in two membranes: the inner, delicate **amnion** and the outer, thicker **chorion**. If two fetuses are developing simultaneously in the same uterus, there are several variants of the placenta and fetal membranes. The most common types are shown in Figure 16–1 and summarized in Table 16–1.

A twin placenta is monochorionic if there is a single chorion, and dichorionic if there are two chorions (which may be secondarily fused where they meet). A monochorionic twin placenta may be either monoamniotic (with one amnion), or diamniotic (with two amnions).

Dizygotic twins have separate placentas, chorions and amnions, but in about

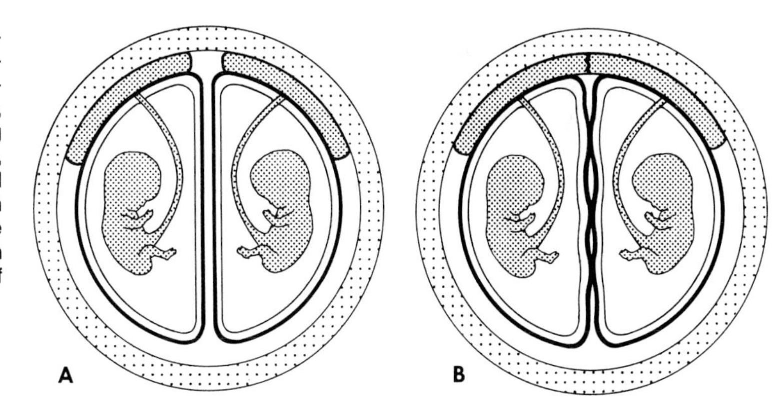

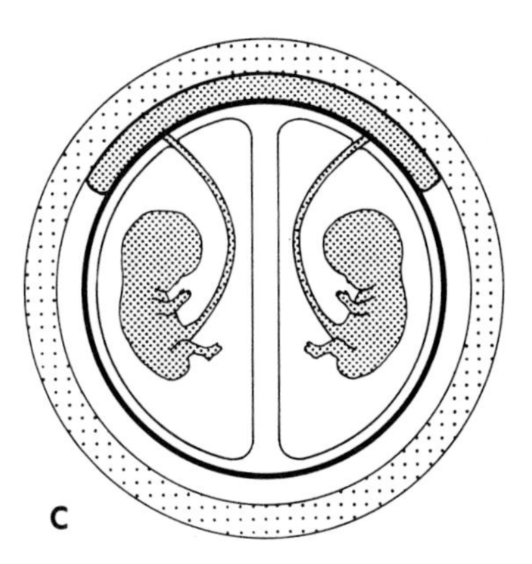

Figure 16–1. Some common arrangements of placentas and fetal membranes in twins. A, Separate placentas and membranes, common in DZ twins but occurring in only 10 percent of MZ twins. B, Separate but secondarily fused placentas and chorions, common in both MZ and DZ twins. C, Single placenta and single chorion with common circulation, diagnostic of MZ twinning.

Table 16–1. TYPES OF PLACENTAS AND FETAL MEMBRANES IN TWINS

	One Chorion		Two Chorions	
	One Amnion	*Two Amnions*	*One Placenta (by secondary fusion)*	*Two Placentas*
MZ	Rare	65 percent	25 percent	10 percent
DZ	—	—	40 percent	60 percent

40 percent of DZ twin pregnancies the two placentas and chorions are secondarily fused and superficially resemble a monochorionic placenta.

Monozygotic twins may have either dichorionic or monochorionic placentas, depending on the time in early embryonic development at which twinning occurred. In about 25 percent of MZ pairs, twinning of the embryo occurs before the third day of development; that is, before the development of the chorion, so that two separate chorions are formed. There may be two separate placentas or, more commonly, a single secondarily fused placenta. Both these types of placentation also occur in DZ twins. In the majority of MZ twins, however, twinning occurs between the third and eighth day of development, by which time the differentiation of the blastocyst has already proceeded too far to allow duplication of the chorion. A monochorionic placenta (which is usually diamniotic) can be regarded as proof of monozygosity.

There are several varieties of monochorionic placenta, depending again on the stage at which twinning of the embryo occurred. Some of the distinguishable types, arranged in order from earlier to later time of twinning, are the following:

1. Each embryo has a separate amnion, and the two fetal circulations remain separate.

2. Each embryo has a separate amnion, but a common fetal circulation has developed by anastomoses of vessels in the placenta. This is the usual type of monochorionic placenta. The anastomoses may sometimes lead to the **transfusion syndrome**, which results in malnutrition of the donor twin, corresponding excessive nutrition in the recipient twin and a large difference in birth weight.

3. Rarely, an even later twinning of the embryo results in a single amniotic sac for both twins. Many monoamniotic twins do not survive.

4. Rarely, still later twinning of the embryo occurs after the formation of the umbilical cord, resulting in conjoined ("Siamese") twins. Conjoined twins are monoamniotic and usually have a common, branched umbilical cord.

Though the fused dichorionic placenta and fetal membranes and the monochorionic type are superficially similar, they can be clearly distinguished if the membrane between the two amniotic cavities is examined. In a dichorionic placenta the membrane is opaque, being composed of two layers of chorion and two layers of amnion, as shown in Figure 16–1B. The two chorions can be readily separated. In contrast, in a monochorionic placenta the two amniotic cavities are separated only by a semitransparent layer composed of two thin amnions, as demonstrated in Figure 16–1C. According to an old tale, if twins are of different sexes (thus DZ) but share a placenta, a heavy curtain is drawn between them to protect their modesty.

In summary, the only kind of placenta that is diagnostic of the type of twinning is monochorionic, which indicates MZ twinning. Separate placentas and membranes or secondarily fused placentas without common circulation can be found with either type of twin. It is helpful to examine and record the nature of the placenta and membranes at like-sexed twin births, since the information about twin zygosity then available may not be easy to obtain in later life, except by extensive tests.

Comparisons of monochorionic and dichorionic MZ twins have so far failed to reveal any striking developmental differences.

GENETIC MARKERS IN TWINS

Monozygotic twins are always alike in sex and in their genetic markers. Dizygotic twins, like ordinary sibs, may be alike or different in these characteristics. For practical purposes, zygosity determination can often be made by careful observation, looking for any differences in hair color and form, iris pattern, ear shape, and so forth. For an objective assessment for research purposes, twins are compared with respect to as many markers as possible. A single difference in any genetic marker proves twins to be DZ. It is impossible to prove monozygosity, since any two children of the same parents resemble one another in many genetic traits; however, it may be possible to show that the probability of monozygosity is very high.

The probability that twins are MZ can best be worked out if the genotypes of the parents and sibs are known. The following example shows how to do the calculation in an oversimplified case. The method of calculation (Bayesian analysis) is described in additional detail in Chapter 14.

	Genotypes			
	Father	*Mother*	*Twin A*	*Twin B*
ABO	*AB*	*OO*	*AO*	*AO*
Rh	R^1r	R^2r	*rr*	*rr*
MNSs	*Ms/Ns*	*MS/MS*	*MS/Ms*	*MS/Ms*
Haptoglobin	2–1	1–1	1–1	1–1

	DZ	MZ
Prior probability	0.70	0.30
Conditional probability:		
That the twins will be alike in sex	0.50	1
That if twin A is *AO*, twin B will be *AO*	0.50	1
That if twin A is *rr*, twin B will be *rr*	0.25	1
That if twin A is *MS/Ms*, twin B will be *MS/Ms*	0.50	1
Joint probability	0.022	0.30
Relative probability	$\frac{0.022}{0.322} = 0.07$	0.93

In this brief example the probability of monozygosity (93 percent) is not high enough to make it safe to conclude that the twins are MZ, but with the numerous additional markers that can now be used it is often possible to raise the probability close to the 100 percent level, or, on the other hand, to show that the twins are DZ. Useful tables of probabilities for situations in which the parental genotypes are unknown are given by Race and Sanger (1975) and Corney and Robson (1975).

DERMATOGLYPHICS

Dermatoglyphics can be helpful in the determination of zygosity, even though the inheritance is multifactorial rather than Mendelian. Tables have been constructed for comparison of a twin pair in terms of the total fingerprint ridge count or the maximal *atd* palmar angle (Maynard-Smith et al., 1961; Corney and Robson, 1975). (Dermatoglyphic terms are defined in Chapter 17.) The probabilities given in these tables can be combined with the probabilities from the other sources just quoted.

Limitations of Twin Studies

The chief drawback of the twin method is that, though it tells something about the strength of the genetic predisposition to develop a disorder, it gives no insight into the genes concerned, their mode of action or their pattern of transmission. Many of the traits in which twin comparison are used are multifactorial. If a trait is determined by a single autosomal dominant gene, it should be half as common in a sib or a DZ co-twin of the proband as in an MZ co-twin; if it is autosomal recessive, it should affect one-quarter as many sibs or DZ co-twins as MZ co-twins. If the proportions vary by much from these ratios, multiple factors (genetic or environmental) are probably involved. For many conditions studied even the MZ concordance rate is well below 50 percent, indicating that environmental factors operative before birth are important in causing them.

Although the twin method assumes that postnatal environment differences are constant for both types of twins, this assumption may be unwarranted. MZ twins, because they are more alike, may seek the same environment and develop along much the same paths. DZ twins, who may even be of different sexes, probably have more different environments than do MZ twins. In many studies a comparison can be made more fairly between MZ twins and like-sexed DZ twins than between all MZ pairs and all DZ pairs. It is noteworthy that DZ twins become less and less alike as they grow older, whereas MZ twins remain remarkably similar throughout life, aging in much the same way and being subject to the same geriatric disorders.

Still another limitation of the twin method is related to bias of ascertainment. Concordant MZ pairs are much more likely to be reported than any other combination. If a twin series is compiled from the literature, it is likely to include a preponderance of this type. Discordant MZ pairs, who may be particularly informative, are unlikely to be reported.

Examples of Twin Studies

A few examples will indicate ways in which twins provide information about medical genetics.

Chromosomal Aberrations. If one member of an MZ twin pair has Down syndrome, the other twin is always affected; if one DZ twin is affected, the other is nearly always normal. This is one of the observations that led to the hypothesis that a chromosomal defect might be the cause of mongolism; it is an example of the way in which a twin study can point to an unusual causative mechanism of disease.

Mutation in Sporadic Cases of Disease. Many cases of Duchenne muscular dystrophy arise as sporadic cases presumably caused by new mutations. The observation of concordance for muscular dystrophy in an MZ twin pair in an otherwise normal family helps to verify the genetic causation of sporadic cases of the disease (Fig. 16–2).

Traits with Complex Inheritance. In Chapter 12 the idea of a heritability index (h^2) was introduced and defined. The index is the proportion of the amount

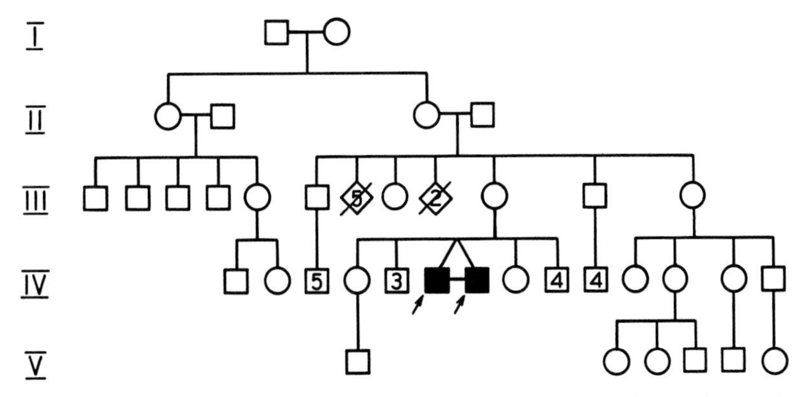

Figure 16–2. A pedigree of Duchenne muscular dystrophy in which only a pair of MZ twins, presumed to be new mutants, are affected. The large number of normal males related to the probands through females (7 brothers, 11 other relatives) makes it unlikely that the mother is a carrier, but the presence of the disease in an MZ pair is evidence of its genetic basis. Redrawn from kindred 31 in Stephens FE, Tyler FH. Studies of disorders of muscle V. The inheritance of childhood progressive muscular dystrophy in 33 kindreds. Am J Hum Genet 1951; 3:111–125.

of variation resulting from genetic differences. When twin data are used, the formula becomes:

$$h^2 = \frac{\text{Variance in DZ pairs} - \text{Variance in MZ pairs}}{\text{Variance in DZ pairs}}$$

As this ratio approaches zero, it indicates that determination of the trait is primarily environmental; as it approaches 1, it indicates primarily genetic determination.

Much of the twin research of the early nineteen-eighties has been epidemiological, concerned with carefully designed comparisons of series of MZ and DZ twin pairs.

Diabetes Mellitus. Twin studies in diabetes mellitus have shown a pronounced difference between the juvenile onset form and the maturity onset form. Concordance in MZ twins is very high (over 90 percent) in the maturity-onset form, but much lower in the juvenile type. Moreover, the juvenile type is associated with specific HLA types, suggesting the possibility of infectious disease as a trigger. (See also Chapter 8.)

Cancer. Estimates of heritability of liability to cancer, based on twin studies, are low both for cancer at all sites and for specific cancers (other than those rare forms known to be determined by single gene mechanisms, such as retinoblastoma).

Behavior Genetics. Studies of the genetics of human behavior make use of twin studies. As an example, Table 16–2 shows some recent data for family resemblances in general intelligence.

For psychiatric diseases, estimates of the heritability of liability to schizophrenia are generally in the 0.6 to 0.7 range. In the affective disorders, when both the unipolar (depressive) and the bipolar (manic-depressive) types are considered, the heritability is even higher, possibly as high as 1.0.

Table 16–2. CORRELATIONS FOR GENERAL INTELLIGENCE WITHIN FAMILIES

Relationship	Correlation
Same individual, repeat test	0.87
MZ twins	0.87
DZ twins	0.62
Sibs	0.41
Parent-child (child reared by parent)	0.35
Parent-child (child reared apart)	0.31
Foster-parent, adopted child	0.16
Spouses	0.29

Data from Henderson ND. Human behavior genetics. Ann Rev Psychol 1982; 33:403–440.

Congenital Malformations. The concordance rates for congenital malformations in MZ and DZ twins indicate the causative role of genetic predisposition. Some typical figures follow:

	Concordance	
	MZ Pairs	*DZ Pairs*
Congenital dislocation of the hip	0.40	0.03
Cleft lip ± cleft palate	0.30	0.05
Club foot	0.23	0.02

Data from Vogel F, Motulsky AG. Human genetics, problems and approaches. New York, Springer-Verlag, 1979.

These few examples show that, for some congenital malformations, the concordance rate is much higher in MZ pairs than in DZ pairs. Twins, particularly MZ twins, also have a higher-than-average risk of certain types of congenital malformation, especially neural tube defects.

Cerebral Palsy. Cerebral palsy is an example of a disorder in which twin concordance rates can be misleading. Cerebral palsy is usually not genetically determined, but results from difficulties at the time of birth, especially prematurity or anoxia. These factors are more prominent in twin births than in single births; therefore a high frequency of twinning, whether MZ or DZ, is found in children with cerebral palsy.

Unusual Types of Twins

Monozygotic Twins with Different Karyotypes (*Heterokaryotic Twins*). Rarely, twins are found to have different karyotypes, though evidence from genetic markers, placenta and physical appearance affirms their MZ origin. Among 12 pairs summarized in one report (Benirschke, 1972), four comprised a normal child and a 21-trisomic Down syndrome member, and the others were 45,X or mosaics with normal or mosaic Turner syndrome twins. In most cases, each twin had a minor population of cultured lymphocytes like the co-twin's major population.

Twins such as these are very rare and of great theoretical interest. Presumably they originate in postzygotic nondisjunction, as do chromosomal mosaics, followed or accompanied by twinning.

Conjoined Twins. Conjoined twins are believed to be MZ twins produced by an incomplete split of the original embryo, occurring relatively late in development (that is, after the eighth day). They are more common than is generally believed, with a frequency of approximately 1 in 400 MZ twin births, or 1 in 120,000 births.

Dizygotic Twins with Different Fathers. Dizygotic twins may have different fathers. One such case has been documented by analysis of the HLA types of the twins, their mother and their putative fathers (Terasaki et al., 1978).

Chimeras. As noted earlier, chimeras are those rare individuals who are composed of a mixture of cells from two separate zygotes. There are two types of naturally occurring chimeras, both exceedingly rare: blood group chimeras comparable to the chimeric heterokaryotic twins described above, and dispermic chimeras.

Blood group chimeras are produced when DZ twins exchange hematopoietic stem cells in utero. The grafted cells are not recognized as foreign, and are retained. Each twin then has two populations of blood cells, one with his own genetic markers, the other with those of the twin.

Dispermic chimeras are thought to develop from fusion of fraternal twin zygotes in very early development. Separate fertilization of the egg and a polar body is one of several other suggested mechanisms. If a dispermic chimera has developed from fused XX and XY zygotes, true hermaphroditism usually results. Indeed, the very first case to be described was found after a deliberate search for a true hermaphrodite with different-colored eyes (Gartler et al., 1962).

GENERAL REFERENCES

Benirschke K. Origin and clinical significance of twinning. Clin Obstet Gynecol 1972; 15:220–235.
Bulmer MG. The biology of twinning in man. Oxford: Clarendon Press, 1970.
Hecht F, Hecht BK. Genetic and related biomedical aspects of twinning. Pediatr Rev 1983; 5:179–183.
Hrubec Z. Robinette D. The study of human twins in medical research. New Engl J Med 1984; 310:435–441.
MacGillivray I, Nylander PPS, Corney G. Human multiple reproduction. London: W. B. Saunders Company, 1975.
Maynard-Smith S, Penrose LS, Smith CAB. Mathematical tables for research workers in human genetics. London: Churchill Livingstone, 1961.
Nance WE, ed. Twin research. Prog Clin Biol Res, Vol. 24, 1978.

PROBLEMS

1. a) A pair of male dichorionic twins each have the following blood group genotypes: *AB*, *Ms/MS*, *CDe/cde*, P_1P_2, Xg^a. The father's genotype is *AO*, *MS/NS*, *cde/cde*, P_1P_1, Xg; the mother's genotype is *BO*, *MS/MS*, *CDe/cDE*, P_1P_2, Xg^a/Xg. What is the probability that the twins are MZ?
   b) Would the following findings increase or decrease the probability of monozygosity?
      1) Total fingerprint ridge count is much higher in one twin that the other. (See Chapter 17.)
      2) A maternal family history of twinning.
      3) A paternal family history of twinning.
      4) Late maternal age.

2. Assume that the incidence of unlike-sexed twins among all twins is 0.46 in a Nigerian

population, 0.35 in a Canadian population and 0.10 in a Japanese population. What is the proportion of MZ twins in each population?

3. These are phenotype data for the twins with two different fathers described by Terasaki and colleagues (see text), as given in their paper.

	ABO	MN	Rh*	Hp	HLA-A	HLA-B	HLA-C
Putative father 1	A	MN	CcDe	2–1	2,3	15	w3
Putative father 2	A	M	cDE	2	2,w24	7,w54	w3
Mother	A	MN	CcDe	2	2,11	w44,27	—
Twin A	A	M	CcDe	2–1	2	w44,15	w3
Twin B	A	MN	CcDEe	2	2,24	w44,w54	w3

*The Rh phenotypes as shown indicate which antigens were detected in the individual's blood. Recall that there is no d antigen, so absence of D is recorded as d. The probable genotypes are PF_1 *CDe/cde*, PF_2 *cDE/cDE*, M *CDe/cde*, Twin A *CDe/cde*, Twin B *CDe/cDE*.

a) Which systems show that the twins were not monozygotic?
b) Which HLA haplotype(s) did the mother give to the twins?
c) Did the mother give the same genes to each twin? Would you expect the maternal contribution to be the same or different? Why?
d) Can either of the putative fathers be excluded as father of either of the twins?
e) Which putative father is responsible for Twin A? for Twin B?
f) Could a different man have fathered both twins?

17

DERMATOGLYPHICS IN MEDICAL GENETICS

Dermatoglyphics are the patterns of the ridged skin of the digits, palms and soles. They are important in medical genetics chiefly because of their diagnostic usefulness in some dysmorphic syndromes, especially Down syndrome. They are also a useful aid in the determination of twin zygosity.

In the early development of the hand and foot, the dermal patterns begin to form at about the 13th week of prenatal life and are essentially complete by the end of the fourth month. Because the patterns remain unchanged thereafter, any abnormalities of the patterns must have had their origin before or during this time span.

The scientific classification of dermatoglyphics was originally proposed by Galton, who was the first to study dermal patterns in families and racial groups. The term dermatoglyphics ("writing on the skin") was coined by Cummins (1961), who noted that the dermal patterns in Down syndrome are characteristic, and different from those of normal individuals. The first dermatoglyphic index for detection of Down syndrome was developed by Walker, who showed that some 70 percent of Down patients could be distinguished from controls on the basis of the dermal patterns alone. Following the discovery of the chromosomal basis of Down syndrome, dermal patterns of patients with other chromosomal disorders were examined and were also found to have distinctive features.

Embryology

While the dermal ridges are being formed, their alignment conforms to the shape of the growing hand or foot. At 10 weeks the fetal hand bears conspicuous volar pads, relatively as large as a cherry on an adult fingertip. At about the 13th week the pads regress; meanwhile, the dermal ridges differentiate in the thickening skin. Foot patterns develop slightly later, but the sequence of events is the same. Growth disturbances at or before this stage can produce abnormal dermatoglyphics, but later ones cannot.

Ridge alignments give the impression that a system of parallel lines has been drawn as economically as possible over an irregular terrain. The topology of ridges on nonflat surfaces has intrigued many scientists, including the mathematician Littlewood (1953), who pointed out that should one comb a spherical dog, at least four loops or two whorls would have to be produced.

Classification of Dermal Patterns

Prints of the digits, palms and soles can be made by one of several standard methods. In our experience, a technique used in many hospitals for identification

Figure 17–1. Three basic fingerprint types: A, arch; B, loop; C, whorl. Ridge counts are made along the indicated lines, excluding the triradius and the pattern core. Here the ridge count is 16 for the loop, 14–10 for the whorl. See text for further details.

of newborns produces excellent results. However, in very young infants, who have soft skin and fine ridges that are usually obscured by dry, scaling epithelium, clear prints may be extremely difficult to obtain. It is much easier to take a print of a child at least a month old than of a newborn.

The dermatoglyphics of importance in medical genetics are fingerprints (rolled prints of the ridged skin of the distal phalanx of each finger), palm prints and prints of the hallucal area of the sole. Rules for the formulation of dermal patterns in these areas are described by Holt (1968).

Pattern combinations and frequencies are more significant than pattern types alone as indicators of abnormal development. In Down syndrome, for example, there is no single dermal pattern that does not also occur in controls, but the combination of a number of patterns, most of which are more common in Down syndrome that in normal persons, is highly specific.

FINGERPRINTS

Fingerprints are classified, according to Galton's system, as **whorls, loops** or **arches**. Examples of each type are shown in Figure 17–1. The classification is made on the basis of the number of triradii: two in a whorl (W), one in a loop and none in an arch (A). A triradius is a point from which three ridge systems course in three different directions, at angles of about 120° (Figure 17–1 *B* and C). Loops are subclassified as **radial** (R) or **ulnar** (U) depending on whether they open to the radial or ulnar side of the finger.

The frequency of the different patterns varies greatly from finger to finger (Table 17–1).

The size of a finger pattern is expressed as the **ridge count**, that is, the number of ridges between the triradius and the pattern core. An arch has a count of zero, since it has no triradius. The line of count for a loop and the two lines for a whorl are shown in Figure 17–1. The **total ridge count** (TRC) of the 10

Table 17–1. PERCENTAGE FREQUENCIES OF DIGITAL PATTERN TYPES ON DIFFERENT DIGITS

	Left					Right				
	5	4	3	2	1	1	2	3	4	5
Whorl	12	37	17	33	31	39	36	18	47	15
Ulnar Loop	85	60	71	36	63	57	31	71	52	84
Radial Loop	<1	<1	3	**19**	<1	<1	**20**	3	<1	<1
Arch	3	3	9	11	6	3	13	6	1	1

Based on prints of 500 normal individuals, chiefly Caucasian.
Data from Walker, 1958.

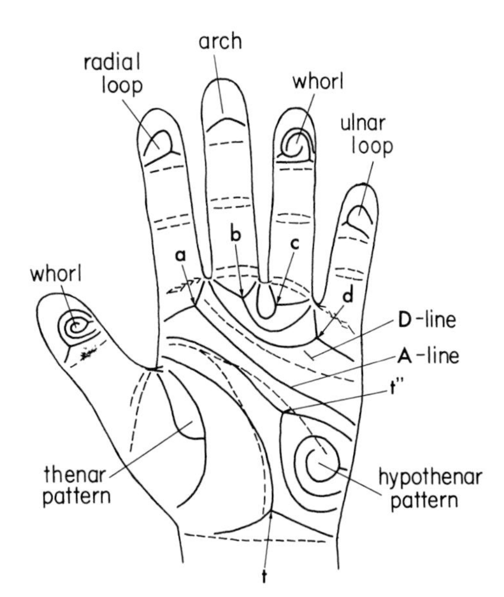

Figure 17–2. The major landmarks of the palm. There are four digital triradii (a,b,c,d) and an axial triradius t (t'' indicates its distal location). Main lines A and D are traced from the corresponding digital triradii. Thenar, hypothenar and digital patterns are also shown. From Smith GF, Berg JM. Down's anomaly, 2nd ed. Edinburgh: Churchill Livingstone, 1976, by permission.

fingers is a useful dermatoglyphic characteristic, which has been shown to be inherited as a multifactorial trait.

PALM PATTERNS

Figure 17–2 shows the chief landmarks of the palm. Palm patterns are defined chiefly by five triradii: four **digital triradii**, near the distal border of the palm, and an **axial triradius**, which is commonly near the base of the palm but sometimes displaced distally, especially in Down syndrome. Interdigital patterns (loops or whorls) may be formed by the recurving of ridges between the digital triradii. Hypothenar or thenar patterns may be present.

The position of the axial triradius is perhaps the single most important feature, because it is distally displaced in many abnormal conditions. Its location may be measured either as a fraction of the total length of the palm, or as the *atd* angle, shown in Figure 17–3.

The **flexion creases**—the heart, head and life lines of palmistry—are not, strictly speaking, dermal ridges, but they are formed at the same time, and may be determined in part by the same forces that affect ridge alignment. A **simian crease** (single transverse crease) in place of the usual two creases, on at least one hand, occurs in 1 percent or more of normal individuals. Simian creases are not unusual in abnormal individuals, such as children with congenital malformations, even when the dermal patterns themselves are not obviously disturbed. In Down syndrome and other chromosomal disorders, single flexion creases are much more common than in controls (Fig. 17–4).

The **Sydney line** is a variant type of palmar crease seen in about 10 percent of normal individuals, in which the proximal crease extends across the palm, like a simian crease, but a distal crease is also present. Various transitional types also occur.

SOLE PATTERNS

Sole (plantar) patterns have been studied less extensively than palm patterns, chiefly because soles are more difficult to print and to classify. Only in the hallucal area have distinctive patterns been described in clinical syndromes. The

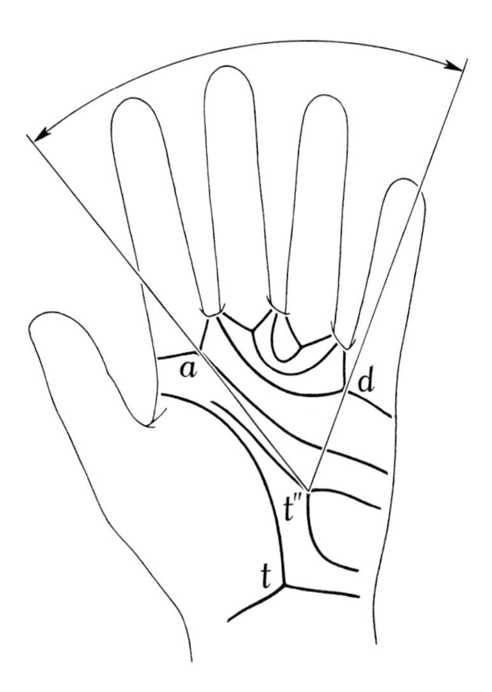

Figure 17–3. Measurement of the *atd* angle of the palm (here 60°). If there is more than one axial triradius, the distal one is used. From Holt SB. The genetics of dermal ridges. Springfield, Illinois: Charles C Thomas, 1968, by permission.

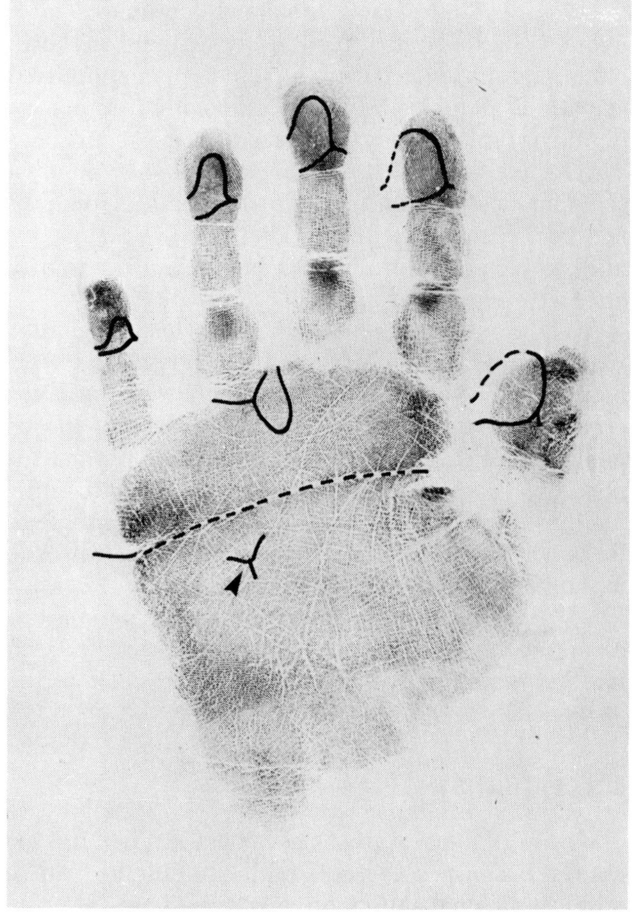

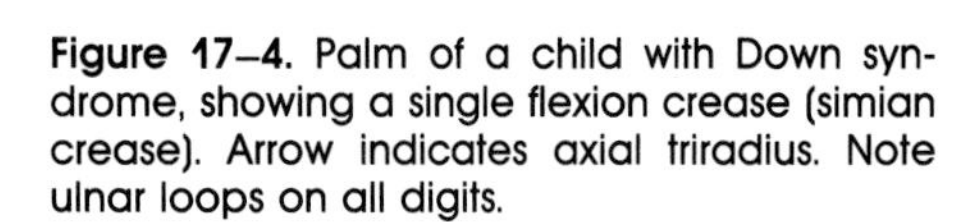
Figure 17–4. Palm of a child with Down syndrome, showing a single flexion crease (simian crease). Arrow indicates axial triradius. Note ulnar loops on all digits.

Figure 17–5. The hallucal tibial arch (A^t), common in Down syndrome but otherwise very rare.

unusual tibial arch (A^t) pattern, which is found in nearly 50 percent of all cases of Down syndrome and very rarely (0.3 percent) in controls, is the single most useful dermal pattern in Down syndrome (Fig. 17–5).

Clinical Applications

CHROMOSOME DISORDERS

In nearly all the chromosome disorders the dermatoglyphic patterns are unusual. This is not surprising because abnormal karyotypes lead to multiple morphological abnormalities, and the dermatoglyphic patterns are influenced by the shape of the underlying structures. Table 17–2 lists some of the chief variants

Table 17–2. CHARACTERISTIC DERMATOGLYPHICS IN SOME CHROMOSOMAL SYNDROMES

	Fingers	Palms	Soles
Down syndrome	Many ulnar loops (usually 10) Radial loop on 4th and/or 5th digits	Single flexion crease in 50% Axial triradius in center of palm in 85%	Tibial arch (50%) or small distal loop (35%)
Trisomy 18	6–10 arches (also on toes) Very low TRC	Single flexion crease	
Trisomy 13	Many arches Low TRC	Axial triradius may be distally displaced Single flexion crease Thenar pattern	Large pattern, fibular arch or tibial loops
45,X	Large loops or whorls High TRC	Axial triradius slightly more distal than in controls	Very large loop or whorl
5p−	Many arches Low TRC	Single flexion crease	Open field
Other syndromes with extra X or Y chromosomes	Many arches Low TRC The more sex chromosomes, the lower the TRC		

Abbreviation: TRC, total ridge count

seen in the classic chromosome syndromes. The differences described are chiefly in the TRC (which may be either higher or lower than normal in different conditions), in the flexion creases of the palm (often reduced to a single crease) and in the position of the axial triradius (usually distally displaced). It must be emphasized that dermal patterns are highly variable, and none of these dermatoglyphic features is in itself abnormal; however, their different frequency in patients and controls and the combination of several different unusual features in a single patient may be distinctive.

Down Syndrome

Figure 17–6 illustrates a set of typical patterns of a Down syndrome patient and a control.

Typically, in Down syndrome the axial triradius is displaced distally to about the center of the palm. The dermal ridges are oriented transversely, rather than obliquely as in most people. In about half the patients there is a single flexion crease on at least one hand.

The characteristic pattern in the hallucal area of the sole has already been mentioned. Over 80 percent of the feet of Down patients have either a tibial arch (see Fig. 17–5) or a very small distal loop, which is an intermediate pattern type closely similar to a tibial arch.

There are a number of methods for analyzing dermal patterns in Down syndrome. Figure 17–7 illustrates one of these, the Walker index (Walker, 1958), based on the ratio of certain finger, palm and sole patterns in a large series of Down patients and controls.

OTHER DISORDERS

Unusual dermatoglyphics have been described in a variety of disorders, including several in which there is no reason to expect dysmorphism. Many of

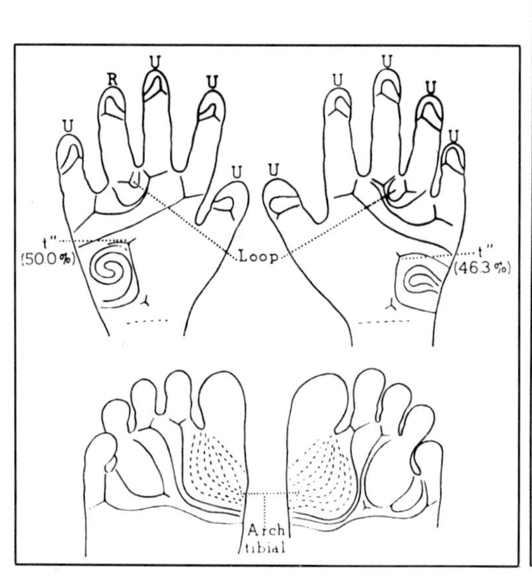

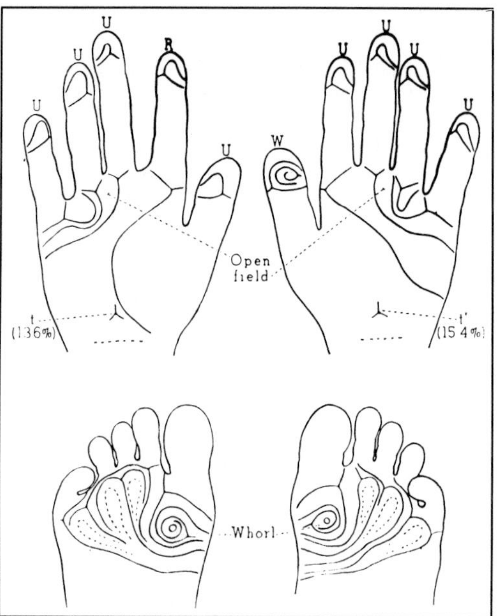

Figure 17–6. Typical dermal patterns in a Down syndrome patient (left) and a control (right). Compare the position of the axial triradii, the hypothenar pattern in the patient but not in the control, the fingerprints (the radial loop on the fourth digit is characteristic of Down patients but rare otherwise) and the sole patterns. For further discussion, see text.

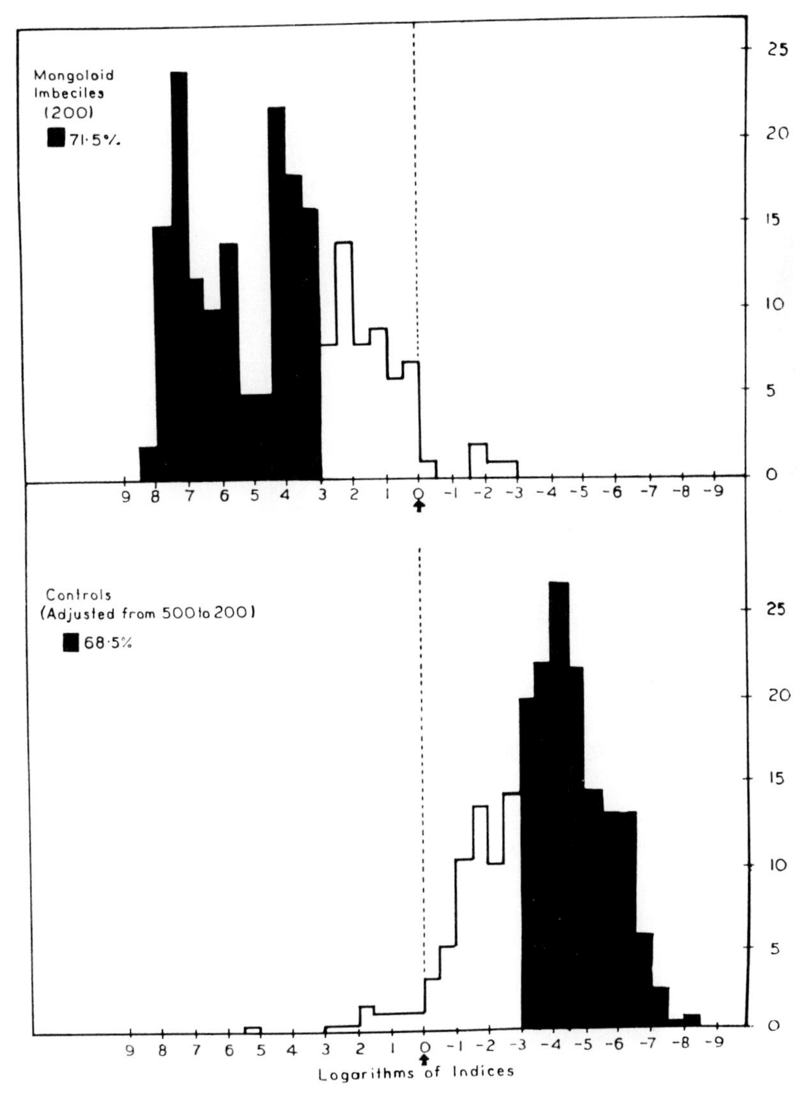

Figure 17–7. Histograms of the Walker index, based on the log of the ratio of dermal pattern frequencies in Down syndrome patients and controls. An index of +3 or higher is found in over 70 percent of Down patients and in almost no controls. From Walker NF. The use of dermal configurations in the diagnosis of mongolism. Pediatr Clin N Am 1958; 5:531–543.

these reports are unreliable and may merely reflect small sample size or inappropriate controls.

Gross distortion of the patterns can occur in association with any limb malformation of early prenatal onset. In arthrogryposis multiplex congenita, for example, joint contractures are present at birth; in some but not all patients the dermatoglyphics are very unusual, with longitudinal orientation of the ridges and other peculiarities. Such findings suggest that the condition was already present, or developing, at the time of ridge differentiation.

Curious digital patterns are seen in digits that have no fingernails, for example, in anonychia, nail-patella syndrome and brachydactyly Type B. Figure 17–8 illustrates the ridge arrangement on a nailless finger. In syndromes characterized by nail dysplasia, the digital patterns are often simple arches.

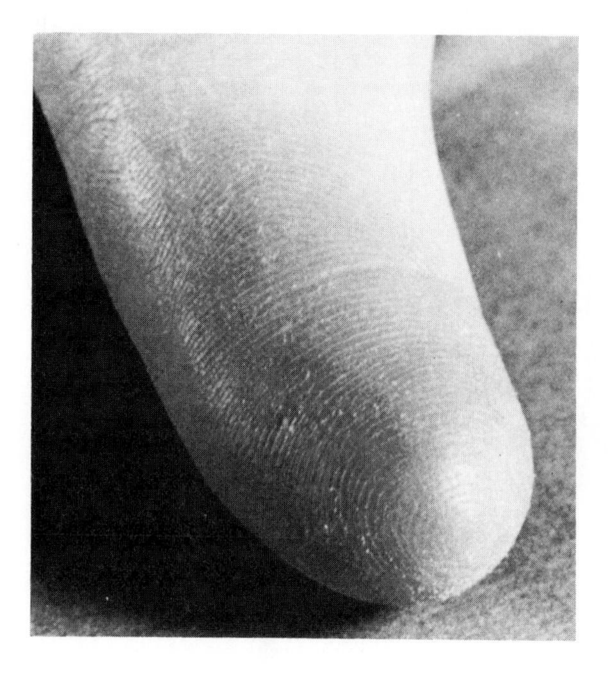

Figure 17–8. The ridge arrangement on a nailless finger in an individual with Type B brachydactyly. Reprinted from Battle HI, Walker NF, Thompson MW. Mackinder's hereditary brachydactyly. Ann Hum Genet 1973; 36:415–424, by permission.

GENERAL REFERENCES

Bartsocas C, ed. Progress in dermatoglyphic research. Prog Clin Biol Res 1982; Vol. 84.
Cummins H, Midlo C. Fingerprints, palms and soles: an introduction to dermatoglyphics. New York: Dover, 1961.
Holt SB. The genetics of dermal ridges. Springfield, Illinois: Charles C Thomas, 1968.
Loesch DZ. Quantitative dermatoglyphics. Classification, genetics, and pathology. Oxford: Oxford University Press, 1983.
Penrose LS. Dermatoglyphics. Sci Am 1969; 221:72–84.
Preus M, Fraser FC. Dermatoglyphics and syndromes. Am J Dis Child 1972; 124:933–943.
Schaumann B, Alter M. Dermatoglyphics in medical disorders. New York: Springer-Verlag, 1976.
Wertelecki W, Plato CC, eds. Dermatoglyphics—fifty years later. Birth Defects 1979; 15(6).

PROBLEMS

Match the following:

A

—total fingerprint ridge count
—45,X Turner syndrome
—very rare except in Down syndrome
—pattern area on ulnar side of palm
—characteristic of trisomy 18
—common in Down syndrome and other chromosomal syndromes
—typically, 5 per palm
—characteristic of palms and soles of embryo
—variant flexion crease
—distally displaced in Down syndrome

B

1) single flexion crease
2) volar pads
3) hallucal tibial arch pattern
4) axial triradius
5) high frequency of arches
6) large fingerprint patterns
7) hypothenar
8) TRC
9) Sydney line
10) thenar
11) triradii

18

PRENATAL DIAGNOSIS

Since 1966, when Steele and Breg demonstrated that amniotic fluid cells could be cultured to reveal the fetal karyotype, prenatal diagnosis has become a major medical genetic service. In earlier chapters it has been mentioned repeatedly in the context of prevention of specific genetic disorders. Here the scope, methodology and implications of prenatal diagnosis will be described in a broader context.

At last count, over 350 centers in 45 countries were known to provide prenatal diagnosis, and the number is increasing. As a general rule, in North American communities about 8 percent of all pregnant women meet the criteria for prenatal diagnosis on the basis of age or other indications (see below); in some centers more than half the eligible pregnancies are monitored. The demand for genetic testing is sure to rise if a screening procedure is developed to identify pregnancies at risk of a chromosomal abnormality in women under 35 years of age, or when techniques for prenatal diagnosis of cystic fibrosis, Duchenne muscular dystrophy and other relatively common genetic disorders are perfected.

In some minds prenatal diagnosis is equated with the issue of abortion, but this is a narrow view of its role. For many families at risk of having a child with a condition that can be diagnosed prenatally, the option of monitoring pregnancies and the availability of elective abortion allow the parents to undertake pregnancies that they would otherwise forego. Only about 2 percent of all pregnancies in which there is prenatal diagnosis are terminated because the fetus has a genetic defect. Much more often, the fetus is found to be unaffected and the pregnancy continues.

Indications for Prenatal Diagnosis

There are widely accepted guidelines for eligibility of pregnant women for prenatal diagnosis. At least theoretically, the guidelines are based on judgement that the risk of an abnormal fetus is at least as great as the risk of the procedure itself. The criteria for pregnancies eligible for prenatal diagnosis include:

1. Maternal age of 35 years or more. In this age range, there is an appreciable risk of Down syndrome or some other trisomy in the child.

2. Previous child with a de novo chromosome abnormality. Here the recurrence risk, at least for trisomies, is about 1 percent irrespective of the mother's age.

3. Presence of a structural chromosomal abnormality in one of the parents. Here the risk of a child with an unbalanced karyotype is theoretically as high as 50 percent.

4. Family history of a neural tube defect. If a first-degree relative of the fetus (parent or sib) has a neural tube defect, the risk to the fetus is in the 2 to 5 percent range. If more remote relatives are affected, the risk to the fetus is less but may still be above the population risk.

5. Family history of some genetic defect, where the risk of a defective fetus is 1 percent or more and the defect might be detectable by prenatal diagnostic techniques.

6. Cases in which some type of population screening has shown that the fetus is at risk of an identifiable defect.

7. X-linked disorders in which there is no specific diagnostic test, but determination of the fetal sex can be used by the parents in deciding whether or not to terminate the pregnancy.

By far the largest proportion of women who receive prenatal diagnosis are of late maternal age (usually defined as age 35 at the estimated date of delivery). Because medical facilities are not available to provide prenatal diagnosis to everyone who wants it, and because the relationship between late maternal age and Down syndrome is widely known, it has been necessary to set a guideline for late maternal age. In some centers, depending on the resources available, this may be higher than the age of 35.

Though the risk of a chromosomally abnormal child is higher among women 35 years of age or older than among younger women, the birth rate is so much higher in the younger age group that many pregnancies in which the fetus has a chromosomal abnormality are not monitored. Methods for identifying younger parents at risk of having chromosomally abnormal children would be very helpful.

Techniques

AMNIOCENTESIS

The term amniocentesis (literally, tapping of the amnion) refers to the procedure of removing a sample of amniotic fluid transabdominally, by syringe. The technique was originally developed by Liley (1962) for therapy of fetuses suffering from the consequences of Rh incompatibility. Ultrasonography is used to determine gestational age, and to outline the position of the fetus and placenta so as to improve the success rate and safety of the procedure. Amniocentesis is usually carried out at a gestational age of 16 to 17 weeks. At an earlier stage the uterus is still too small and inaccessible; at a later stage there may be insufficient time to complete testing before the pregnancy has progressed past the age at which termination can be performed, even if the fetus is found to be abnormal. Amniocentesis is performed as an out-patient procedure and does not normally require local anesthesia. The risks of the procedure seem very small, though there is a slight risk of inducing miscarriage (about 1 in 200). Maternal infection is an unlikely complication. If the mother is Rh-negative, Rh immune globulin should be administered to prevent Rh sensitization (see Chapter 9).

CHORIONIC VILLOUS SAMPLING (CVS)

Fetal tissue for analysis can be aspirated from the villous area of the chorion transcervically by catheter at a gestational age of 9 to 12 weeks. The technique is still under development and is currently available in only a limited number of centers. Its chief advantages are the following:

1. Fetal testing at 9 to 12 weeks is more acceptable to mothers than later testing.

2. Chorionic villous cells are rapidly dividing, and there may be enough mitotic cells in a sample for chromosome analysis without cell culture; thus, a prompt result is often possible.

3. Termination of pregnancy, if required, is a safer and simpler procedure in the first than in the second trimester.

The safety of CVS has not yet been fully evaluated; it appears to be acceptable, though the miscarriage rate may be higher than that for amniocentesis. Because half of all spontaneous first-trimester abortions are chromosomally abnormal, the fetus may be found by CVS to be chromosomally abnormal in some pregnancies that would eventually have terminated spontaneously; in other words, the frequency of chromosomal abnormalities in fetuses tested by CVS may be appreciably higher than the frequency in fetuses tested by amniocentesis or than the frequency at birth.

ULTRASONOGRAPHY

In addition to its use in prenatal diagnosis as a necessary tool in either amniocentesis or chorionic villous sampling, ultrasound is also used for detection of some congenital malformations—especially neural tube defects, but also skeletal dysplasias, some anomalies of the gastrointestinal and urinary tracts, and various congenital heart defects. Diagnosis of multiple pregnancy and determination of fetal growth rate can also be accomplished ultrasonographically. Fetal blood sampling or skin biopsy can now be performed under ultrasonic guidance rather than by fetoscopy (see below).

Ultrasound instrumentation has improved greatly since ultrasonography was first introduced. At present the method of choice is real-time scanning. In addition to advances in instrumentation, the expertise of the ultrasonographer is an important factor in the ability to diagnose congenital malformations.

FETOSCOPY

Fetoscopy, which allows direct visualization of a second-trimester fetus by means of a fiberoptic endoscope, is a valuable adjunct to other prenatal diagnostic techniques in some cases, especially for the detection of external anatomical defects and for taking direct samples of fetal blood, skin or even liver. Only a limited number of centers perform fetoscopy at present. The fetal loss rate of about 5 percent is higher than that for amniocentesis.

PRENATAL CHROMOSOME ANALYSIS

In addition to late maternal age and the birth of a previous child with a de novo chromosomal abnormality, there are other, less common indications for analysis. These include mosaicism or a balanced chromosomal rearrangement in one parent or the risk of the fragile X syndrome or one of the rare chromosomal breakage syndromes in the fetus. Fetal sex determination for X-linked disorders is another indication. Moreover, chromosome analysis is usually routinely performed on all amniotic fluid samples, even when the indication for amniocentesis is the risk of a non-chromosomal defect such as a neural tube disorder.

Chromosome analysis is by far the most time-consuming part of the prenatal diagnosis process; in many centers the cytogenetic capacity of the program is the factor that limits the number of amniocenteses that can be performed. Thus any prenatal diagnosis program requires communication and cooperation between clinicians and cytogenetic laboratories.

Problems in Cytogenetic Analysis

There are certain problems in fetal chromosome analysis that should be kept in mind by the patient as well as by her physician:

1. **Culture failure.** A culture may not grow, or may grow too poorly for

adequate analysis. In prenatal diagnosis, there may not be time to repeat the amniocentesis and culture a second sample. If a fetal blood sample can then be obtained by fetoscopy, it can be cultured and a report can be made available within a few days.

2. **Maternal cell contamination.** Undetected maternal cell contamination may lead to misdiagnosis of a male fetus as female, or to the prediction of XX/XY mosaicism in the fetus.

3. **Mosaicism.** When two (or more) cell lines are found in cultured fetal cells, there may be serious problems in interpretation, both as to whether the fetus is truly mosaic and as to the clinical significance of the observed type of mosaicism.

Cytogeneticists distinguish between **true mosaicism** and **pseudomosaicism,** in which the abnormal cell line is not present in the fetus but has arisen as an artifact during culture or through maternal cell contamination. If the abnormal cells are confined to a single culture vessel, the probable interpretation is pseudomosaicism. If several cultures show both the normal and the abnormal cell line, the fetus may be a true mosaic. In either case, the finding must be discussed with the parents and interpreted for them as accurately as possible; if termination of pregnancy ensues, an effort must be made to verify the finding.

True mosaicism is known for the common chromosomal syndromes, and may sometimes be associated with milder expression of the phenotype. However, the proportion of abnormal cells present in an amniotic fluid cell culture may not reflect the proportion in the fetus (or the proportion present in the fetus during differentiation, some weeks earlier). Moreover, the abnormal cell line might be derived from extra-embryonic tissues and not from the fetus itself.

The uncertainty related to the interpretation of mosaicism is one of the major problems in prenatal diagnosis. It is important that the parents be warned in advance of the possibility that the chromosomal findings may be difficult to interpret.

Alpha-Fetoprotein (AFP)

Alpha-fetoprotein is a protein of unknown function, specific to the fetus, synthesized in the yolk sac, the gastrointestinal tract and, especially, the liver. It is passed into the amniotic fluid in the fetal urine, reaching its peak concentration there at about 12 to 14 weeks' gestation. The concentration in adult serum is normally extremely low, but rises in pregnant women. It has been known since 1972 that increased levels of AFP are found in the amniotic fluid of fetuses with anencephaly or open spina bifida (Brock and Sutcliffe, 1972). This observation has led to the wide use of assay of AFP in amniotic fluid for prenatal diagnosis of neural tube defects, and of AFP in maternal serum (MSAFP) as a screening test. (See also Chapters 12 and 14.)

AMNIOTIC FLUID ALPHA-FETOPROTEIN (AFAFP)

The presence of a neural tube defect in a first-degree relative (parent or sib), a second-degree relative or even a third-degree relative of a fetus is considered to be an indication for AFAFP assay of the amniotic fluid, though there is little evidence for an increased risk unless the affected relative is a parent or sib, or there is more than one affected relative.

For AFP determination, the amniotic fluid obtained by amniocentesis at 16 weeks' gestation is usually measured by immunoassay. Because the assay is

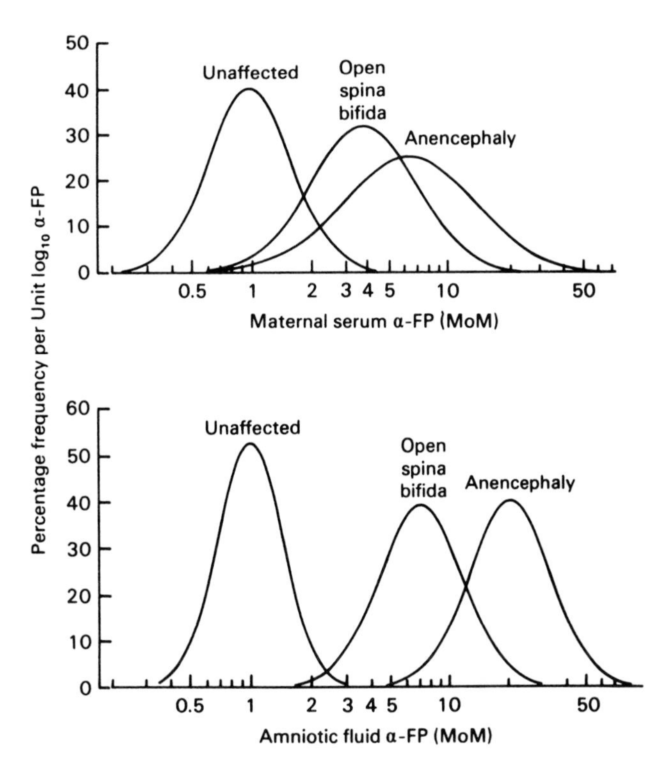

Figure 18–1. Alpha-fetoprotein levels in maternal serum (above) and amniotic fluid (below), expressed in multiples of the median value (MoM). From Wald NJ. UK collaborative study, Fourth Report. In: J Epidemiol Community Health 1982; 35:87–95.

relatively simple and inexpensive, usually all amniotic fluids are assayed, whether or not the indication for amniocentesis was the risk of a neural tube defect. Because there is an overlap between the distributions of AFP levels in normal pregnancies and those with a neural tube defect, the assay carries a small risk of misinterpretation; for example, among measurements more than three multiples of the median above the median about 1 percent are false positives, and among those below that cutoff about 1 percent are false negatives (Fig. 18–1). However, when AFP assay is used in conjunction with ultrasonography, close to 98 percent of all cases of open spina bifida and virtually all anencephalics can be identified by 18 weeks' gestation.

Problems in Amniotic Fluid AFP Assay

A number of technical and diagnostic problems may lead to elevated AFP in a fetus without a neural tube defect:

1. Contamination of the sample by fetal blood in which the concentration of AFP is higher than in amniotic fluid.
2. Twin pregnancy, in which a high level of AFP in one amniotic sac can diffuse to the other.
3. Fetal death.
4. Certain abnormalities, including omphalocele and at least one form of congenital nephrosis.

MATERNAL SERUM ALPHA-FETOPROTEIN (MSAFP)

When the level of AFP is elevated in the amniotic fluid, a less striking increase in AFP level is usually also present in the maternal serum. This observation has been exploited by the use of maternal serum AFP measurement as a population screening test for neural tube defects. Since more than 90 percent of all babies with neural tube defects are born to parents with no family history of such disorders, a screening test that could be given to all pregnant women rather than only to those known to be at risk could lead to a significant reduction in the occurrence of neural tube defects.

Several pilot programs for maternal serum AFP screening have now been performed. In these studies, MSAFP is assayed at 16 to 18 weeks' gestation. Since the level of MSAFP is rising rapidly during this stage of pregnancy, determination of precise gestational age by ultrasound is essential.

Because there is considerable overlap between the MSAFP range in normal fetuses and those with a neural tube defect, a proportion of affected fetuses with open spina bifida are missed (Fig. 18–1).

Elevated MSAFP levels are found in twin pregnancies, in missed or threatened abortions, and in a number of other conditions as well as in neural tube defects. Women with elevated MSAFP are referred for ultrasound examination and possibly for amniocentesis in order to diagnose the specific cause of the abnormally high level and to allow intervention if it is indicated. The most common cause of an MSAFP level that is higher than expected is underestimated gestational age. Many elevated MSAFP levels are simply false positives in normal fetuses.

Biochemical Disorders

Many rare biochemical disorders can be diagnosed by detection of a specific biochemical defect in cultured amniotic fluid cells (Table 18–1). Most such conditions are extremely rare, and the knowledge that a given pregnancy is at risk for a specific biochemical defect comes either from population screening for heterozygotes (as in Tay-Sachs disease) or from a history of a previous affected child. Most biochemical disorders are autosomal recessive, and thus have a 25 percent recurrence risk, far higher than the recurrence risk for most chromosomal defects or for neural tube defects.

Recombinant DNA Technology in Prenatal Diagnosis

The use of recombinant DNA methods for prenatal diagnosis has been described in several earlier chapters, but is mentioned here for completeness. Over the next few years this area is expected to expand very rapidly and to become widely applied as a prenatal diagnostic method, perhaps replacing currently used techniques.

Table 18–1. METABOLIC DISORDERS DETECTABLE IN CULTURED AMNIOTIC FLUID CELLS (partial list)

Adenosine deaminase deficiency	Lesch-Nyhan syndrome
Arginosuccinic aciduria	Maple syrup urine disease
Fabry disease	Menkes disease
G_{M1} gangliosidosis	Metachromatic leukocystrophy
G_{M2} gangliosidosis (Tay-Sachs disease)	Methylmalonic aciduria
(Sandhoff disease)	Mucopolysaccharidoses
Hypercholesterolemia, familial	Niemann-Pick disease
Krabbe disease	Propionic acidemia

Data from Patrick AD. Inherited metabolic disorders. Br Med Bull 1983; 39:378–385, and Stephenson SR, Weaver DD. Prenatal diagnosis—a compilation of diagnosed conditions. Am J Obstet Gynecol 1981; 141:319–343.

The Organization of Prenatal Diagnosis Programs

Prenatal diagnosis is an area that requires the combined skills of obstetricians, laboratory scientists and geneticists. The organization of these elements into a high-quality prenatal diagnosis program requires careful attention to lines of responsibility and communication among the professionals themselves and between professional and patient.

There are several ways of organizing a prenatal genetic program. For example, pregnant women may be referred directly by a family physician or obstetrician to a central clinic. Depending on whether the referral is strictly for late maternal age or for some other indication, a decision is made as to whether genetic counseling will be provided by the obstetrician who is to perform the amniocentesis, or whether the patient will receive a genetic assessment (usually including family history and confirmation from medical records and often also requiring clinical examination and chromosome analysis or other tests), plus genetic counseling from a member of the genetics staff.

A major concern in any prenatal diagnosis program is time. Time is required for the obstetrician or geneticist to discuss the genetic risks with the patient during her appointment. Time is also needed to evaluate any unusual genetic information that comes to light, to arrange for any necessary investigation and to allow the pregnant woman and her partner to consider their options before the deadlines of 16 weeks' gestation for amniocentesis and (usually) 20 weeks' gestation for termination of pregnancy, if it is indicated. Ideally, genetic counseling should precede pregnancy; otherwise, it should be provided as early in pregnancy as possible.

Termination of Pregnancy

Few issues of the present day are as hotly debated as abortion, but despite many legal restrictions it has come into widespread use in recent years. Without a means of legal termination of pregnancy when the fetus is found to be seriously abnormal, amniocentesis would not have developed into the widely accepted procedure that it has become.

Termination of pregnancy in the first trimester is usually performed by dilatation of the cervix and evacuation of the uterus, a minor surgical procedure. Termination at 17 weeks' gestation or later is more complicated; abortion is induced by infusion of hypertonic sodium chloride solution or, more frequently, prostaglandin into the amniotic sac, thus bringing about expulsion of the uterine contents within about 24 hours. It is important to attempt to confirm the presence in the fetus of the defect for which termination was performed. This is usually possible for chromosomal defects, but is less likely to be successful for anatomical abnormalities.

GENERAL REFERENCES

Ferguson-Smith M, ed. Prenatal genetic diagnosis. Br Med Bull 1983; 39:301–404.
Harrison MR, Golbus MS, Filly RA. The unborn patient. Orlando: Grune and Stratton Inc., 1984.
Milunsky A. Genetic disorders and the fetus. New York: Plenum Press, 1979.
Rodeck CH, Nicolaides KH. Prenatal diagnosis. Chichester: John Wiley and Sons, 1984.

PROBLEMS

General Review, Part I

1. Bob and Mary are a pair of twins. Their ABO blood groups are unusual; Bob's red cells are all type AB whereas Mary has 90 percent type O and 10 type AB cells. Could Mary have an AB child? Explain.

2. Laura, who has normal parents, has Christmas disease (hemophilia B), an X-linked recessive condition. Give three possible explanations.

3. Which of the following explanations best accounts for the maintenance of the Hb S polymorphism in West Africa?
   a) Recurrent mutation of β^A to β^S.
   b) Selective advantage of $\beta^A\beta^S$ heterozygotes over $\beta^S\beta^S$ homozygotes.
   c) Selective advantage of $\beta^A\beta^S$ heterozygotes over $\beta^A\beta^A$ homozygotes.
   d) Selection favoring both homozygotes over the heterozygote.
   e) Selection favoring the heterozygote over both homozygotes.

4. Name two ways in which chiasmata are fundamental to the genetic process.

5. When co-cultivated, fibroblasts from a Hurler syndrome patient and a Scheie syndrome patient will not mutually correct for the intracellular accumulation of mucopolysaccharides, as happens when fibroblasts from a Hurler syndrome patient and a Hunter syndrome patient are co-cultivated. Explain.

6. A woman who has a brother and a son both affected with Duchenne muscular dystrophy asks for genetic counseling concerning her risk of another Duchenne child. What other information would you require? What is your estimate of the risk for her next son?

7. The loci for the deutan type of red-green color blindness and G6PD are both near the tip of the long arm of the X chromosome. In a certain family the father is color-blind and G6PD-A, and the mother is not color-blind and is G6PD-AB. The couple has a total of 20 children.
   Sons: 5 color-blind, G6PD-A
   1 color-blind, G6PD-B
   4 not color-blind, G6PD-B
   Daughters: 3 color-blind, G6PD-A
   1 not color-blind, G6PD-A
   6 not color-blind, G6PD-AB
   a) What is the mother's genotype at each locus? What is the linkage phase of the genes?
   b) Which children and what proportion of the total are recombinants?
   c) What estimate of the distance between the loci does this family provide?

8. A certain gene F has two mutant alleles f^1 and f^2, each of which when homozygous causes a genetic disorder. If f^1f^1 patients occur with a frequency of 1/360,000 and f^2f^2 patients occur with a frequency of 1/160,000, what is the frequency of patients with the f^1f^2 genotype?

9. Which of the following is NOT a known teratogen in man? Radiation, phenytoin (Dilantin), Coumadin (Warfarin), cyclamates, alcohol, rubella.

10. How frequently does meiotic recombination occur between:
    a) two loci on chromosome 2 located 8 centimorgans apart?
    b) two loci near the tips of Xp and Xq?
    c) two loci, one of which is on chromosome 4 and the other on chromosome 7?

11. A couple who have had a child with PKU ask for prenatal diagnosis. You use the Southern blotting procedure and a cDNA probe for phenylalanine hydroxylase

associated with fragment lengths of 4.2 and 3.8 respectively. The family data show: Father 4.2/3.8 Mother 4.2/3.8

Patient 3.8 Sister 4.2 Fetus 4.2/3.8

a) Which fragment length in each parent is associated with the PKU allele?
b) Is the fetus affected, unaffected or a carrier?
c) What is the sister's probability of being a carrier?

12. In another family, the data are as follows:

Father 4.2/3.8 Mother 4.2

2 affected sons, both 4.2 Normal daughter 4.2/3.8 Fetus 4.2

a) What is the genotype of the normal daughter?
b) What is the genotype and phenotype of the fetus?

13. Name and briefly discuss three situations in which a couple, both of whom are normal, could have only retarded children.

14. a) What method is used for population screening of pregnant women for neural tube defects in their offspring?
b) When false-positive and false-negative values are encountered in screening, what are their causes? What steps can be taken to reduce or eliminate them?

15. For three different autosomal recessive diseases, the frequency of each is A: 0.001, B: 0.00005, C: 0.000001. For each, indicate:
a) the gene frequency
b) the carrier frequency
c) the proportion of marriages that can produce an affected child.

19

GENETIC COUNSELING

Genetic counseling, as defined by Harper (1984), is "the process by which patients or relatives at risk of a disorder that may be hereditary are advised of the consequences of the disorder, the probability of developing and transmitting it, and the ways in which this may be prevented or ameliorated." In this chapter, the general principles and procedures basic to genetic counseling are brought together. Table 19–1 summarizes the major steps in the process.

The unique feature of genetic disorders, as compared to other medical problems, is their tendency to recur within families. For a patient with any disorder, genetic or otherwise, clinical management includes:

1. Making a diagnosis.
2. Advising the patient or the patient's family of the consequences of the disorder.
3. Providing treatment and support, either directly or by referral.

If the disorder is genetic, there are additional responsibilities: to inform the patient or family of the probability that the disorder will recur in other family members (usually called the recurrence risk) and of the methods available to prevent recurrence or at least to reduce the risk. This is the aspect of genetic counseling that clearly differentiates it from other forms of clinical management.

The Genetic Counseling Process

The most common situation in genetic counseling is that the consultands—the persons receiving genetic counseling—are parents who have had a child with a genetic defect and want to know about the defect and its implications for their subsequent children. In this as in any genetic counseling situation the first step is to try to reach a firm and correct diagnosis, without which all the other components of the process may be inaccurate.

Unfortunately, this essential first requirement cannot always be met, especially if the affected family members are deceased and there is no autopsy information. If parents who have had a stillborn child or an infant who died soon after birth request genetic counseling before undertaking another pregnancy, lack of information about the previous child may make their request impossible to fulfill. For any child who is stillborn or dies in the neonatal period, the following tests, at a minimum, should be performed:

1. Chromosome analysis
2. Photographs
3. X-rays
4. Autopsy

Information from these tests can often help form a diagnosis, or allow some possible causes of the stillbirth to be ruled out.

Table 19–1. STAGES IN GENETIC COUNSELING

Pre-Counseling Assessment	**Diagnosis** Family history Confirmation of family history information Special tests (chromosomes, etc.)
Recurrence Risk Estimation	**Based on pedigree analysis, medical literature, test results**
Communication	**Nature and consequences of the disorder** Recurrence risk Treatment, if available Preventive measures (prenatal diagnosis, artificial insemination by donor, etc.)
Follow-up	**Written reports to referring physician and consultand** Referral to appropriate health care agencies, self-help organizations, psychotherapist, etc., as required

SPECIAL GENETIC COUNSELING PROBLEMS

A variety of other situations are occasionally seen in genetics clinics. Some of these are listed and briefly discussed below; their genetic aspects are presented more fully in earlier chapters.

Consanguinity. Consanguineous couples may request information about their risks, either before marriage or before reproduction. The genetic aspects of consanguinity have been discussed earlier (Chapter 15). The family history will show the exact relationship of the couple and might indicate some genetic problems in the kindred that could pose an increased risk for their child. Consanguineous couples can usually be reassured that their risk of a child with a genetic defect is only slightly increased. However, if they actually do produce such a child, they should return for a further genetic assessment since, if the child's disorder is autosomal recessive, the parents are proved to be carriers of the same recessive gene, probably inherited from one of their common ancestors.

Post-Screening Genetic Counseling. If a person is identified through a screening program as being at risk of either developing a genetic disorder or having an affected child, genetic counseling should be provided to make sure that the genetic risk, its implications, and ways of preventing or ameliorating the harmful consequences are understood.

Genetic Counseling After Prenatal Diagnosis has Shown that a Fetus is Abnormal or at Risk of Being Abnormal. This type of genetic counseling has the great disadvantage of being provided when the parents are extremely anxious and when a prompt decision must be made about continuation or termination of the pregnancy. To some extent this disadvantage can be minimized if the parents are fully informed of the possibility of an unfavorable or questionable outcome before prenatal diagnosis is attempted.

Teratogenic Exposure. In recent years there has been an increase in the number of inquiries directed to genetic centers about exposure of a mother in an early stage of pregnancy to possible teratogenic agents. In fact, in some areas "teratogen hotlines" have been set up to meet this need. Though information about teratogenic risk is gradually increasing, all too often hard data are not available to help women exposed to possible teratogens to reach decisions about possible termination of pregnancy.

Pre-Adoption Counseling. Occasionally a couple planning to adopt a child, or an agency with a child to be placed for adoption, requests genetic counseling.

There is usually little genetically significant family history information about children available for adoption, but occasionally an adoptable child is the consequence of incest or has a family history of a serious genetic disorder. In such cases, careful clinical assessment may clarify the child's condition and potential.

FAMILY HISTORY

In Chapter 1 the importance of family history was emphasized and its main elements described. The family history should include the most accurate information possible about medical problems in the family, particularly about the condition for which genetic counseling has been requested.

Confirmation of Family History Information

Wherever possible the details recorded in the family history should be confirmed from medical records. There may be difficulties in accomplishing this objective. Because consent for the release of medical records must be obtained from the person concerned or the next of kin, obtaining the records can be time-consuming and may be impossible. Moreover, the patient or parents referred for counseling may be reluctant to involve other family members or even to have it known within the family that they are concerned about the risk of a genetic disorder in their progeny.

Old family photographs are often helpful, especially when there is a suspected dysmorphic syndrome in a family member who is not available for examination.

GENETIC TESTS

While the family history information is being collected and reviewed, it may become obvious that certain family members should receive genetic tests; for example, a relative with Down syndrome may not previously have had a chromosome analysis, or female relatives of a boy with Duchenne muscular dystrophy may not have had carrier tests. Because the results of such tests can profoundly affect the recurrence risk estimations, they should be carried out before genetic counseling is given.

RECURRENCE RISK ESTIMATION

The estimation of recurrence risk is the central problem of genetic counseling. Ideally it is based on knowledge of the genetic nature of the disorder and on the pedigree of the particular family; however, in the absence of precise information it must rely on the experience of the geneticist concerned and of others who have had to deal with the particular problem, and may then be little more than an informed guess. Obviously, the geneticist's degree of uncertainty about the risk estimate should be shared with the family.

Baseline Risks

Many people are unaware that in any pregnancy there is an appreciable risk of an unfavorable outcome. Table 19–2 summarizes the baseline risks that any

Table 19–2. BASELINE RISKS OF ABNORMALITIES IN THE GENERAL POPULATION

Risk of a child being born with some birth defect	1 in 30
Risk of a child being born with some serious physical or mental handicap	1 in 50
Risk of a child of first-cousin parents being born with some serious physical or mental handicap	1 in 20
Risk of any pregnancy ending in a spontaneous abortion	1 in 8
Risk of stillbirth (North America)	1 in 125
Risk of perinatal death (North America)	1 in 150
Risk of death in first year of life after first week (North America)	1 in 200
Risk that couple will be infertile	1 in 10

Based on Harper PS. Practical genetic counselling, 2nd ed. Bristol: Wright, and Littleton, Wright, PSG, 1984; and Ontario Vital Statistics, 1982.

couple faces in any pregnancy. Recurrence risks for specific genetic defects that have already appeared in a family are in addition to these risks.

Mendelian Risks

When a disorder is known to have single-gene inheritance, its recurrence risk for specific family members can usually be determined readily. A few considerations should be noted:

1. Many autosomal dominants, in particular, are variable in expression or may even exhibit lack of penetrance. The parents may be more interested in the risk of a severe expression of the gene than in the 50 percent risk of its transmission. The importance to the parents of knowing the "worst-case scenario" is well recognized (Lippman-Hand and Fraser, 1979).

2. The Mendelian risk estimates can sometimes be modified, especially in X-linked recessive pedigrees, by means of Bayesian analysis of the pedigree, as described in Chapter 14.

3. As noted above, genetic tests can often allow much improved accuracy in risk estimation by clarifying the genotype or phenotype of certain at-risk individuals.

Empiric Risks

For multifactorial and most chromosomal disorders, recurrence risk estimates are empiric, that is, based on experience rather than on knowledge of the underlying mechanism. As noted above, for many conditions the data base from which to judge the recurrence risk is still inadequate. Eventually this might be remedied if recurrence risk data for large populations could be evaluated, centrally stored and made available to genetics clinics with computer assistance.

COMMUNICATION

The diagnosis is clear, the medical and family data have been collected, and the recurrence risk has been estimated as accurately as possible. The next step is to inform the consultands about the condition, its consequences, its recurrence risk, and possible ways to deal with the problems that result from it, and to help them to arrive at a decision.

A common problem at this stage is how to transmit information, which is often quite complicated and expressed in terms of probabilities, in a form that can be clearly understood and remembered by the individuals counseled. Although there are exceptions, most persons have little knowledge of either the medical or the genetic aspects of the disorder for which they are receiving counseling. Adequate time must be allowed for explanation and for discussion of the reproductive decisions that could be made on the basis of the information. A written report summarizing the information provided is helpful to the consultands, reinforcing their recollection of the interview.

Timing

The usefulness of genetic counseling depends partly on its timing. Ideally, for example, it should not be attempted soon after a serious diagnosis has been made in a family member, or soon after loss of a child. On the other hand, it should be provided before the next pregnancy, or, if the consultand is already pregnant, as early as possible in the pregnancy.

Some people at risk of transmitting a disease are identifiable by carrier tests; for example, all sisters of boys with Duchenne muscular dystrophy are at risk of being carriers, and some can be confirmed by biochemical tests. In such cases genetic counseling is ineffectual, however, until the girl is old enough to appreciate the risk and its significance for her reproductive future. Moreover, because there is often a possibility of improved methods of prenatal diagnosis, genetic counseling can sometimes be more precise if it is postponed until needed. However, delay increases the risk of seeing a carrier for the first time when she is already pregnant.

Treatment

Although there is a persistent impression that to label a disease as genetic is to label it as incurable, actually many genetic disorders can be treated with reasonable success. Many defective genes are defective only in certain environments, and environmental modification can often ameliorate gene expression. Whether or not direct replacement of harmful genes becomes possible in the immediate future, there are many other means of clinical management of genetic defects.

Preventive Measures

The chief purpose of clinical genetics is prevention, and the chief preventive measure available to parents at risk of having a child with a genetic defect is prenatal diagnosis (see Chapter 18). However, prenatal diagnosis is by no means a universal solution. There are many disorders for which prenatal diagnosis cannot be provided, and many patients who are reluctant to terminate a pregnancy even if the fetus is demonstrably abnormal.

Other preventive measures may be appropriate for certain families:

1. If the decision of the paents is not to have (further) children, contraception or sterilization may be their choice, and they may need information about the possible procedures or appropriate referral.

2. Some parents choose to adopt a child. At present, because of contraception, elective abortion and the decision of many single mothers to keep their children, adoption is not always feasible.

3. Artificial insemination by donor (AID) is an option that some parents choose in preference to adoption. It is appropriate if the father has or is at risk for an autosomal dominant defect, or if both parents are known carriers of an autosomal recessive condition. Obviously it is not indicated when the mother carries an autosomal dominant or X-linked trait. By the same token, in vitro fertilization of a donor egg by the partner's sperm, followed by embryo transfer, may eventually become a practical option if the mother has an autosomal

dominant gene, carries a recessive gene that her partner also carries, or is a carrier of an X-linked disorder.

4. Sex selection—the ability to conceive a child of the desired sex—may become a practical possibility. Many women who carry harmful X-linked genes would choose this option if it were available.

5. Special tests (such as chromosome analysis, biochemical carrier tests or molecular genetic analysis) will sometimes provide reassurance to parents that they are not at risk, and in other cases will point to the need for preventive measures such as those outlined above.

6. If the parents decide to terminate a pregnancy, provision of information and support is an appropriate part of their genetic counseling.

THE FOLLOW-UP TO GENETIC COUNSELING

The experience of attending a genetics clinic is new and strange to most of the individuals counseled. In order to make sure they understand and remember the information they are given, it is often helpful to arrange one or more follow-up interviews. A letter to the family should review the medical and family information, the genetic risk, the alternative decisions that might be made, and, if a decision has been reached, the reasoning behind it.

Many families with genetic disorders are seen in genetics clinics repeatedly over a period of years, when new family members develop the family disease or want information to assist them in reproductive planning.

Finally, genetic disorders can lead to serious personal and family problems. Though of course this is also true of nongenetic disorders, the concern generated by knowledge that the condition might recur, plus the necessity of making reproductive decisions, can lead to special problems for persons with genetic disorders or for parents of a genetically abnormal child. Many people have the strength to deal personally with such problems; they prefer receiving even bad news to remaining ignorant, and make their own decisions on the basis of the most complete and accurate information they can obtain. Others require more support, and some may even need referral for psychotherapy. The many disease-oriented self-support groups that have been formed in recent years are helpful to some families.

WHEN IS REFERRAL FOR GENETIC COUNSELING INDICATED?

Broadly speaking, genetic counseling is simply one aspect of the overall medical management of a patient, and much genetic counseling can and should be done by primary care physicians. However, most genetic disorders are individually so rare that few physicians have the opportunity to gain wide experience in providing appropriate genetic advice for them. Moreover, the lengthy process of obtaining and verifying medical and family information fits poorly into the normal pattern of medical practice.

Today most medical schools and teaching hospitals, and some other hospitals, have medical genetics units to which patients can be referred for assistance with diagnosis, prenatal evaluation, genetic or cytogenetic testing and other aspects of genetic counseling. Outreach clinics are also available in many communities.

Screening for Genetic Disease

Genetic screening, unknown until a few years ago, is now an important part of public health programs. Population screening originally had the objective of identifying newborns with genetic disorders that could be treated if diagnosed promptly. The prototype disorder is phenylketonuria, but a number of other disorders can also be diagnosed in newborns if urine is screened for abnormal metabolites, or blood for amino acids. Recently hypothyroidism, which is not usually genetic but is treatable, has been added to the list of conditions suitable for routine neonatal screening. Heterozygote screening programs have also been introduced. Perhaps as a consequence of overenthusiasm and haste in putting some programs into operation prematurely, not all have been successful; some have generated public concern about the legal and ethical aspects of screening, which has been called "an arguably unique intrusion into the domain of personal freedom" (Reilly, 1975), but also "a major philosophical advance in medicine" (Levy, 1973).

NEWBORN SCREENING PROGRAMS

The principles of an adequate newborn screening program are obvious. The condition should be clearly defined and treatable and should have a reasonable frequency in the population; the screening test itself should be one that can be done rapidly and inexpensively on a large scale; the test should yield few false positives and, ideally, no false negatives; follow-up for definitive diagnosis and initiation of treatment should be well organized and prompt.

If the condition does not meet these requirements, problems may arise. The question of validity of the test results is very important; false positives cause great concern to the parents, and false negatives vitiate the whole reason for the project.

Not everyone agrees that untreatable conditions should not be screened, basing their argument on the need for early identification so that genetic counseling can be provided and births of additional affected children in the family can be prevented. This argument is made in particular for Duchenne muscular dystrophy (Scheuerbrandt and Mortier, 1984). Identification of DMD can be made by a screening test in newborns, whereas the disease may not be evident clinically until another affected boy has been born. The benefit of early diagnosis for genetic counseling must be balanced, however, against the consideration of whether it is humane to inform the parents of the child's problem so far in advance of its manifestation.

HETEROZYGOTE SCREENING

For heterozygote screening to be feasible, three minimal criteria must be met: (1) occurrence of the disorder chiefly in a specific population group; (2) availability of a test suitable for mass screening; and (3) possibility of prenatal diagnosis. Three disorders that meet these criteria are Tay-Sachs disease, sickle cell disease and β thalassemia.

Tay-Sachs Disease

Tay-Sachs disease has a high incidence in Ashkenazi Jews. Since 1969, when mass screening for carriers of the disease began, over 300,000 individuals have been screened. In North America, the incidence of Tay-Sachs disease has already dropped 65 to 85 percent as a result of screening followed by prenatal diagnosis.

In Quebec, through the Quebec Network of Genetic Medicine, a screening program for Tay-Sachs disease at the high school level has demonstrated that carrier testing of high school students is acceptable to the screened population and is an efficient screening procedure (Zeesman et al., 1984).

Sickle Cell Disease

Sickle cell disease occurs almost exclusively in blacks, among whom the carrier frequency is as high as 45 percent in some parts of Africa and about 8 percent in the United States.

Early attempts to screen the American black population for sickle cell trait were poorly accepted, partly because identification as a carrier was seen as stigmatization and partly because at that time no method of prenatal diagnosis was available. Efforts are now being made to reinstitute screening programs, supported by information and genetic counseling. As noted earlier (Chapter 5), prenatal diagnosis can now be offered to carrier couples.

Beta Thalassemia

Prevention of β thalassemia by heterozygote screening and prenatal diagnosis has already led to a precipitous drop in the incidence of the disease in Cyprus and Sardinia. In Quebec, a program for β thalassemia prevention (providing educational information and reproductive counseling as well as heterozygote screening) in the community at large and among high school students of Italian and Greek ethnic background has proved to be acceptable, efficient and cost-effective (Scriver et al., 1984).

The success of carrier screening programs for Tay-Sachs disease and β thalassemia and the relative failure of early attempts at screening for sickle cell trait indicate that community consultation, availability of genetic counseling and acceptable methods of prenatal diagnosis are requirements that must be fulfilled if any such program is to be effective.

Screening for Maternal Serum Alpha-fetoprotein (MSAFP)

The usefulness of maternal serum α-fetoprotein screening in the detection of neural tube defects has been discussed in previous chapters. A second possible benefit of MSAFP screening is that unusually low values may indicate that the fetus has Down syndrome (Cuckle et al., 1984).

Other Possible Screening Programs

It is theoretically possible to screen adult populations to identify members who are at risk for one of several kinds of disorders that could be forestalled by diet, by avoidance of smoking, or by other health-promoting measures that are matters of life style rather than of medicine. For example, population screening for hyperlipoproteinemia would allow identification of people at high risk of coronary artery disease, and population screening for α_1-antitrypsin deficiency would identify not only homozygotes susceptible to severe lung and liver disease, but also heterozygotes whose lung function might be impaired by smoking or other environmental hazards. Public health officials who promote health programs aimed at the whole population might have better success if they were able to concentrate on the identification and persuasion of high-risk groups. This is a consideration for the future.

LEGAL ASPECTS

In a legal case that has attracted considerable interest, parents who had been tested because they were Ashkenazi Jews and might be carriers of Tay-Sachs disease were informed that they were both noncarriers, but later had a Tay-Sachs child. The testing laboratory had previously been warned that its technical methods were inaccurate. In this case, the court recognized the child's claim for "wrongful life," a claim by the child that it should never have been allowed to be born.

Several other cases involving Tay-Sachs disease and thalassemia have led to malpractice suits (Shaw, 1984). It is clear that according to established standards of medical care, at least in the United States, a physician may now be considered negligent if he or she fails to take a history including ethnic and family information, fails to advise the parents of their risk and fails to offer prenatal diagnosis when it is possible. It is also clear that laboratories performing screening tests are considered negligent if a test result is shown to be a false negative.

THE NEGATIVE IMPLICATIONS OF SCREENING PROGRAMS

The objective of genetic screening is to improve the public health; the insidious aspects that have since been pointed out by lawyers and philosophers come as something of a surprise to those who had instituted the programs for what they saw as altruistic reasons. Invasion of privacy is regarded by some as a major problem. Concern is also expressed about stigmatization of individuals as abnormal on the basis of the tests; about whether informed consent is obtained, or whether compulsion, overt or implied, is exercised; about "the right not to know" of one's inborn handicaps; about whether the data will be assembled in data banks that are not totally confidential; and so on. These considerations must enter into the planning of future screening programs.

Concluding Comments

Medical genetics already has many useful applications to clinical practice, and promises to become even more important as the new genetics, based on analysis of disease at the molecular level, moves out of the research laboratories and into the mainstream of health services. Molecular genetics has the potential to provide fresh information on the molecular basis of many diseases, on sophisticated methods of prenatal diagnosis and eventually, perhaps, even on gene therapy.

There are many problems to be solved if the remarkable research advances of the last few years are to contribute appropriately to the management and, especially, to the prevention of genetic disease. In genetics, perhaps more than in most other areas of medicine, there is a broad spectrum of knowledge, ranging from molecular biology to clinical application. Medical research is becoming more and more complex, and very few people are competent in both basic research and clinical application (Littlefield, 1984). Ways of bridging the gap must be found.

The ethical issues surrounding genetic disease have created much interest and discussion. A major topic of concern has been prenatal diagnosis, with its implicit link to selective abortion. The philosophical question of what is to be regarded as "normal" also arises, and there is concern about where society's apparent intolerance of its abnormal members might lead. There is a basic dilemma: how to manage to work vigorously toward the prevention of genetic diseases, and at the same time to provide humane supportive treatment to defective persons and their families (Callahan, 1983).

In spite of all that genetic research and its clinical applications can provide, genetic disorders will not disappear, since recurrent mutation will continue to introduce new defective genes. However, in time the incidence of many genetic disorders could be strikingly reduced. Those most likely to become rare are, of course, the late-onset autosomal dominants (such as Huntington disease), the recessives for which heterozygotes can be detected by screening, and those that can be identified prenatally. Disorders in which a high proportion of the patients are new mutants, such as Duchenne muscular dystrophy and achondroplasia, are less likely to show a large drop in frequency.

Genetic counseling, screening and prenatal diagnosis are important to the practice of genetics, but are by no means the whole of it. Because genes interact with environment, a large part of the task of medical genetics is to find ways of manipulating the environment in order to allow individuals with defective genes to be viable. Will this lead to "genetic pollution"? This is a hard question and one that will have to be resolved, but it is not a new issue; medical ethics has always been on the side of the welfare of the individual if it conflicts with the welfare of society as a whole. The application of genetic knowledge to the improvement of human health is the ultimate goal of genetics in medicine.

GENERAL REFERENCES

Childs B, Simopoulos AP, eds. Genetic screening: programs, principles and research. Washington: National Academy of Sciences, 1975.

Emery AEH, Pullen I, eds. Psychological aspects of genetic counselling. London: Academic Press, 1984.

Fuhrmann W, Vogel F. Genetic counseling, 3rd ed. New York: Springer-Verlag, 1983.

Harper PS. Practical genetic counseling, 2nd ed. Bristol: Wright and Littleton: Wright, PSG, 1984.

Kelly P. Dealing with dilemma: a manual for genetic counselors. New York: Springer-Verlag, 1977.

Lubs HA, De La Cruz F, eds. Genetic counseling. New York: Raven Press, 1977.

Murphy EA, Chase GA. Principles of genetic counseling. Baltimore: Johns Hopkins University Press, 1975.

Stevenson AC, Davison BCC. Genetic counseling, 2nd ed. London: Heinemann, 1976.

PROBLEMS

General Review, Part II

1. Assume that a neonatal serum screening test for cystic fibrosis (CF) is developed. The test identifies 60 percent of all CF patients but also gives abnormal results in 30 percent of controls. The frequency of CF is approximately 1 in 2000 births.
   a) Determine the specificity and sensitivity of the test. What would be the predictive value of an abnormal test in a child in the general population? In a sib of a CF child?
   b) In what way(s) does this test fail to meet the standards for an acceptable neonatal screening test?

2. You are asked to see an infant with multiple malformations and failure to thrive. You arrange for chromosome analysis. The laboratory reports that one chromosome 18 has a deletion of the long arm at band 21.
   a) Using Figure 2–3 as a guide, sketch this abnormal chromosome.

      Your next step is to arrange for chromosome analysis of both parents.

      The laboratory report states that the mother's karyotype is 46,XX, but the father's has a reciprocal translocation between band q21 of chromosome 18 and band q23 of chromosome 8.
   b) Sketch the father's derivative chromosomes.
   c) Which of his chromosomes, 8, der (8), 18 and der (18), did the father transmit to the baby?
   d) Is the father's karyotype balanced or unbalanced?
   e) Is the baby's karyotype balanced or unbalanced?

f) Why is the baby abnormal?
g) What genetic counseling would you provide?

3. Beta thalassemia is an extremely heterogeneous condition, in which β globin chain production may be absent (β^0 thalassemia) or reduced ($\beta+$ thalassemia). Which type of thalassemia would you expect to result from each of the following mechanisms?
   a) Deletion of a large segment of the β gene.
   b) Frameshift mutation or nonsense mutation within the gene that leads to premature termination of a globin chain.
   c) Missense mutation in promoter region upstream of the protein coding sequences.
   d) Mutation in splice region that results in abnormal processing of primary RNA transcript.
   e) Heterozygosity for a β thalassemia allele.

4. You review a grant application in which the researcher plans to study the incidence of Down syndrome at birth in one of the following ways:
   1) Review of the number of cases karyotyped per year at a major pediatric hospital.
   2) Review of number of cases reported in physicians' birth notifications.
   3) Review of number of cases identified by prenatal diagnosis.
   4) Analysis of chromosomes of 55,000 newborns.
   Briefly comment on each proposed method of ascertainment.

5. Name two disorders that are congenital but not genetic, two that are genetic but not congenital, and two that are both.

6. A curiosity that has not been mentioned previously in this book is the type of disorder that is thought to be caused by mutation in a mitochondrial gene. The mitochondrial chromosome is circular, and contains 16,569 base pairs. It has been completely sequenced (Anderson et al., 1981). Though some enzymes functional in mitochondria are coded in nuclear DNA, several respiratory enzymes of mitochondria are known to be coded in the mitochondrial DNA itself.

   Almost all mitochondria are maternally inherited, passing from mother to child in the cytoplasm of the ovum. Two disorders, Leber optic atrophy and mitochondrial cytopathy, may be transmitted by mitochondria.

   Leber optic atrophy is a disorder in which the pattern of transmission is not Mendelian. Male-to-male transmission probably does not occur, but all sisters of affected males seem to be carriers. In mitochondrial cytopathy (a generalized disorder, probably heterogeneous, in which mitochondrial abnormality is a prominent feature), exclusively maternal inheritance has been described in a majority of families (Egger and Wilson, 1983).
   a) Sketch a pedigree with 3 generations and 10 or more members, showing the features to be expected in inheritance of a mitochondrial disease.
   b) Why could some families with mitochondrial cytopathy show Mendelian transmission while others do not?

7. Many but not all people, after eating asparagus, detect a pungent odor in the urine. It had been thought that excretion of an abnormal urinary component by some individuals but not others after eating asparagus was a polymorphic hereditary trait, "excretor" being dominant to "nonexcretor." However, the polymorphism now seems not to be in excretion of the odoriferous substance but in ability to smell it (Lison et al., 1980).

   In this study, Lison and colleagues found that 30 of 307 persons tested were "smellers" who could identify the characteristic odor even at very low concentrations, whereas the remainder were "nonsmellers" who could detect it only at high concentrations, if at all.

   You undertake a family study to determine whether this polymorphism can be explained on a single-gene basis and, if so, which allele is recessive. Here are your findings.

Parents	Children	
	Smellers	*Nonsmellers*
Smeller × smeller	7	2
Smeller × nonsmeller	80	85
Nonsmeller × nonsmeller	–	812

What are your conclusions about the inheritance of this polymorphism? What are the (approximate) frequencies of the two alleles and of the three genotypes?

8. Hemophilia A is thought to have a fitness (f) of 0.70; in other words, hemophiliacs have 70 percent of the normal number of offspring.

   Emery (1976) notes that the incidence of hemophilia in males (I) equals the chance of receiving a new mutation from one of the two parents (2μ) + half the frequency of heterozygous women ($\frac{1}{2}H$), and the incidence of carrier females is $2\mu + \frac{1}{2}H$ + the chance of inheriting the gene from an affected father, which is the incidence of affected males multiplied by the fitness factor. In brief:

$$I = \mu + \tfrac{1}{2}H$$
$$H = 2\mu + \tfrac{1}{2}H + If$$

   a) For hemophilia A, what is the population incidence of carrier females? of affected males? (Answer in terms of multiples of the mutation rate).
   b) For Duchenne muscular dystrophy (or any other genetically lethal X-linked trait), what is the population incidence of carrier females? Of affected males?
   c) As far as is known, the X-linked blood types Xg(a+) and Xg(a−) each have a fitness of 1. What are the relative proportions of heterozygous women and Xg(a−) men, if Xg(a−) men have a population incidence of 0.33? (See data in Chapter 15.)
   d) Looking back at (a) above, if a woman has a son with hemophilia A, what is the relative risk that she is a carrier? If she is a carrier, what is the chance that the gene is a new mutation? That she inherited it from her mother? That she inherited it from an affected father?

9. A problem in genetic counseling is that, through no fault of their own, people seeking counseling are often poorly informed about genetics. How would you respond to each of the following (all genuine questions, comments or requests, directed to medical geneticists at one major genetic clinic)?
   a) Are the chromosomes fully developed in a 16-week fetus?
   b) My doctor would like me to have my genes counted.
   c) I'm not worried about this pregnancy because the first-born is never affected in our family.
   d) Genetic disease? Isn't that like V.D. and those things?
   e) Why do you say my baby's disorder is dominant? My physician says it must be very recessive, as he has seen only one similar case.
   f) My sister and I are both carriers (of Duchenne muscular dystrophy), but her test result was higher than mine, so she is more of a carrier than I am.
   g) Will I get to be less of a carrier as I get older?
   h) Will you please send me all the information you have on genetics?

10. Physicians also make errors in interpretation of medical genetics. The following are actual comments from letters referring patients or families for genetic assessment and management. Have you any comments on these comments?
    a) Testing for carriers of Tay-Sachs disease is nonsense. I've never seen a case in my life. Why don't they test for carriers of Down syndrome instead?
    b) Familial polyposis of the colon (an autosomal dominant condition), if not treated surgically, will progress eventually to malignancy. If I have a patient who is an affected 30-year-old male, I recommend postponement of surgery until he has had a chance to reproduce, since the operation carries a 30 percent risk of impotence.
    c) I don't know whether this patient with osteogenesis imperfecta is a new mutant, or whether he has a dominant gene that he will transmit to his children.
    d) I'd like to refer a little boy with Hurler syndrome. I know it can't be Hunter syndrome because he isn't dwarfed.
    e) A report from the Chromosome Lab says the child has trisomy 21. At this stage they are unable to say whether there is any extra problem such as deletion of chromosome 21.
    f) It is my impression that Huntington disease can skip one or two generations.
    g) I suspect that thalassemia is much like the Rh problem and the baby's risk will be dependent on whether her husband is homozygotic or heterozygotic.
    h) Certainly if she is carrying a gene which is lethal to male fetuses she has a 50 percent chance of producing a healthy female child, and if she can at least be encouraged to keep trying she will eventually have a family of girls.

GLOSSARY

Acentric A chromosome fragment lacking a centromere.

Acrocentric A chromosome with the centromere near one end. In man the acrocentric chromosomes have satellited short arms that carry genes for ribosomal RNA.

Allele Alleles are alternative forms of a gene. If there are more than two alleles at a given locus, they are called multiple alleles or members of an allelic series.

Allograft A graft in which donor and host are members of the same species but not genetically identical.

Amino acids The building blocks of protein, for which DNA carries the genetic code. The amino acids are listed in Table 3–1.

Amniocentesis Needle puncture of the uterus and amniotic cavity through the abdominal wall of a pregnant woman to obtain amniotic fluid.

Amplification The production of many copies of a region of DNA.

Anaphase The phase of mitosis or meiosis at which the chromosomes leave the equatorial plate and pass to the poles of the cell.

Aneuploid A chromosome number that is not an exact multiple of the haploid number, or an individual with an aneuploid chromosome number.

Antibody An immunoglobulin molecule, formed by immune-competent cells in response to an antigen stimulus, that reacts specifically with that antigen.

Anticipation The term used to denote the progressively earlier appearance and increased severity of a disease in successive generations. It is thought to result from bias of ascertainment.

Antigen Any macromolecule that can elicit antibody formation from immune-competent cells and react specifically with the antibody so produced.

Ascertainment The method of selection of individuals for inclusion in a genetic study.

Association The occurrence together in a population of two or more phenotypic characteristics more often than expected by chance. Not to be confused with linkage.

Assortative mating Selection of a mate with preference for a particular genotype, that is, nonrandom mating. May be positive (preference for a mate of the same genotype) or negative (preference for a mate of a different genotype).

Assortment The random distribution of different combinations of the parental chromosomes to the gametes. Nonallelic genes assort independently to the gametes, unless they are linked.

Autograft A graft of the host's own tissue.

Autoimmunity The ability to form antibodies against one's own antigens, resulting in autoimmune disease.

Autoradiography A technique for detecting radioactively labeled molecules (by their effect of creating an image on photographic films) in a cell or tissue.

Autosome Any chromosome other than the sex chromosomes. Man has 22 pairs of autosomes.

B cells Small lymphocytes that respond to antigenic stimulation by producing humoral antibodies.

Backcross Mating of a heterozygote to a recessive homozygote (*Aa* × *aa*), in which the progeny (½ *Aa*, ½ *aa*) reveal the genotype of the heterozygous parent. The double backcross mating (*AaBb* × *aabb*) is particularly useful for linkage analysis in family studies.

Bacteriophage A virus that infects bacteria.

Banding A technique of staining chromosomes in a characteristic pattern of cross bands. See G bands, Q bands.

Barr body The sex chromatin as seen in female somatic cells. Named for Murray Barr, who with his student E. G. Bertram first described sexual dimorphism in somatic cells.

Base analogue A chemical that may act as a mutagen because its molecular structure mimics that of a DNA base.

Base pair (bp) In nucleic acids, adenine must always pair with thymine (or, in RNA, with uracil) and guanine with cytosine. The specificity of base pairing is fundamental to DNA replication and to its transcription into RNA.

Bivalent A pair of homologous chromosomes in association as seen at metaphase of the first meiotic division.

Blood group A genetically determined antigen on a red cell. The antigens formed by a set of allelic genes make up a blood group system.

Burden In medical genetics the burden is the total impact of a genetic disorder on the patient, his family and society as a whole.

Cancer A disease characterized by uncontrolled invasive cell multiplication.

Carcinogen An agent that induces cancer.

Carrier An individual who is heterozygous for a normal gene and an abnormal gene that is not expressed phenotypically, though it may be detectable by appropriate laboratory tests.

CAT box Refers to a DNA sequence of unknown function often found in eukaryotic genes about 70 to 80 bases upstream from the start of transcription.

cDNA Complementary DNA or copy DNA, synthetic DNA transcribed from a specific RNA through the action of the enzyme reverse transcriptase.

Centimorgan (cM) In linkage, a unit that represents 1 percent recombination. Also called a *map unit*.

Centriole One of the pair of organelles that form the points of focus of the spindle during cell division. The centrioles lie together outside the nuclear membrane at prophase and migrate to opposite poles of the cell during cell division.

Centromere The heterochromatic region within a chromosome by which the chromatids are held together and to which the kinetochore is attached. Also called the *primary constriction*.

Chiasma Literally, a chiasma is a cross. The term refers to the crossing of chromatid strands of homologous chromosomes, seen at diplotene of the first meiotic division. Chiasmata are evidence of interchange of chromosomal material (crossovers) between members of a chromosome pair.

Chimera An individual composed of cells derived from different zygotes. In human genetics the term is used especially with reference to blood chimerism, which results from exchange of hematopoietic stem cells in utero by dizygotic twins, who then continue to form blood cells of both types. Dispermic chimerism, in which two separate zygotes are fused into one individual, is much rarer. Not to be confused with mosaicism.

Chorionic villous sampling (CVS) A procedure used for prenatal diagnosis at 9 to 12 weeks' gestation. Fetal tissue for analysis is aspirated by catheter through the cervix from the villous area of the chorion (chorion frondosum), under ultrasonic guidance.

Chromatid A chromosome in a dividing cell can be seen to consist of two parallel strands held together at the centromere. Each strand is a chromatid. A chromosome replicates during the DNA synthesis stage of the cell cycle, and is then composed of two chromatids until the next mitotic division, at which each chromatid becomes a chromosome of a daughter cell.

Chromatin The nucleoprotein fibers of which chromosomes are composed.

Chromomere A densely coiled region of chromatin on a chromosome. Chromomeres give the extended chromosome a beaded appearance, especially obvious at meiotic prophase.

Chromosomal aberration An abnormality of chromosome number or structure.

Chromosomal satellite A small mass of chromatin at the end of the short arm of each chromatid of an acrocentric chromosome. Not to be confused with *satellite DNA*.

Clone A cell line derived by mitosis from a single ancestral diploid cell.

Codominant If both alleles of a pair are expressed in the heterozygote, the alleles (or the traits determined by them) are codominant.

Codon A triplet of three bases in a DNA or RNA molecule, specifying a single amino acid.

Coefficient of inbreeding (F) The probability that an individual has received both alleles of a pair from an identical ancestral source. Also, the proportion of loci at which he is homozygous by descent.

Coefficient of relationship The probability that two persons have inherited a certain gene from a common ancestor or the proportion of all their genes that have been inherited from common ancestors.

Colinearity Term used to describe the parallel relationship between the base sequence of the DNA of a gene (or the RNA transcribed from it) and the amino acid sequence of the corresponding polypeptide.

Complementation In genetics, the ability of two different genetic defects to correct for one another, thus demonstrating that the defects are not in the same gene. See text.

Compound A genotype in which the two alleles are different mutations from the wild type, or an individual who carries two different mutant alleles at a locus.

Concordant In human genetics, a twin pair in which both members exhibit a certain trait.

Congenital Present at birth; not necessarily genetic.

Consanguinity Relationship by descent from a common ancestor.

Consultand The individual referred for genetic counseling.

Crossover or crossing over Exchange of genetic material between members of a chromosome pair. The chiasmata seen at diplotene are the physical evidence of crossing over.

Cytogenetics The study of the relationship of the microscopic appearance of chromosomes and their behavior during cell division to the genotype and phenotype of the individual.

Cytoplasmic inheritance Inheritance through genes present in cytoplasmic organelles such as mitochondria.

Deletion Usually, a chromosomal aberration in which a portion of a chromosome is lost. May also refer to loss of any DNA segment, as in some thalassemia genes.

Denaturation Conversion of DNA from the double-stranded to the single-stranded state, usually accomplished by heating to destroy chemical bonds involved in base pairing.

Deoxyribonucleic acid See *DNA*.

Dermatoglyphics The patterns of the ridged skin of the palms, soles and digits.

Dicentric A structurally abnormal chromosome with two centromeres.

Dictyotene The stage of the first meiotic division in which a human oocyte remains from late fetal life until ovulation.

Diploid The number of chromosomes in most somatic cells, which is double the number found in the gametes. In man, the diploid chromosome number is 46.

Discordant In human genetics, a twin pair in which one member shows a certain trait and the other does not. See *concordant*.

Dispermy Fertilization of a duplicated egg nucleus, or egg and polar body, by two sperm. See *chimera*.

Divergence Change in DNA sequence between two related genes or changes in amino acid sequence between two related proteins due to mutation. The accumulated differences may be expressed as a percentage of the entire sequence.

Dizygotic twins Twins produced by two separate ova, separately fertilized. Fraternal twins.

DNA (Deoxyribonucleic acid) The nucleic acid of the chromosomes. See text.

Domain A region of the amino acid sequence of a protein that can be equated with a particular function, or a corresponding segment of a gene.

Dominant A gene is dominant if its expression is the same in heterozygotes as in homozygotes. Modifications of this definition are discussed in the text.

Dosage compensation Refers to the fact that the amount of product of an X-linked gene is equivalent whether one X chromosome or two (or, rarely, more than two) are present.

Duplication Presence of a segment of a chromosome in duplicate. Duplication may involve whole genes, series of genes or only part of a gene.

Dysmorphism Morphological developmental abnormality, as seen in many syndromes of genetic or environmental etiology.

Empiric risk Probability that a trait will occur or recur in a family based on past experience rather than on knowledge of the causative mechanism.

Endonuclease An enzyme that cleaves bonds within a DNA or RNA strand.

Eukaryote A unicellular or multicellular organism in which the cells have a nucleus with a nuclear membrane and other specialized characteristics. See *prokaryote*. For further discussion, see text.

Exon Any segment of an interrupted gene that is represented in the mature messenger RNA product and thus codes for protein.

Expressivity The extent to which a genetic defect is expressed. If there is variable expressivity, the trait may vary in expression from mild to severe but is never completely unexpressed in individuals who have the corresponding genotype.

F See *coefficient of inbreeding*.

F_1 ("F one") The first-generation progeny of a mating.

Familial Refers to any trait that is more common in relatives of an affected individual than in the general population.

Fetoscopy A technique for direct visualization of the fetus, used in prenatal diagnosis.

Fingerprint 1. The pattern of the ridged skin on the distal phalanx of a finger. 2. A method of combining electrophoresis and chromatography to separate the components of a protein such as hemoglobin.

Fitness The probability of transmitting one's genes to the next generation and having them survive in that generation to be passed on to the next, relative to the average probability for the population.

Flanking sequence A region of a gene preceding or following the transcribed region.

Forme fruste A mild expression of a genetic trait, of no clinical significance.

Founder effect A high frequency of a mutant gene in a population founded by a small ancestral group when one or more of the founders was a carrier of the mutant gene.

Frameshift mutation A mutation involving a deletion or insertion that is not an exact multiple of 3 bp and therefore changes the reading frame of the gene.

G bands The dark and light cross-bands seen in chromosomes after treatment with trypsin and Giemsa stain.

Gamete A reproductive cell (ovum or sperm) with the haploid chromosome number.

Gene A segment of a DNA molecule involved in the synthesis of a polypeptide chain or RNA molecule and carrying the code for that polypeptide or RNA. See text.

Gene flow Gradual diffusion of genes from one population to another, by migration and mating.

Gene map A representation of the human karyotype showing the chromosomal locations of the genes that have been mapped. See text.

Gene pool All the genes present at a given locus in the population.

Genes in common Those genes inherited by two individuals from a common ancestral source.

Genetic Determined by genes. Not to be confused with *congenital.*

Genetic code The base triplets that specify the 20 different amino acids. See Table 3–1.

Genetic counseling Provision of information to patients or relatives at risk of a disorder that may be genetic, concerning the consequences of the disorder, the probability of developing and/or transmitting it, and ways in which it may be prevented or ameliorated.

Genetic death Failure of a mutation to be passed on to the next generation because of its damaging phenotypic effects.

Genetic drift Random fluctuation of gene frequencies in small populations.

Genetic lethal A gene or genetically determined trait that leads to failure to reproduce in affected individuals.

Genetic load The sum total of death and defect caused by mutant genes.

Genetic marker A trait can be used as a genetic marker in studies of cell lines, individuals, families and populations if it is genetically determined, can be accurately classified, has a simple unequivocal pattern of inheritance and has heritable variations common enough to allow it to be classified as genetic polymorphism.

Genetic screening Testing on a population basis to identify individuals at risk of having a specific disorder, or of producing a child with a specific genetic disorder. Screening tests normally are applied only when some method of treatment or intervention is available.

Genome The full complement of DNA, either haploid or diploid.

Genotype 1. The genetic constitution (genome). 2. More specifically, the alleles present at one locus.

Germ line The cell line from which gametes are derived.

Haploid The chromosome number of a normal gamete, with only one member of each chromosome pair. In humans, the haploid number is 23.

Haplotype 1. A group of alleles from closely linked loci, usually inherited as a unit. Example: the HLA complex. 2. A set of restriction fragment sites closely linked to one another and to a gene of interest.

Hardy-Weinberg law The law that relates gene frequency to genotype frequency. See text.

Hemizygous A term that applies to the genes on the X chromosome in a male. Since males have only one X, they are hemizygous, not homozygous or heterozygous, with respect to X-lined genes.

Heritability A statistical measure of the degree to which a trait is genetically determined. See text.

Heterogeneity If a phenotype (or very similar phenotypes) can be produced by different genetic mechanisms, the phenotype is genetically heterogeneous.

Heterokaryon A cell with two separate nuclei, formed by fusion of two genetically different cells.

Heteromorphism A normal variant of a chromosome; sometimes called *polymorphism.*

Heteroploid Any chromosome number other than the normal.

Heterozygote An individual who has two different alleles, one of which is a normal allele, at a given locus on a pair of homologous chromosomes. May also refer to a carrier of a balanced chromosome rearrangement. Adjective: heterozygous.

Histocompatibility A host will accept a particular graft only if it is histocompatible, that is, if the graft contains no antigens that the host lacks.

Histones Proteins associated with DNA in the chromosomes, rich in basic amino acids (lysine or arginine) and virtually invariant throughout eukaryote evolution.

Holandric Pattern of inheritance of genes on the Y chromosomes from a father to all his sons but none of his daughters.

Homologous chromosomes A pair of chromosomes, one from each parent, having the same gene loci in the same order.

Homozygote An individual possessing a pair of identical alleles at a given locus on a pair of homologous chromosomes. Adjective: homozygous.

Housekeeping genes Genes (theoretically) expressed in all cells because they provide basic functions.

Hybrid cell A cell formed by fusion of two cells of different origin in which the two nuclei have merged into one. Can be cloned to produce hybrid cell lines.

Hybridization In molecular genetics, pairing of an RNA and a DNA strand or of two different DNA strands.

Hybridoma Cells formed by fusion of mutant myeloma cells and cells of another type, such as spleen cells from an immunized mouse. The hybrid cells grow in culture and secrete a single antibody specific for the antigen used to immunize the mouse. Such monoclonal antibodies have many scientific uses.

Immune reaction The specific reaction between antigen and antibody.

Immunological homeostasis The characteristic condition of a normal adult, who has certain antigens and the ability to react to antigens by producing antibodies, but does not produce antibodies in response to his own antigens.

Immunological tolerance Inability to respond to a specific antigen, resulting from previous exposure to that antigen, especially during embryonic life.

Inborn error A genetically determined biochemical disorder in which a specific enzyme defect produces a metabolic block that may have pathological consequences.

Inbreeding The mating of closely related individuals. The progeny of close relatives are said to be inbred.

Index case See *proband*.

Insertion A structural chromosomal aberration in which part of an arm of one chromosome is inserted into the arm of a nonhomologous chromosome.

Interphase The part of the cell cycle between two successive cell divisions.

Intervening sequence See *intron*.

Intron A segment of a gene that is initially transcribed but is then removed from within the primary transcript by splicing together the sequences on either side of it.

Inversion A chromosomal aberration in which a segment of a chromosome is reversed end to end. See *paracentric inversion* and *pericentric inversion* in text.

Isochromosome An abnormal chromosome with duplication of one arm (forming two arms of equal length that bear the same loci in reverse sequence), and with deletion of the other arm of the normal chromosome.

Isograft A tissue graft between two individuals who have identical genotypes.

Isolate A subpopulation in which matings take place exclusively with other members of the same subpopulation.

Isolated Refers to an individual who is the only affected member of his or her family, either by chance or through new mutation.

Isozymes Multiple molecular forms of an enzyme, which may be in the same or different tissues within the same individual.

Karyotype The chromosome set. The term is also used for a photomicrograph of the chromosomes of an individual, arranged in the standard classification, or for the process of preparing such a photomicrograph.

kb (kilobase) A unit of 1000 bases in DNA or RNA.

Kindred An extended family.

Kinetochore A structure beside the centromere to which the spindle fibers are attached.

Lethal equivalent A gene carried in the heterozygous state which, if homozygous, would be lethal. Also, a combination of two genes in the heterozygous state each of which, if homozygous, would cause the death of 50 per cent of homozygotes; any equivalent combination.

Linkage Linked genes have their loci within measurable distance of one another on the same chromosome. See *synteny*.

Linkage disequilibrium The tendency of two linked alleles to occur together on the same chromosome more frequently than would be expected by chance.

Locus The position of a gene on a chromosome. Different forms of the gene (alleles) may occupy the locus.

Lyon hypothesis (Lyonization) See *X inactivation*.

Manifesting heterozygote A female heterozygous for an X-linked disorder in whom, because of X inactivation, the trait is expressed clinically with approximately the same degree of severity as in hemizygous affected males.

Map unit A measure of distance between two loci on a chromosome based on the percentage frequency of recombination between them; one map unit is equivalent to one centimorgan or one percent recombination.

Meiosis The special type of cell division occurring in the germ cells by which gametes containing the haploid chromosome number are produced from diploid cells. Two meiotic divisions occur, meiosis I and meiosis II. Reduction in number takes place during meiosis I. To be distinguished from *mitosis*.

Meiotic drive Term used for the situation in which, in heterozygotes, one allele is significantly more likely than the other to be transmitted to the progeny; that is, prezygotic selection is operating in favor of one allele at the expense of the other.

Messenger RNA (mRNA) An RNA transcribed from the DNA of a gene, forming the template from which a protein is translated.

Metaphase The stage of mitosis or meiosis in which the chromosomes have reached their maximum condensation and are lined up on the equatorial plane of the cell, attached to the spindle fibers. This is the stage at which chromosomes are most easily studied.

Missense mutation Mutation that changes a codon specific for one amino acid to specify another amino acid.

Mitochondrial DNA The DNA in the circular chromosome of the mitochondria, cytoplasmic organelles which possess their own unique DNA. Mitochondrial DNA is present in many copies per cell, is maternally inherited, and evolves 5 to 10 times as rapidly as genomic DNA.

Mitosis Somatic cell division resulting in the formation of two cells, each with the same chromosome complement as the parent cell. Distinguish from *meiosis*.

Monoclonal antibody See *hybridoma*.

Monosomy A condition in which one chromosome of a pair is missing, as in 45,X Turner syndrome. Partial monosomy may occur.

Monozygotic twins Twins derived from a single fertilized ovum. Identical twins.

Mosaic An individual or tissue with at least two cell lines differing in genotype or karyotype, derived from a single zygote. Distinguish from *chimera*.

Multifactorial Determined by multiple factors, genetic and possibly also nongenetic, each with only a minor effect. See also *polygenic*.

Multiplex Refers to a pedigree in which there is more than one case of a particular disease.

Mutagen An agent that increases the mutation rate by causing changes in DNA.

Mutant 1. A gene in which a mutation has occurred. 2. An individual carrying such gene.

Mutation Any permanent heritable change in the sequence of genomic DNA.

Usually defined as a change in a single gene (point mutation), although the term is sometimes used more broadly for a structural chromosomal change.

Mutation rate The rate at which mutations occur at a given locus, expressed as mutations per gamete per locus per generation.

Myeloma Tumor arising from a clone of plasma cells and producing a single antibody.

Nondisjunction The failure of two members of a chromosome pair to disjoin during anaphase of cell division, so that both pass to the same daughter cell.

Nonsense mutation Mutation to a chain-termination codon.

Nuclease. See *endonuclease*.

Nucleosome Primary repeating unit of DNA structure in chromatin. See text.

Nucleotide A molecule composed of a nitrogenous base, a 5-carbon sugar and a phosphate group. A nucleic acid is a polymer of many nucleotides.

Oncogene One of a small number of normal genes of vertebrates that have been preserved throughout evolution and show increased activity in various types of human cancer.

p 1. The short arm of a chromosome (from the French *petit*). 2. In population genetics, often used to indicate the frequency of the more common allele of a pair.

Palindrome A phrase that reads the same forward and backward. The codes of many restriction sites are palindromes.

Pedigree In medical genetics, a diagram of a family history indicating the family members, their relationship to the proband and their status with respect to a particular hereditary condition.

Penetrance The frequency of expression of a genotype. When it is less than 100 percent, the trait is said to exhibit reduced penetrance or lack of penetrance. In an individual who has a genotype that characteristically produces an abnormal phenotype but is phenotypically normal, the trait is said to be nonpenetrant.

Pharmacogenetics The area of biochemical genetics concerned with drug responses and their genetically controlled variations.

Phenocopy A mimic of a phenotype that is usually determined by a specific genotype, produced instead by the interaction of some environmental factor with a different genotype.

Phenotype The observable physical, biochemical and physiological characteristics of an individual, as determined by his genotype and the environment in which he develops. Also, in a more limited sense, the expression of some particular gene or genes.

Philadelphia chromosome (Ph^1) The structurally abnormal chromosome 22 typically occurring in a proportion of the bone marrow cells in most patients with chronic myelogenous leukemia. The abnormality is a reciprocal translocation between the distal portion of the long arm of chromosome 22 and the distal portion of the long arm of chromosome 9.

Pleiotropy If a single gene or gene pair produces multiple effects, it is said to exhibit pleiotropy (or to have pleiotropic effects).

Polygenic Determined by many genes at different loci, with small additive effects. Also termed quantitative. To be distinguished from *multifactorial*, in which environmental as well as genetic factors may be involved.

Polymorphism 1. The occurrence together in a population of two or more genetically determined alternative phenotypes, each with appreciable frequency. Arbitrarily, if the rarer allele has a frequency of at least 0.01, so that the heterozygote frequency is at least 0.02, the locus is considered to be polymorphic. 2. In molecular genetics, a restriction fragment length polymorphism (RFLP) is a polymorphism in DNA sequence that can be detected on the basis of differences in the length of DNA fragments produced by digestion with a specific restriction enzyme. See also *heteromorphism*.

Polypeptide A chain of amino acids, held together by peptide bonds between the amino group of one and the carboxyl group of an adjoining one. A protein molecule may be composed of a single polypeptide chain or of two or more identical or different polypeptides.

Polyploid Any multiple of the basic haploid chromosome number, other than the diploid number; thus $3n$, $4n$ and so forth.

Primary constriction Centromere of a chromosome.

Primary transcript The first RNA transcript of a gene, containing introns as well as exons.

Proband The family member through whom the family is ascertained. Also called index case (if affected) or propositus.

Probe In molecular genetics, a radioactive DNA or RNA sequence used to detect the presence of a complementary sequence by molecular hybridization.

Prokaryote A simple unicellular organism, such as a bacterium, lacking a separate nucleus and simpler than eukaryotic cells in other ways. See *eukaryote*.

Prophase The first stage of cell division, during which the chromosomes become visible as discrete structures and subsequently thicken and shorten. Prophase of the first meiotic division is further characterized by pairing (synapsis) of homologous chromosomes.

Propositus See *proband*.

Pseudogene An inactive gene within a gene family, derived by mutation of an ancestral active gene. An evolutionary relic.

q 1. The long arm of a chromosome. 2. In population genetics, often used to indicate the frequency of the rarer allele of a pair.

Q bands The pattern of bright and dim cross-bands seen on chromosomes under fluorescent light after quinacrine staining.

Quasicontinuous variation The type of variation shown by a multifactorial trait, determined by an underlying continuous distribution, that has a threshold effect and thus appears to be either present or absent, that is, discontinuous.

Quasidominant The pattern of inheritance produced by the mating of a recessive heterozygote, so that recessively affected members appear in two or more successive generations.

Random mating Selection of a mate without regard to the genotype of the mate. In a randomly mating population, the frequencies of the various matings are determined solely by the frequencies of the genes concerned.

Reading frame One of the three possible ways of reading a nucleotide sequence as a series of triplets. An *open reading frame* contains no termination codons, and thus is potentially translatable into protein.

Recessive A trait or gene that is expressed only in homozygotes. For further discussion, see text.

Recombinant An individual who has a new combination of genes not found together in either parent. Usually applied to linked genes.

Recombinant DNA Artificially synthesized DNA in which a gene or part of a gene from one organism is inserted into the genome of another.

Recombination The formation of new combinations of linked genes by crossing over between their loci.

Recurrence risk The probability that a genetic disorder present in one or more members of a family will recur in another member of the same or a subsequent generation.

Reduction division The first meiotic division, so called because at this stage the chromosome number per cell is reduced from diploid to haploid.

Restriction enzyme A nuclease that recognizes specific sequences in DNA and cleaves the DNA strand at those points. Used in recombinant DNA technology.

Restriction map A linear array of sites on DNA cleaved by various restriction enzymes.

Reverse transcriptase An enzyme that catalyzes the synthesis of DNA on an RNA template.

RFLP Restriction fragment length polymorphism. See *polymorphism*.

Ribonucleic acid See *RNA*.

Ribosomes Cytoplasmic organelles composed of ribosomal RNA and protein, on which polypeptide synthesis from messenger RNA occurs.

Ring chromosome A structurally abnormal chromosome in which the end of each arm has been deleted and the broken arms have reunited in ring formation.

RNA (ribonucleic acid) A nucleic acid formed upon a DNA template, containing ribose instead of deoxyribose. Messenger RNA (mRNA) is the template upon which polypeptides are synthesized. Transfer RNA (tRNA), in cooperation with the ribosomes, brings activated amino acids into position along the mRNA template. Ribosomal RNA (rRNA), a component of the ribosomes, functions as a nonspecific site of polypeptide synthesis. See text for further details.

Robertsonian translocation A translocation between two acrocentric chromosomes by fusion at the centromere and loss of the short arms.

Satellite DNA DNA containing many tandem repeats of a short basic repeating unit. Not to be confused with *chromosomal satellite*.

Secondary constriction A narrowed heterochromatic region of a chromosome. The secondary constrictions of satellited chromosomes contain genes coding for ribosomal RNA.

Segregation In genetics, the separation of allelic genes at meiosis. Since allelic genes occupy the same locus on homologous chromosomes, they pass to different gametes; that is, they segregate.

Selection In population genetics, the operation of forces that determine the relative fitness of a genotype in the population, thus affecting the frequency of the gene concerned.

Sensitivity In assessment of test results, the sensitivity of the test is the proportion of all affected individuals who have an abnormal test result.

Sex chromatin A chromatin mass in the nucleus of interphase cells of females of most mammalian species, including humans. It represents a single X chromosome that is inactive in the metabolism of the cell. Normal females have sex chromatin, and thus are chromatin-positive; normal males lack it, and thus are chromatin-negative. Synonym: *Barr body*.

Sex chromosomes Chromosomes responsible for sex determination. In humans, XX in female, XY in male.

Sex-influenced A trait that is not X-linked in its pattern of inheritance but is expressed differently, either in degree or in frequency, in males and females.

Sex-limited A trait that is expressed in only one sex though the gene determining it is not X-linked.

Sex-linked See *X linkage*.

Sib, sibling Brother or sister. In precise usage, a sibling is a younger brother or sister.

Silent gene A mutant gene that has no detectable phenotypic effect.

Simplex In human genetics, used to describe a family history with only one affected member.

Sister chromatid exchange Exchange of segments of DNA between sister chromatids. Occurs with particularly high frequency in patients with Bloom syndrome.

Solenoid A coil of wire wound around a hollow core. Used in cytogenetics to describe the coiled structure into which nucleosomes are wound during chromatin condensation.

Somatic cell genetics The study of genetic phenomena in cultured somatic cells.

Somatic mutation A mutation occurring in a somatic cell rather than in the germ line.

Southern blot A technique for transferring DNA fragments separated by agarose gel electrophoresis to a nitrocellulose filter, on which specific DNA fragments can then be detected by their hybridization to radioactive probes. Devised by Edwin Southern.

Specificity In assessment of test results, the specificity of the test is the proportion of all normal individuals who have normal test results.

Spindle Intracellular microtubules involved in the organization of the chromosomes on the metaphase plate and their segregation at anaphase.

Structural gene A gene coding for any RNA or protein product other than a regulator.

Synapsis Close pairing of homologous chromosomes in prophase of the first meiotic division.

Syndrome A characteristic overall pattern of anomalies, presumed to be causally related.

Synteny Presence together on the same chromosome of two or more gene loci, whether or not they are close enough together for linkage to be demonstrated. Adjective: syntenic.

T cells Small lymphocytes committed by the influence of the thymus gland to be responsible for cell-mediated response to antigens.

Tandem repeats Multiple copies of the same DNA sequence arranged in direct succession along a chromosome.

TATA box A conserved DNA sequence about 25 bp upstream from the startpoint of the coding region of genes, apparently involved in the initiation of transcription.

Telophase The stage of cell division that begins when the daughter chromosomes reach the poles of the dividing cell and lasts until the two daughter cells take on the appearance of interphase cells.

Teratogen An agent that produces or raises the incidence of congenital malformations.

Termination codon One of the three codons UAG, UAA or UGA that cause termination of protein synthesis. Also called a nonsense codon.

Transcription The synthesis of RNA on a DNA template.

Translation The synthesis of a polypeptide with an amino acid sequence derived from the codon sequence of a corresponding messenger RNA. See text for details.

Translocation The transfer of a segment of one chromosome to another chromosome. If two nonhomologous chromosomes exchange pieces, the translocation is reciprocal. See also *Robertsonian translocation*.

Triplet In molecular genetics, a unit of three DNA or RNA bases, coding for a specific amino acid. A codon.

Triploid A cell with three of each chromosome, or an individual made up of such cells.

Triradius In dermatoglyphics, a point from which the dermal ridges course in three directions at angles of approximately 120°.

Trisomy The state of having three representatives of a given chromosome instead of the usual pair, as in trisomy 21 (Down syndrome).

Ultrasonography A technique in which high-frequency sound waves are used to outline internal body structures.

Vector In cloning, the plasmid or phage used to carry the cloned DNA segment.

Wild type Term used especially in experimental genetics to indicate the normal allele (often symbolized as +) or the normal phenotype.

Xenograft Graft from a donor of one species to a host of a different species.

X inactivation Inactivation of genes on one X chromosome in early embryonic life in somatic cells of female mammals. See text.

X linkage Genes on the X chromosome, or traits determined by such genes, are X-linked.

Zygosity Twins may be either monozygotic (MZ) or dizygotic (DZ). To determine whether a certain twin pair is MZ or DZ is to determine their zygosity.

REFERENCES

Aird I, Bentall HH, Roberts JAF. A relationship between cancer of the stomach and the ABO blood groups. Br Med J 1953; 1:799–801.
Allderdice PW, Browne N, Murphy DP. Chromosome 3 duplication q21 → qter deletion p25 → pter syndrome in children of carriers of a pericentric inversion inv (3) (p25q21). Am J Hum Genet 1975; 27:699–718.
Ames BN, Durston WE, Yamasaki E, Lee FD. Carcinogens are mutagens: a simple test system combining liver homogenates for activation and bacteria for detection. Proc Natl Acad Sci USA 1973; 70:2281–2285.
Anderson S, Bankier AJ, Barrel BG, et al. Sequence and organization of the human mitochondrial genome. Nature 1981; 290:457–465.
Angell RR, Aitken RJ, vanLook PFA, et al. Chromosome abnormalities in human embryos after *in vitro* fertilization. Nature 1981; 303:336–338.
Auerbach C. A possible case of delayed mutation in man. Ann Hum Genet 1956; 20:266–269.
Baird PA, McGillivray B. Children of incest. J Pediatr 1982; 101:854–857.
Bakker E, Hofker MH, Goor N, et al. Prenatal diagnosis and carrier detection of Duchenne muscular dystrophy with closely linked RFLPs. Lancet 1985; 1:655–658.
Bateson W, Punnett RC. Experimental studies in the physiology of heredity. Reports 2, 3, 4 to the Evolution Committee of the Royal Society. Reprinted in Peters JA, ed. Classic papers in genetics. Englewood Cliffs, New Jersey: Prentice-Hall, 1959.
Beadle GW, Tatum EL. Genetic control of biochemical reactions in Neurospora. Proc Natl Acad Sci USA 1941; 27:499–506.
Benirschke K. Origin and clinical significance of twinning. Clin Obstet Gynecol 1972; 15:220–235.
Bodmer WF, Cavalli-Sforza LL. Genetics, evolution, and man. San Francisco: W.H. Freeman, 1976.
Bonaiti-Pellié C, Smith C. Risk tables for genetic counselling in some common congenital malformations. J Med Genet 1974; 11:374–377.
Botstein D, White RL, Skolnick M, Davis RW. Construction of a genetic linkage map using restriction fragment length polymorphisms. Am J Hum Genet 1980; 32:314–331.
Brock DJH, Sutcliffe RG. Alpha-fetoprotein in the antenatal diagnosis of anencephaly and spina bifida. Lancet 1972; 2:197–199.
Brown, MS, Goldstein JL. Familial hypercholesterolemia: defective binding of lipoproteins to cultured fibroblasts associated with impaired regulation of 3-hydroxy-3-methylglutaryl coenzyme A reductase activity. Proc Natl Acad Sci USA 1974; 71:788–792.
Byers PH, Bonadio JF. The molecular basis of clinical heterogeneity in osteogenesis imperfecta: mutations in type 1 collagen genes have different effects on collagen processing. In Butterworth's International Medical Reviews, Pediatrics, Vol. 5. Genetic and metabolic diseases, Scriver CR, Lloyd J, eds. London: Butterworths, 1985.
Callahan D. The meaning and significance of genetic disease. Philosophical perspectives. In Silber T, ed. Ethical issues in the treatment of children and adolescents. Thorofare, New Jersey: Slack Inc, 1983.
Carr DH. Chromosomes and abortion. Adv Hum Genet 1971; 2:201–258.
Carr DH, Gedeon M. Population genetics of human abortuses. In Hook EB, Porter IH, eds. Population cytogenetics: studies in humans. New York: Academic Press, 1977.
Carter CO. Genetics of common disorders. Br Med Bull 1969; 25:52–57.
Carter CO. Genetics of common single malformations. Br Med Bull 1976; 32:21–26.
Cavenee WK, Dryja TP, Phillips RA, et al. Expression of recessive alleles by chromosomal mechanisms in retinoblastoma. Nature 1983; 305:779–784.
Childs B. Genetics in the medical curriculum. Am J Med Genet 1982; 13:319–324.
Clarke CA, Donohoe WTA, McConnell RB, et al. Further experimental studies on the prevention of Rh haemolytic disease. Br Med J 1963; 1:979–984.
Clow CL, Scriver CR. Knowledge about and attitudes toward genetic screening among high school students: the Tay-Sachs experience. Pediatrics 1977; 59:86–91.
Comings DE. Mechanisms of chromosome banding and implications for chromosome structure. Ann Rev Genet 1978; 12:25–46.
Cori GT, Cori CF. Glucose-6-phosphatase of liver in glycogen storage disease. J Biol Chem 1952; 199:661–667.
Corney G, Robson EB. Types of twinning and determination of zygosity. In MacGillivray I, Nylander

PPS, Corney G, eds. Human multiple reproduction. Philadelphia: W.B. Saunders Company, 1975.
Creagan RP, Ruddle FH. New approaches to human gene mapping by somatic cell genetics. In Yunis JJ, ed. Molecular structure of human chromosomes. New York: Academic Press, 1977.
Crow JF. Mutation in man. Prog Med Genet 1961; 1:1–26.
Cuckle HS, Wald NJ, Lindenbaum RH. Maternal serum alpha-fetoprotein measurement: a screening test for Down syndrome. Lancet 1984; 1:926–929.
Darlington GJ, Bernhard HP, Ruddle FH. Human serum albumen phenotype activation in mouse hepatoma–human leukocyte cell hybrids. Science 1974; 185:859–862.
Davidson RG, Nitowsky HM, Childs B. Demonstration of two populations of cells in the human female heterozygous for glucose-6-phosphate dehydrogenase variants. Proc Natl Acad Sci USA 1963; 50:481–485.
Davies KE, Pearson PL, Harper PS, et al. Linkage analysis of two cloned DNA sequences flanking the Duchenne muscular dystrophy locus on the short arm of the human X chromosome. Nucl Acids Res 1983; 11:2303–2312.
Davis RM. Localization of male-determining factors in man: a thorough review of structural anomalies of the Y chromosome. J Med Genet 1981; 18:161–195.
Donahue RP, Bias WB, Renwick JH, McKusick VA. Probable assignment of the Duffy blood group locus to chromosome 1 in man. Proc Natl Acad Sci USA 1968; 61:949–955.
Duckworth-Rysiecki G, Cornish K, Clarke CA, Buchwald M. Identification of two complementation groups in Fanconi anemia. Somatic Cell Mol Genet 1985; 11:35–41.
Edwards JH. The simulation of mendelism. Acta Genet 1960; 10:63–70.
Edwards JH, Harnden DG, Cameron AH, et al. A new trisomic syndrome. Lancet 1960; 1:787–790.
Edwards RG. Test-tube babies, 1981. Nature 1981; 293:253–256.
Edwards RG, Steptoe PC. Current status of in-vitro fertilisation and implantation of human embryos. Lancet 1983; 2:1265–1269.
Egger J, Wilson J. Mitochondrial inheritance in a mitochondrially mediated disease. New Engl J Med 1983; 309:142–146.
Epstein CJ, Motulsky AG. Evolutionary origins of human proteins. Prog Med Genet 1965; 4:85–127.
Fehilly CB, Willadsen SM, Tucker EM. Interspecific chimaerism between sheep and goat. Nature 1984; 307:634–636.
Ferguson-Smith MA, ed. Early prenatal diagnosis. Br Med Bull 1983; 39:301–404.
Fisher RA. The correlation between relatives on the supposition of mendelian inheritance. Trans Roy Soc Edinburgh 1918; 52:399–433.
Fogh-Andersen, P. Inheritance of hare-lip and cleft-palate. Op Dom Biol Hered Hum Kbh 4. Copenhagen: Munksgaard, 1942.
Fölling A. Über Ausscheidung von Phenylbrenztraubensäure in den Harn als Stoffwechselanomalie in Verbindung mit Imbezillität. Hoppe-Seyler's Z Physiol Chem 1934; 227:169–176.
Ford CE, Jones K, Polani P, et al. A sex chromosome anomaly in a case of gonadal dysgenesis (Turner's syndrome). Lancet 1959; 1:711–713.
Fraser FC. The multifactorial/threshold concept—uses and misuses. Teratology 1976; 14:267–280.
Fraser FC. Evolution of a palatable multifactorial threshold model. Am J Hum Genet 1980; 32:796–813.
Fraser GR. The causes of profound deafness in childhood. Baltimore: Johns Hopkins University Press, 1976.
Fratantoni JC, Hall CW, Neufeld EF. Hurler and Hunter syndromes: mutual correction of the defect in cultured fibroblasts. Science 1968; 162:570–572.
Freda VJ, Gorman JG, Pollack W. Successful prevention of experimental Rh sensitization in man with an anti-Rh gamma-2 globulin antibody preparation: a preliminary report. Transfusion 1964; 4:26–32.
Gallico GG, O'Connor NE, Compton CC, et al. Permanent coverage of large burn wounds with autologous cultured human epithelium. New Engl J Med 1984; 311:448–451.
Garrod AE. The incidence of alkaptonuria: a study in chemical individuality. Lancet 1902; 2:1616–1620.
Gartler SM, Waxman SH, Giblett ER. An XX/XY human hermaphrodite resulting from double fertilization. Proc Natl Acad Sci USA 1962; 48:332–335.
Giannelli F, Choo KH, Winship PR, et al. Characterisation and use of an intragenic polymorphic marker for detection of carriers of haemophilia B (factor IX deficiency). Lancet 1984; 1:239–241.
Giblett ER. Erythrocyte antigens and antibodies. In: Williams WJ, Beutler E, Erslev AJ, Lichtman MA, eds. Hematology, 3rd ed. New York: McGraw-Hill, 1983.
Giblett ER, Anderson JE, Cohen F, et al. Adenosine deaminase deficiency in two patients with severely impaired cellular immunity. Lancet 1972; 2:1067–1069.
Gravel RA, Lam KF, Scully KJ, Hsia YE. Genetic complementation of propionyl co-A carboxylase deficiency in cultured human fibroblasts. Am J Hum Genet 1977; 29:378–388.
Grüneberg H. Genetical studies on the skeleton of the mouse. IV. Quasicontinuous variation. J Genet 1952; 51:95–114.
Gusella JF, Wexler NS, Conneally PM, et al. A polymorphic DNA marker genetically linked to Huntington's disease. Nature 1983; 306:234–238.
Haldane JBS. The rate of spontaneous mutation of a human gene. J Genet 1935; 31:317–326.
Hansmann I. Factors and mechanisms involved in nondisjunction and X-chromosome loss. In: Sandberg AA, ed. Cytogenetics of the mammalian X chromosome. Part A. Basic mechanisms of X chromosome behavior. New York: Liss, 1983.

Harper PS. Practical genetic counseling, 2nd ed. Bristol: Wright; Littleton: Wright, PSG, 1984.
Harris H, Hopkinson DA. Average heterozygosity per locus: an estimate based on enzyme polymorphism. Ann Hum Genet 1972; 36:9–20.
Hassold T, Chiu D, Yamane JA. Parental origin of autosomal trisomies. Ann Hum Genet 1984; 48:129–144.
Hassold T, Matsuyama A. Origin of trisomies in human spontaneous abortions. Hum Genet 1979; 46:285–294.
Hassold T, Quillen SD, Yamane JA. Sex ratio in spontaneous abortions. Ann Hum Genet 1983; 47:39–47.
Hayden MR. Huntington's chorea. New York: Springer-Verlag, 1981.
Herrmann J, Opitz JM. Delayed mutation as a cause of genetic disease in man: achondroplasia and the Wiedemann-Beckwith syndrome. In: Nichols WW, Murphy DG, eds. Regulation of cell proliferation and differentiation. New York: Plenum Press, 1977.
Hirschhorn K, Hirschhorn R. Immunodeficiency disorders. In: Emery AEH, Rimoin DL, eds. Principles and practice of medical genetics. Edinburgh: Churchill Livingstone, 1983.
Holt SB. The genetics of dermal ridges. Springfield, Illinois: Charles C Thomas, 1968.
Hulten M. Chiasma distribution at diakinesis in the normal human male. Hereditas 1974; 76:55–78.
Ingram VM. A specific chemical difference between the globins of normal human and sickle cell anaemia haemoglobin. Nature 1956; 178:792–794.
Ingram VM. Separation of the peptide chains of human globin. Nature 1959; 183:1795–1798.
Ingram VM. The hemoglobins in genetics and evolution. New York: Columbia University Press, 1963.
Jacobs PA, Price WH, Court-Brown WM, et al. Chromosome studies on men in a maximum security hospital. Ann Hum Genet 1968; 31:339–358.
Jacobs PA, Strong JA. A case of human intersexuality having a possible XXY sex-determining mechanism. Nature 1959; 183:302–303.
Jervis GA. Phenylpyruvic oligophrenia: deficiency of phenylalanine oxidizing system. Proc Soc Exp Biol Med 1953; 82:514–515.
Kalter H, Warkany J. Congenital malformations. Etiological factors and their role in prevention. New Engl J Med 1983; 308:424–431, 491–497.
Kan YW, Dozy AM. Polymorphism of DNA sequence adjacent to human beta-globin structural gene: relationship to sickle mutation. Proc Natl Acad Sci USA 1978; 75:5631–5635.
Karp LE. Older fathers and genetic mutations. Am J Med Genet 1980; 7:405–406.
Kimura M. The neutral theory of molecular evolution. New York: Cambridge University Press, 1983.
Kingston HM, Sarfarazi M, Thomas NST, Harper PS. Localization of the Becker muscular dystrophy gene on the short arm of the X chromosome by linkage to cloned DNA sequences. Hum Genet 1984; 67:6–17.
Klinefelter HF, Reifenstein EC, Albright F. Gynecomastia, aspermatogenesis without aleydigism and increased excretion of follicle-stimulating hormone. J Clin Endocrinol 1942; 2:615–627.
La Du BN, Zannoni VG, Laster L, Seegmiller JE. The nature of the defect in tyrosine metabolism in alkaptonuria. J Biol Chem 1958; 230:251–260.
Laberge C. Hereditary tyrosinemia in a French Canadian isolate. Am J Hum Genet 1969; 21:36–45.
Land H, Parada LF, Weinberg RA. Cellular oncogenes and multistep carcinogenesis. Science 1983; 222:771–778.
Landsteiner K. Über Agglutinationserbscheinungen normalen menschlicken Blutes. Wien klin Wschr 1901; 14:1132–1134.
Landsteiner K, Levine P. A new agglutinable factor differentiating individual human bloods. Proc Soc Exp Biol N Y 1927; 24:600–602.
Latt SA. Microfluorimetric detection of deoxyribonucleic acid replication in human metaphase chromosomes. Proc Natl Acad Sci USA 1973; 70:3395–3399.
Lavett DK. Secondary structure and intron-promoter homology in globin switching. Am J Hum Genet 1984; 36:338–345.
Ledbetter DH, Mascarello JT, Riccardi VM, et al. Chromosome 15 abnormalities and the Prader-Willi syndrome: a follow-up report of 40 cases. Am J Hum Genet 1982; 34:278–285.
Leder P, Batty J, Lenoir G, et al. Translocations among antibody genes in human cancer. Science 1983; 222:765–770.
Lejeune J, Gautier M, Turpin R. Étude des chromosomes somatiques de neuf enfants mongoliens. C R Acad Sci Paris 1959; 248:1721–1722.
Lejeune J, Lafourcade J, Bergen R, et al. Trois cas de deletion partielle du bras court du chromosome 5. C R Acad Sci Paris 1963; 257:3098–3102.
Lesch M, Nyhan WL. A familial disorder of uric acid metabolism and central nervous system function. Am J Med 1964; 36:561–570.
Levy HL. Genetic screening. Adv Hum Genet 1973; 4:1–104.
Ley TJ, De Simone J, Anagnou NP, et al. 5-Azacytidine selectively increases γ-globin synthesis in a patient with β^+ thalassemia. New Engl J Med 1982; 307:1469–1475.
Li CC, Mantel N. A simple method of estimating the segregation ratio under complete ascertainment. Am J Hum Genet 1968; 20:61–81.
Liley AW. Intrauterine transfusion of foetus in haemolytic disease. Br Med J 1963; 2:1107–1109.
Lippman-Hand A, Fraser FC. Genetic counseling—the postcounseling period: I. Parents' perceptions of uncertainty. Am J Med Genet 1979; 4:51–71.
Lippman-Hand A, Fraser FC. Genetic counseling—the postcounseling period: II. Making reproductive choices. Am J Med Genet 1979; 4:73–87.

Lison M, Blondheim SH, Melmed RN. A polymorphism of the ability to smell urinary metabolites of asparagus. Br Med J 1980; 281:1676–1678.
Littlefield JW. On the difficulty of combining basic research and patient care. Am J Hum Genet 1984; 36:731–735.
Littlewood JE. A mathematician's miscellany. London: Methuen, 1953.
Lyon MF. Sex chromatin and gene action in the mammalian X-chromosome. Am J Hum Genet 1962; 14:135–148.
Maeda N, Yang F, Barnett DR, et al. Duplication within the haptoglobin *Hp*[2] gene. Nature 1984; 309:131–135.
Maynard-Smith S, Penrose LS, Smith CAB. Mathematical tables for research workers in human genetics. London: Churchill, 1961.
Mayr E. Animal species and evolution. Cambridge: Harvard University Press, 1963.
McInnes RR, Shih V, Chilton S. Interallelic complementation in an inborn error of metabolism: genetic heterogeneity in argininosuccinate lyase deficiency. Proc Natl Acad Sci USA 1984; 81:4480–4484.
McKusick VA. The anatomy of the human genome. Am J Med 1980; 69:267–276.
Meinecke-Tillmann S, Meinecke B. Experimental chimaeras—removal of reproductive barrier between sheep and goat. Nature 1984; 307:637–638.
Melnick M, Bixler D, Fogh-Andersen P, Conneally M. Cleft lip ± cleft palate: an overview of the literature and an analysis of Danish cases born between 1941 and 1968. Am J Med Genet 1980; 6:83–97.
Mikkelsen M, Poulsen H, Grinsted J, Lange A. Non-disjunction in trisomy 21: study of chromosomal heteromorphisms in 110 families. Ann Hum Genet 1980; 44:17–28.
Miller E, Hare JW, Cloherty JP, et al. Elevated maternal hemoglobin A1c in early pregnancy and major congenital anomalies in infants of diabetic mothers. N Engl J Med 1981; 304: 1331–1334.
Morton NE. Sequential tests for the detection of linkage. Am J Hum Genet 1955; 7:277–318.
Morton NE. Genetic tests under incomplete ascertainment. Am J Hum Genet 1959; 11:1–16.
Muller HJ. Artificial transmutation of the gene. Science 1927; 66:84–87.
Murphree AL, Benedict WF. Retinoblastoma—clues to human oncogenesis. Science 1984; 223:1028–1033.
Neel JV. The inheritance of sickle cell anemia. Science 1949; 110:64–66.
Neel JV. Editorial. Genetic epidemiology 1984; 1:5–6.
Neel JV, Schull WJ. Human heredity. Chicago: University of Chicago Press, 1954.
Neufeld EF. Cell mixing and its sequelae. Am J Hum Genet 1983; 35:1081–1085.
Neufeld EF, Fratantoni JC. Inborn errors of mucopolysaccharide metabolism. Science 1970; 169:141–146.
Nora JJ. Multifactorial inheritance hypothesis for the etiology of congenital heart diseases. The genetic-environmental interaction. Circulation 1968; 38:604–617.
Nora JJ, Randall MPH, Lortscher H, et al. Genetic-epidemiologic study of early-onset ischemic heart disease. Circulation 1980; 61:503–508.
Nowell PC, Hungerford DA. Chromosome studies on normal and leukemic human leukocytes. J Natl Cancer Inst 1960; 29:911–931.
Nyhan WL, Bakay B, Connor JD, et al. Hemizygous expression of glucose-6-phosphate dehydrogenase in erythrocytes of heterozygotes for the Lesch-Nyhan syndrome. Proc Natl Acad Sci USA 1970; 65:214–218.
Ohno S. Major sex-determining genes. Berlin: Springer-Verlag, 1979.
Opitz JM. "Unstable premutations" in achondroplasia: penetrance vs phenotrance. Am J Med Genet 1984; 19:251–254.
Palmiter RD, Brinster RL, Hammer RE, et al. Dramatic growth of mice that develop from eggs injected with metallothionein–growth hormone fusion genes. Nature 1982; 300:611–615.
Parisi P, Gatti M, Prinzi G, Caperna G. Familial incidence of twinning. Nature 1983; 304:626–628.
Patau K, Smith DW, Therman E, et al. Multiple congenital anomaly caused by an extra autosome. Lancet 1960; 1:790–793.
Pauling L, Itano HA, Singer SJ, Wells IC. Sickle cell anemia, a molecular disease. Science 1949; 110:543–548.
Peake IR, Furlong BL, Bloom AL. Carrier detection by direct gene analysis in a family with haemophilia B (Factor IX deficiency). Lancet 1984; 1:242–243.
Pinsky L, Kaufman M, Killinger DW, et al. Human minimal androgen insensitivity with normal dihydrotestosterone-binding capacity in cultured genital skin fibroblasts: evidence for an androgen-selective qualitative abnormality of the receptor. Am J Hum Genet 1984; 36:965–978.
Preston AE, Barr A. The plasma concentration of factor VIII in the normal population. II. The effects of age, sex and blood group. Br J Haematol 1964; 10:238–245.
Prockop DJ. Osteogenesis imperfecta: phenotypic heterogeneity, protein suicide, short and long collagen. Am J Hum Genet 1984; 36:499–505.
Race RR, Sanger R. Blood groups in man, 6th ed. Oxford: Blackwell Scientific Publications, 1975.
Reed TE. Caucasian genes in American negroes. Science 1969; 165:762–7678.
Reilly P. Genetic screening legislation. Adv Hum Genet 1975; 5:310–376.
Reiser CA, Pauli RM, Hall JG. Achondroplasia: unexpected familial recurrence. Am J Med Genet 1984; 19:245–250.

Renwick JH. Nail-patella syndrome: evidence for modification by alleles at the main locus. Ann Hum Genet 1956; 21:159–169.
Reynolds GA, Basu SK, Osborne TF, et al. HMG CoA reductase: a negatively regulated gene with unusual promoter and 5′ untranslated regions. Cell 1984; 38:275–285.
Riccardi VM, Dobson CE, Charkraborty R, Bontke C. The pathophysiology of neurofibromatosis: IX. Paternal age as a factor in the origin of new mutations. Am J Med Genet 1984; 18:169–176.
Rowley JD. A new consistent chromosomal abnormality in chronic myelogenous leukemia identified by quinacrine fluorescence and Giemsa staining. Nature 1973; 243:290–293.
Rowley JD. Human oncogene locations and chromosome aberrations. Nature 1983; 301:290–291.
Russell ES. Hereditary anemias of the mouse: a review for geneticists. Adv Genet 1979; 20:357–459.
Russell ES. Developmental studies of mouse hereditary anemias. Am J Med Genet 1984; 18:621–641.
Russell WL, Russell LB, Cupp MB. Dependence of mutation frequency on radiation dose in female mice. Proc Natl Acad Sci USA 1959; 45:18–23.
Sanger R, Race RR. The Lutheran-secretor linkage in man: support for Mohr's findings. Heredity 1958; 12:513–520.
Scheuerbrandt G, Mortier W. Voluntary newborn screening for Duchenne muscular dystrophy: a nationwide pilot program in West Germany. In: Serratrice G, Cros D, Desnuelle C, et al., eds. Neuromuscular diseases. New York: Raven Press, 1984.
Schull WJ, Otake M, Neel JV. Genetic effects of the atomic bombs: a reappraisal. Science 1981; 213:1220–1227.
Scriver CR, Bardanis M, Cartier L, et al. β-Thalassemia disease prevention: genetic medicine applied. Am J Hum Genet 1984; 36:1024–1038.
Shapiro LJ, Mohandas T. DNA methylation and the control of gene expression on the human X chromosome. Cold Spring Harbor Symp Quant Biol 1982; 47:631–637.
Shaw MW. Presidential address. To be or not to be? That is the question. Am J Hum Genet 1984; 36:1–9.
Sherman SL, Morton NE, Jacobs PA, Turner G. The marker (X) syndrome: a cytogenetic and genetic analysis. Ann Hum Genet 1984; 48:21–37.
Shows TB, Alper CA, Bottsman D, et al. International system for human gene nomenclature (ISGN 1979). Cytogenet Cell Genet 1979; 25:96–116.
Shows TB, McAlpine PJ, Miller RL. The 1983 catalogue of mapped human markers and report of the nomenclature committee. Seventh International Workshop on Human Gene Mapping. Cytogenet Cell Genet 1984; 37:340–393. Simultaneous publication in Birth Defects 1984; 20(2).
Sillence DO, Barlow KK, Garber AP, et al. Osteogenesis imperfecta type II. Delineation of the phenotype with reference to genetic heterogeneity. Am J Med Genet 1984; 17:407–423.
Sillence DO, Rimoin DL. Danks DM. Clinical variability in osteogenesis imperfecta—variable expressivity or genetic heterogeneity. Birth Defects 1979; 15(5B):112–129.
Skolnick MH, Willard HF, Menlove LA. Report of the committee on human gene mapping by recombinant DNA techniques. In: Sparkes RS, Berg K, Evans JH, Klinger HP, eds. Human gene mapping 7. Seventh international workshop on human gene mapping. Cytogenet Cell Genet 37, 1984. Simultaneous publication in Birth Defects 1984; 20(2).
Smith CAB. Linkage scores and corrections in simple two- and three-generation families. Ann Hum Genet 1968; 32:127–150.
Smith SM, Penrose LS, Smith CAB. Mathematical tables for research workers in human genetics. London: Churchill, 1961.
Smithells RW, Sheppard S, Schorah CJ, et al. Possible prevention of neural-tube defects by periconceptional vitamin supplementation. Lancet 1980; 1:339–340.
Smithies O, Connell GE, Dixon GH. Chromosomal rearrangements and the evolution of haptoglobin genes. Nature 1962; 196:232–236.
Solomon E, Bodmer WF. Evolution of sickle variant gene. Lancet 1979; 1:923.
Southern EM. Detection of specific sequences among DNA fragments separated by gel electrophoresis. J Mol Biol 1975; 98:503–517.
Steele MW, Breg WR. Chromosome analysis of human amniotic fluid cells. Lancet 1966; 1:383–385.
Stewart DA, ed. Children with sex chromosome aneuploidy: follow-up studies. Birth Defects 1982; 18(4).
Stewart TA, Mintz B. Successive generations of mice produced from an established culture of euploid teratocarcinoma cells. Proc Natl Acad Sci USA 1981; 78:6314–6318.
Stoll C, Roth MP, Bigel P. A re-examination of paternal age effect on the occurrence of new mutants for achondroplasia. Prog Clin Biol Res 1982; 104:419–426.
Taylor JH, Woods PS, Hughes WL. The organization and duplication of chromosomes as revealed by autoradiographic studies using tritium-labeled thymidine. Proc Natl Acad Sci USA 1957; 43:122–128.
Terasaki PI, Gjertson D, Bernoco D, et al. Twins with two different fathers identified by HLA. New Engl J Med 1978; 299:590–592.
Therman E, Denniston C, Serto GE, Ulber M. X chromosome constitution and the human female phenotype. Hum Genet 1980; 54:133–143.
Tjio HJ, Levan A. The chromosome number of man. Hereditas 1956; 42:1–6.
Trounson A, Mohr L. Human pregnancy following cryopreservation, thawing and transfer of an eight-cell embryo. Nature 1983; 305:707–709.
Turner G, Jacobs P. Marker (X)-linked mental retardation. Adv Hum Genet 1983; 13:83–112.

Turner HH. A syndrome of infantilism, congenital webbed neck and cubitus valgus. Endocrinology 1938; 23:566–574.
Uchida IA. Maternal radiation and trisomy 21. In: Hook EB, Porter IA, eds. Population genetics: studies in humans. New York: Academic Press, 1977.
Uchida IA, Lin CC. Identification of triploid genome by fluorescence microscopy. Science 1972; 176:304–305.
Vogel F. 1970. ABO blood groups and disease. Am J Hum Genet 1970; 22:464–475.
Waardenburg PJ. Das menschliche Auge und seine Erbanlagen. The Hague: Nijhoff, 1932.
Wald NJ. Neural-tube defects and vitamins: the need for a randomized clinical trial. Br J Obstet Gynecol 1984; 91:516–523.
Walker NF. The use of dermal configuration in the diagnosis of mongolism. Pediat Clin N Am 1958; 5:531–543.
Watkins WM. Biochemistry and genetics of the ABO, Lewis, and P blood group systems. Adv Hum Genet 1980; 10:1–136.
Watson JD, Crick FHC. Molecular structure of nucleic acids—a structure for deoxyribose nucleic acid. Nature 1953; 171:737–738.
White C, Wyshak G. Inheritance in human dizygotic twinning. New Engl J Med 1964; 271:1003–1005.
Woo SLC, Lidsky AS, Güttler F, et al. Cloned human phenylalanine hydroxylase gene allows prenatal diagnosis and carrier detection of classical phenylketonuria. Nature 1983; 306:151–155.
Woolf CM. Congenital cleft lip: a genetic study of 496 propositi. J Med Genet 1971; 8:65–71.
Wright S. An analysis of variability in numbers of digits in an inbred strain of guinea pigs. Genetics 1934; 19:506–536.
Wyrobek AJ, Bruce WR. The induction of sperm shape abnormalities in mice and humans. In: Hollaender A, de Serres FJ, eds. Chemical mutagens: principles and methods for their detection. Vol 5. New York: Plenum Press, 1978.
Yunis JJ. The chromosomal basis of human neoplasia. Science 1983; 221:227–236.
Zavala C, Morton NE, Rao DC, et al. Complex segregation analysis of diabetes mellitus. Hum Hered 1979; 29:325–333.
Zeesman S, Clow CL, Cartier L, Scriver CR. A private view of heterozygosity: eight year follow-up on carriers of the Tay-Sachs gene detected by high school screening. Am J Med Genet 1984; 18:769–778.
Zenzes MT, Reed TE. Variability in serologically detected male antigen titre and some resulting problems: a critical review. Hum Genet 1984; 66:103–109.

ANSWERS

CHAPTER 2

1. a) *A* and *a* b) 1) At first meiotic division 2) At second meiotic division
2. a) 2 b) 32 c) 2^n
3. Two of the 2^{23} possible combinations; one of each of the two parental combinations.
4. Segregation of alleles at anaphase of the first or second meiotic division; random recombination of nonalleles by the random assortment of nonhomologous chromosomes at anaphase of the first meiotic division.
5. 250–500; 60–125; 35–70.
6. Meiosis I; primary spermatocyte.

CHAPTER 3

1. I. point mutation, UAU to UGU changes tyr to cys.

   II. frameshift mutation, deletion of the first nucleotide of the third codon.

   III. frameshift mutation, insertion of G after the first codon.
2. A. 5,9,3,2,6,8,7,10,—,4,1.

CHAPTER 4

1. This question gives the student practice in composing a pedigree chart and some insight into the difficulty of obtaining complete and accurate family history information.
2. b) 1/36 c) 1/4
3. b) They are homozygous for the hemophilia gene. (Their mother is an obligate carrier.)
   c) 100 percent d) 50 percent
4. b) Only X-linked dominant inheritance is ruled out.
   c) Probably X-linked recessive, though in view of the consanguinity autosomal recessive inheritance cannot be ruled out.
   d) Very low, as low as the population risk if this is X-linked.
5. b) Cannot estimate because genetic pattern is not clear; could be anything from very low to 50 percent for Hedy.
   c) Now appears to be autosomal dominant, so risk for Hedy is 1/2, and for her child 1/4.
   d) In view of her age, Hedy appears to have escaped the gene.
6. Risk that Linda's father, who is still unaffected at age 40+, will eventually develop HD is about 1/3. Thus risk for Linda is about 1/6 or less as she is still unaffected, and risk for her child is about 1/12 or less.
7. *d*, limb-girdle muscular dystrophy; X*cb*, color blindness gene on X chromosome. Father *dd* X*cb*Y. Mother *Dd* XX*cb*. Daughter *dd* X*cb* X*cb*.
8. a) *DdEe* b) 1/12

9. a) D, dentinogenesis imperfecta; d, normal; T, taster; t, nontaster
   $DdTt$ X $ddTt$
   b) 1/2 (daughter) × 1/2 (dd) × 1/4 (tt) = 1/16

10. a) 1/2 × 0.8 = 0.4 b) 1/6 × 1/2 × 0.8 = 0.07 (See also Chapter 14.)

CHAPTER 5

1. These two mutations affect different globin chains. The expected offspring are 1/4 normal, 1/4 Hb M Saskatoon heterozygotes with methemoglobinemia, 1/4 Hb M Boston heterozygotes with methemoglobinemia, 1/4 double heterozygotes with four hemoglobin types: normal, both types of Hb M and a type with an abnormality in each chain. In the double heterozygotes, the clinical consequences are unknown—probably more severe methemoglobinemia.

2. 2/3 × 2/3 × 1/4 = 1/9

3. 1/4

4. a) Patient and uncle hemizygous for Hunter gene, mother heterozygous, father's genotype irrelevant but 2/3 chance of being heterozygous for Scheie syndrome.
   b) Patient a Hurler-Scheie compound, mother Hurler heterozygote, father Scheie heterozygote, maternal uncle Hurler homozygote.

5. a) Theoretically, no effect because Tay-Sachs patients never reproduce.
   b) Theoretically, this would raise the frequency of the Tay-Sachs gene by increasing the proportion of heterozygotes in the offspring.

6. 6, 5, 8, 1, 9, 2, 3, 4, 11, 10

CHAPTER 6

1. a) A mosaic karyotype with a normal male cell line and a cell line with an extra chromosome 21. Unbalanced.
   b) An unbalanced male karyotype with a missing chromosome 14 and an additional translocation chromosome composed of the part of chromosome 14 between band p11 and the end of the long arm and the part of chromosome 21 between band p11 and the end of the long arm. The short arms of chromosome 14 and 21 have been lost.
   c) A balanced female karyotype in which the segment of chromosome 2 extending from q22 to q32 has been inverted and inserted into a break at band p14 of chromosome 5.
   d) An unbalanced male karyotype, in which one chromosome 7 is missing and has been replaced by a derivative chromosome 7, where there has been a reciprocal translocation between chromosomes 7 and 11. Breakage and reunion have occurred between bands 7q36 and 11q21. The segments distal to these bands have been exchanged between the two chromosomes. The der(7) chromosome has been inherited from the patient's mother.
   e) A complex balanced translocation in a female. Breakage and reunion have occurred in chromosomes 2, 5 and 7 at bands 2p21, 5q23 and 7q22 respectively. The segment of chromosome 2 distal to 2p21 has been translocated to chromosome 5 at 5q23, the segment of chromosome 5 distal to 5q23 has been translocated to chromosome 7 at 7q22, and the segment of chromosome 7 distal to 7q22 has been translocated to chromosome 2 at 2p21.

2. a) Ovum or sperm has 2n chromosome number because of failure of normal meiotic division.
   b) Two sperm may fertilize same egg. Fertilization of ovum by diploid sperm is most common type.

3. This is probably the result of an early postzygotic loss of chromosome 21 in a 47,XY,+21 embryo, accompanied by twinning of the original embryo. Loss of the 46,XY, line in one twin and of the 47,XY,+21 line in the other twin might have followed. However, each twin might be a mosaic with some cells of the other line.

4. See text for retinoblastoma, Prader-Willi syndrome, and the aniridia–Wilms tumor association.

5. 88-96 percent.

6. About 8 percent.

7. This question is for class discussion. Consider the risk of a 21q22q unbalanced translocation in offspring of a male or female heterozygote, the need for an approach to the family and how this can be arranged, the physician's responsibility to inform the family, etc.

CHAPTER 7

1. Apparently from the mother, since the father has given her his X with his color-blindness gene.

2. The error must have been in the paternal gamete, since the daughter received neither the paternal X nor the paternal Y.

3. a) X*cb*X*cb* Y
   b) Father XY, mother XX*cb*; probably nondisjunction at meiosis II in mother.

4. a) Theoretically, 1/2 X and 1/2 XX.
   b) 1/4 each XX, XY, XXX, XXY.

5. a) Paternal, meiosis I.
   b) Mother, nondisjunction at either meiosis I or meiosis II.

6. No. The nondisjunctional event that produces XXY can occur in either parent, and at either meiotic division in the female or at meiosis I in the male; however, XYY can arise by nondisjunction in the male at meiosis II only.

7. a) A female karyotype in which one X is replaced by an isochromosome, made up of two entire long arms of the X chromosome separated by a centromere. Unbalanced.
   b) A balanced female karyotype in which a reciprocal translocation has occurred between an X chromosome and a chromosome 21. Breakage and reunion have taken place at band p21 of the X chromosome and band p12 of chromosome 21, with exchange of the distal segments of the short arms of these chromosomes.

CHAPTER 8

1. MZ twin, DZ twin and sib, father and mother, cousin, unrelated person.

2. a) 2.6 b) 109 c) 20

CHAPTER 9

1. a) A, B or AB
   b) All men of group O, that is, approximately 46 percent of a typical North American population.

2. a) No
   b) 1, 2 and 4 can be excluded, but not 3.
   c) Father must have an *O* and an *M* gene.

3. AB child to A × AB parents.
   B child to B × B parents.
   The two A children cannot be definitely assigned without further tests.
   The O × O parents have not found their child.

4. a) H, A, B b) M, N, S, s

CHAPTER 10

1. a) Three
   b) A, C, F; B, E; D
   c) Mutations at a minimum of three loci can produce the phenotype.

2. Tabulate observations:

	1	2	3	4	7	8	9	10	13	19	20	21	X
I				+	+	+		+			+		+
II	+		+		+							+	+
III							+		+	+		+	+
IV	+			+				+		+			+
V		+	+						+		+		+

Conclusions:
Enzyme A has its locus on chromosome 2.
Enzyme B has its locus on the X chromosome.
Enzyme C has its locus on chromosome 19.
Enzyme D cannot be assigned to any of these 13 chromosomes.

3. 3, 7, 8, 5, 6, 4, 10, 9, 11, 2

CHAPTER 11

1. Question on Figure 11–7
   a) Phase-known pedigree; in II–2, A is in coupling with the nail-patella gene. III–6 is a recombinant.
   b) 6NR, 1R LOD Score = 0.857 at recombination frequency θ of 0.15.

2. a and b are in disequilibrium; c and d are in equilibrium.

3. At equilibrium HLA type A1B8 should have a frequency of $0.17 \times 0.11 = 0.02$. Similarly, the frequency of A3B7 should be $0.13 \times 0.11 = 0.014$.

4. a) and b) The frequencies are the same (0.018). Prove this to yourself by calculating the frequency of each possible combination per chromosome, then of each diploid combination. This is equivalent to showing that the two loci are in equilibrium.
   c) Changing the recombination frequency does not change the frequencies of the combinations, unless special circumstances such as recent admixture or mutation, or selective advantage, are acting to maintain certain combinations at a high frequency.

CHAPTER 12

1. a) Autosomal dominant with reduced penetrance (because risk drops by half with each step of more remote relationship).
   b) Any other evidence for autosomal dominance and against multifactorial inheritance, such as twin data (concordance about 20 percent in MZ twins, 10 percent in DZ twins) or equal risk for relatives of probands of either sex.

2. c) 12 to 15 percent; a) about 5 percent; b) a little less than 5 percent; e) of the order of 1 percent; the fetus has an affected second-degree relative and an affected third-degree relative; d) well below 1 percent; fetus has a third-degree affected relative only. Probably all except d would qualify.

3. Brother 1 in 5, sister 1 in 14, as for other first-degree relatives (Table 12–4).

4. Rules are in text. If the malformation is autosomal recessive, the risk for sibs is much higher than the risk for parents, consanguinity of parents does not affect risk for sibs, and the concordance rate is four times as high in MZ as in DZ twins.

5. For X linkage, other affected members are in the maternal line; sex ratio is $q{:}q^2$ for X-linked recessive, $q{:}2pq+q^2$ for X-linked dominants and so forth.

CHAPTER 13

1. Mosaic
2. Teratocarcinoma
3. Malformation
4. Syndrome
5. Chimera
6. Implantation
7. Pleiotropy
8. Teratogen
9. Sequence
10. Deformation

CHAPTER 14

1. a) $(1/2)^6 = 1/64$ b) $20\ (1/2)^3(1/2)^3 = 5/16$

2. a) 1/2 b) $(1/2)^4 = 1/16$ c) $1 - (1/2)^4 = 15/16$

3. a) 1/4 b) $(3/4)^5 = 243/1024$ c) $10(3/4)^3(1/4)^2 = 270/1024$

4. a) $(1/2)^3 = 1/8$ b) $(1/2)^3 = 1/8$ c) $3(1/2)(1/2)^2 = 3/8$

5. a) 57.1 percent b) 62.5 percent

6. Figure 14–2. Calculation of probability that IV–2 is a carrier. I–1 is a carrier.

	Carrier	Not a carrier
II–1		
Prior probability	1/2	1/2
Conditional probability (all 4 sons unaffected)	(1/2) 4	1
Joint probability	1/32	1/2
Relative probability	1/17	16/17
III–1		
Prior probability	1/34	33/34
Conditional probability (all 4 sons unaffected)	(1/2) 4	1
Joint probability	1/544	33/34
Relative probability	1/529	528/529
IV–1		
Prior probability	1/1058	1057/1058
Conditional probability (1 son unaffected)	1/2	1
Joint probability	1/2116	1057/1058 = 2114/2116
Relative probability	1/2115	

7.

	Fetus normal	Fetus with NTD
Prior probability (fetus has affected first degree relative)	0.95	0.05
Conditional probability of elevated AFP	0.05	0.80
Joint probability	0.0475	0.04
Relative probability	$\frac{0.0475}{0.0475 + 0.04} = 0.54$	0.46

8. a) 0.016 b) About 0.009—reduced by almost half.

9. Read off from the graph in Figure 4–25 the approximate probability of being normal or affected at the ages mentioned. One way of doing this problem:

	Huntington disease (*Hh*)	Unaffected (*hh*)
Proband		
Prior risk	0.5	0.5
Conditional risk of normal phenotype at age 50	0.4	1
Conditional risk of 2 sons aged 30 and 28 with normal phenotype (if he is *Hh*, each son could be either *hh* [1/2] or *Hh* but not yet manifesting [1/2 × 2/3]; total chance for each son 0.83)	$(.83)^2$	1
Conditional risk of normal grandson aged 5 is almost 1 and can be ignored	——	——
	0.14	0.5
Relative risk	$\frac{0.14}{0.64} = 0.22$	0.78

The man has a 0.78 chance of never developing Huntington disease.

10. a) X linkage: $(1/2)^4 \times (1)^4 = 0.063$
    Autosomal recessive: $4p^3q \times p^4 = 0.007$
    X linkage is more likely.
    b) X linkage: not really consistent if any females are affected, unless the mother is affected and the trait is dominant.
    Probability = 0
    Autosomal recessive: $(4p^3q)(6p^2q^2) = 0.09$
    Autosomal recessive is more likely.

CHAPTER 15

1. a) *A* 0.9, *a* 0.1 b) Same c) $(0.18)^2 = 0.03$
2. a) The frequency of carrier × carrier matings is expected to be about 0.001 b) the frequency of affected offspring will be about 1 in 3300.
3. 36 percent A, 64 percent O.
4. 1/8
5. 3/8
6. a) 1/20,000 b) 1/100; 2/100 c) 1/10,000; 1/5,000
7. In equilibrium: e only
8. a) $aa = 0.09$ $a = \sqrt{0.09} = 0.3$ $A = 1 - 0.3 = 0.7$
   Equilibrium *AA* 0.49, *Aa* 0.42, *aa* 0.09
   b) $(2 \times 0.33) + (1 \times 0.34) = 0.50$ *M*, 0.50 N
   M 0.25, MN 0.50, N 0.25
   c) M 0.25, MN 0.50, N 0.25
   d) *AA* 0.64, *Aa* 0.32, *aa* 0.04
9. a) As in other X-linked lethals, theoretically 1/3 are new mutations.
   b) Since a proportion (theoretically 1/3) of the surviving babies would be heterozygotes, the gene frequency would rise and the proportion of new mutants among all cases of the disease would decline. (Keep in mind that we have no idea whether reproductive compensation has been operative in the past in such families.)
10. a) Site 1: $2(0.74)(0.26) = 0.38$
    Site 2: $2(0.81)(0.19) = 0.31$
    Sites 1 and 2 simultaneously: $(0.38 \times 0.31) = 0.12$
    b) 0.62 not heterozygous at site 1
    0.69 not heterozygous at site 2
    $0.62 \times 0.69 = 0.43$ not heterozygous at either site
    c) Yes

CHAPTER 16

1. a) 0.985
   b) 1, 2 and 4 would decrease the probability of monozygosity; 3 is thought to be irrelevant.
2. Nigerian 0.08, Canadian 0.30, Japanese 0.80
3. a) All except the ABO system, which is not informative here.
   b) Apparently A2, Bw44—to each twin.
   c) She appears to have given M to twin A and N to twin B. Biologically, 2 separate ova are expected. An identical maternal contribution to each twin is very unlikely.
   d) PF1 is excluded as father of Twin B by the HLA system. Similarly PF2 is excluded as father of Twin A by the HLA system and by the Hp system. (For further discussion, refer to the original paper.)
   e) PF1: Twin A; PF2: Twin B
   f) A man who could father both twins would have to be of the following types: MN or M, Ee in the Rh system, Hp 2–1, and Aw24-Bw54-Cw3 and A2 (or −)-B15-Cw3. Only 1 in 140,000 men in the population has this genotype.

CHAPTER 17

A. 8, 6, 3, 7, 5, 1, 11, 2, 9, 4

CHAPTER 18

1. No. She is a blood group chimera whose "true" blood type is O, thus she could not have an AB child.

2. 1) An X-chromosome abnormality such as 45,X or an X/autosome translocation in a carrier.
   2) Non-paternity or new mutation in the father; Laura has received the gene from both her parents.
   3) Laura is a carrier in whom by chance X inactivation has overwhelmingly favored the normal X.

3. e

4. 1) They are involved in genetic recombination and thus in increasing the amount of shuffling of genetic information from generation to generation.
   2) They maintain bivalent association in meiosis I and thus are important in normal chromosome segregation.

5. In Hurler and Scheie syndromes, the same enzyme (α-L-iduronidase) is defective. A different enzyme (iduronate sulfatase) is defective in Hunter syndrome.

6. She is an obligate carrier. No carrier tests or further pedigree data are necessary. The risk to her next son is 0.50.

7. a) Mother *CBD* Gd^A/ + Gd^B
   b) The color-blind, G6PD-B son and the daughter with normal color vision who is G6PD-A are the recombinants.
   c) About 10 centimorgans (2/20).

8. Gene f^1: frequency = 1/600
   Gene f^2: frequency = 1/400
   f^1f^2 = 2(1/600) (1/400) = 1/120000

Note that the compound is more common in the population than either homozygote.

9. All except cyclamates are known teratogens.

10. a) 0.08; b) 0.50; c)0.50
    (Keep in mind that 50 percent is the maximum recombination frequency.)

11. a) Each has the PKU allele in the 3.8 fragment.
    b) Carrier.
    c) Not a carrier.

12. Here only the father is heterozygous for the polymorphism. He has given both affected sons the 4.2 marker, and has given the daughter the 3.8.
    a) The daughter could be either homozygous normal or heterozygous.
    b) The father has given the fetus the 4.2 marker, so it is at least heterozygous. There is a 50 percent chance it has received the mother's PKU allele and is affected.

13. 1) One parent a carrier of 21q21q translocation. Assuming normal fertilization, a child could only have a single chromosome 21 or an additional chromosome 21.
    2) Mother affected with a metabolic disorder such as PKU.
    3) Mother must use a medication that is teratogenic.

14. For answer, see text (Chapter 18).

15.

	A	B	C
a)	0.03	0.007	0.001
b)	0.06	0.014	0.002
c)	$(0.06)^2$	$(0.014)^2$	$(0.002)^2$

Marriages of a recessive × heterozygote or recessive × recessive are so rare that they are ignored in this answer. These frequencies roughly fit the following diseases:
A. Tay-Sachs disease in Ashkenazi Jews
B. Phenylketonuria in the Italian population
C. Familial hypercholesterolemia, in which the *heterozygotes* have premature coronary artery disease.

CHAPTER 19

1. a) Specificity 0.70; sensitivity 0.60. The predictive value of an abnormal test is about 0.001 in the general child population and 0.33 in sibs of patients.
   b) See text for details. In particular, there are too many false negatives with this test.

2. c) Father gave baby 8 and der (18).
   d) Balanced
   e) Unbalanced
   f) Baby is hemizygous for large segment of chromosome 18. He lacks about 30 percent of one chromosome 18, which normally contains about 3 percent of the genome. How many genes are in 30 percent of 3 percent of 50,000 or more genes? A special problem about deletions is that they allow recessives on the normal chromosome of the pair to be expressed. As a rule, deletions are more severe than duplications.
   g) Genetic counseling should include discussion of the option of prenatal diagnosis in any subsequent pregnancy. In addition, the father's parents (and, if indicated, his sibs and other relatives) should be encouraged to have chromosome analysis.

3. Keep in mind that knowledge of the β globin gene and its flanking sequences is still incomplete.
   a) β^0; deletion is a rare cause of β thalassemia overall, but a particular deletion is common in one Indian form. This deletion involves the 3′ end of the gene and extends past the end of the gene.
   b) β^0; no normal β globin chains are formed.
   c) β^+; these mutations appear to reduce the rate of β globin chain production but not to stop it completely.
   d) Various splice mutations (mutations in the splice regions that lead to abnormal processing of RNA) can lead to either type of β thalassemia, depending on the consequences of the molecular change.
   e) In a heterozygote for a β^0 allele, only half the normal amount of β globin chains is produced; thus this is a form of β^+ thalassemia.

4. The first is obviously inadequate because you would not know the denominator. The second is poor because not all Down children are recognized and reported at birth; a Down syndrome infant might be unreported, or might be reported as, for example, a case of congenital heart defect rather than a case of Down syndrome. The third is inappropriate because the women who receive prenatal diagnosis are not a random sample of women of child-bearing age; in particular, they are older. A chromosome study of a large number of newborns is required.

5. Congenital, not genetic: fetal alcohol syndrome, rubella syndrome, etc. Genetic, not congenital: Huntington disease, muscular dystrophies, etc. Both genetic and congenital: Down syndrome, achondroplasia, etc.

6. a) Pedigree should show the features mentioned: affected males have normal children only, and all their sisters are carriers. Note that you cannot distinguish a *single* pedigree of this type from an X-linked pedigree, but a collection of similar pedigrees is convincing.
   b) As noted, some mitochondrial enzymes are coded in nuclear DNA whereas others are encoded in mitochondrial DNA itself.

7. The family data suggest that nonsmellers are homozygous recessive. On this basis, the gene frequencies from the data of Lison and colleagues are smeller 0.05, nonsmeller 0.95. Among the smellers, most (over 97 percent) are heterozygous.

8. a) Female 18μ; males 10μ.
   b) Females 4μ; males 3μ.
   c) Gene frequencies are approximately Xg^a 0.67, Xg 0.33. Thus carrier females 0.44, Xg(a−) males 0.33, or 4:3.
   d) About 0.90. (Work this out using Bayesian probability.) Referring to the equation $H = 2\mu + \frac{1}{2}H + If = 18\mu$, the probability that she is a new mutant is 2/18 (0.11), the probability that she is heterozygous by maternal descent is 0.50, and the probability that she had an affected father is 0.39.

9 and 10. No specific answers are provided here. A major purpose of these questions is to help you to recognize and handle common misconceptions about medical genetics.

NAME INDEX

SUBJECT INDEX

Page numbers in *italic* type indicate illustrations; page numbers followed by t refer to tables.

Pseudomosaicism, 294